STUDENT MATHEMATICAL LIBRARY
IAS/PARK CITY MATHEMATICAL SUBSERIES
Volume 51

Differential Equations, Mechanics, and Computation

Richard S. Palais
Robert A. Palais

American Mathematical Society, Providence, Rhode Island
Institute for Advanced Study, Princeton, New Jersey

Editorial Board of the Student Mathematical Library
Gerald B. Folland   Brad G. Osgood (Chair)
Robin Forman   Michael Starbird

Series Editor for the Park City Mathematics Institute
John Polking

2000 *Mathematics Subject Classification.* Primary 34–01, 65–01, 70–01.

For additional information and updates on this book, visit
www.ams.org/bookpages/stml-51

Library of Congress Cataloging-in-Publication Data
Palais, Richard S.
  Differential equations, mechanics, and computation / Richard S. Palais, Robert A. Palais.
       p. cm. — (Student mathematical library ; 51. IAS/Park City mathematical subseries)
  Includes bibliographical references and index.
  ISBN 978-0-8218-2138-1 (alk. paper)
   1. Evolution equations. 2. Differential equations, Linear—Numerical solutions. 3. Differential equations—Numerical solutions. 4. Mechanics. I. Palais, Robert Andrew. II. Title.

QA371.P34   2009
518'.6—dc22   2009011294

Copying and reprinting. Individual readers of this publication, and nonprofit libraries acting for them, are permitted to make fair use of the material, such as to copy a chapter for use in teaching or research. Permission is granted to quote brief passages from this publication in reviews, provided the customary acknowledgment of the source is given.

Republication, systematic copying, or multiple reproduction of any material in this publication is permitted only under license from the American Mathematical Society. Requests for such permission should be addressed to the Acquisitions Department, American Mathematical Society, 201 Charles Street, Providence, Rhode Island 02904-2294 USA. Requests can also be made by e-mail to reprint-permission@ams.org.

© 2009 by the American Mathematical Society. All rights reserved.
    The American Mathematical Society retains all rights
    except those granted to the United States Government.
              Printed in the United States of America.

∞ The paper used in this book is acid-free and falls within the guidelines
            established to ensure permanence and durability.
        Visit the AMS home page at http://www.ams.org/

             10 9 8 7 6 5 4 3 2 1   14 13 12 11 10 09

We dedicate this book to our families, teachers, and students and to the memories of George Mackey and Andrew Gleason, the mentors of both authors, for teaching us how to learn and how to teach.

Contents

IAS/Park City Mathematics Institute	ix
Preface	xi
Acknowledgments	xiii
Introduction	1
Chapter 1. Differential Equations and Their Solutions	5
1.1. First-Order ODE: Existence and Uniqueness	5
1.2. Euler's Method	16
1.3. Stationary Points and Closed Orbits	19
1.4. Continuity with Respect to Initial Conditions	22
1.5. Chaos—Or a Butterfly Spoils Laplace's Dream	25
1.6. Analytic ODE and Their Solutions	31
1.7. Invariance Properties of Flows	33
Chapter 2. Linear Differential Equations	37
2.1. First-Order Linear ODE	37
2.2. Nonautonomous First-Order Linear ODE	48
2.3. Coupled and Uncoupled Harmonic Operators	50
2.4. Inhomogeneous Linear Differential Equations	52
2.5. Asymptotic Stability of Nonlinear ODE	53
2.6. Forced Harmonic Oscillators	55
2.7. Exponential Growth and Ecological Models	56

Chapter 3. Second-Order ODE and the Calculus of Variations — 63

- 3.1. Tangent Vectors and the Tangent Bundle — 63
- 3.2. Second-Order Differential Equations — 66
- 3.3. The Calculus of Variations — 68
- 3.4. The Euler-Lagrange Equations — 70
- 3.5. Conservation Laws for Euler-Lagrange Equations — 73
- 3.6. Two Classic Examples — 75
- 3.7. Derivation of the Euler-Lagrange Equations — 78
- 3.8. More General Variations — 80
- 3.9. The Theorem of E. Noether — 81
- 3.10. Lagrangians Defining the Same Functionals — 82
- 3.11. Riemannian Metrics and Geodesics — 85
- 3.12. A Preview of Classical Mechanics — 86

Chapter 4. Newtonian Mechanics — 91

- 4.1. Introduction — 91
- 4.2. Newton's Laws of Motion — 92
- 4.3. Newtonian Kinematics — 96
- 4.4. Classical Mechanics as a Physical Theory — 99
- 4.5. Potential Functions and Conservation of Energy — 106
- 4.6. One-Dimensional Systems — 111
- 4.7. The Third Law and Conservation Principles — 118
- 4.8. Synthesis and Analysis of Newtonian Systems — 122
- 4.9. Linear Systems and Harmonic Oscillators — 124
- 4.10. Small Oscillations about Equilibrium — 126

Chapter 5. Numerical Methods — 133

- 5.1. Introduction — 133
- 5.2. Fundamental Examples and Their Behavior — 144

Contents

5.3. Summary of Method Behavior on Model Problems	169
5.4. Paired Methods: Error, Step-Size, Order Control	177
5.5. Behavior of Example Methods on a Model 2×2 System	180
5.6. Stiff Systems and the Method of Lines	187
5.7. Convergence Analysis: Euler's Method	213
Appendix A. Linear Algebra and Analysis	225
A.1. Metric and Normed Spaces	225
A.2. Inner-Product Spaces	227
Appendix B. The Magic of Iteration	233
B.1. The Banach Contraction Principle	233
B.2. Newton's Method	238
B.3. The Inverse Function Theorem	240
B.4. The Existence and Uniqueness Theorem for ODE	241
Appendix C. Vector Fields as Differential Operators	243
Appendix D. Coordinate Systems and Canonical Forms	247
D.1. Local Coordinates	247
D.2. Some Canonical Forms	250
Appendix E. Parametrized Curves and Arclength	255
Appendix F. Smoothness with Respect to Initial Conditions	257
Appendix G. Canonical Form for Linear Operators	259
G.1. The Spectral Theorem	259
Appendix H. Runge-Kutta Methods	263

Appendix I. Multistep Methods	281
Appendix J. Iterative Interpolation and Its Error	303
Bibliography	307
Index	311

IAS/Park City Mathematics Institute

The IAS/Park City Mathematics Institute (PCMI) was founded in 1991 as part of the "Regional Geometry Institute" initiative of the National Science Foundation. In mid-1993 the program found an institutional home at the Institute for Advanced Study (IAS) in Princeton, New Jersey. The PCMI continues to hold summer programs in Park City, Utah.

The IAS/Park City Mathematics Institute encourages both research and education in mathematics and fosters interaction between the two. The three-week summer institute offers programs for researchers and postdoctoral scholars, graduate students, undergraduate students, high school teachers, mathematics education researchers, and undergraduate faculty. One of PCMI's main goals is to make all of the participants aware of the total spectrum of activities that occur in mathematics education and research: we wish to involve professional mathematicians in education and to bring modern concepts in mathematics to the attention of educators. To that end the summer institute features general sessions designed to encourage interaction among the various groups. In-year activities at sites around the country form an integral part of the High School Teacher Program.

Each summer a different topic is chosen as the focus of the Research Program and Graduate Summer School. Activities in the Undergraduate Program deal with this topic as well. Lecture notes from the Graduate Summer School are published each year in the IAS/Park City Mathematics Series. Course materials from the Undergraduate Program, such as the current volume, are now being published as part of the IAS/Park City Mathematical Subseries in the Student Mathematical Library. We are happy to make available more of the excellent resources which have been developed as part of the PCMI.

John Polking, Series Editor

April 13, 2009

Preface

> "I have attempted to deliver [these lectures] in a spirit that should be recommended to all students embarking on the writing of their PH.D. thesis: Imagine that you are explaining your ideas to your former smart, but ignorant, self, at the beginning of your studies!"
>
> —Richard P. Feynman

This quote of Dick Feynman expresses well both our goal in writing this book and the style in which we have tried to present the material—this is how we would have liked someone to explain things to us when we were first learning about differential equations and mechanics.

Richard Palais and Bob Palais

Acknowledgments

Our thanks go to the many friends and colleagues who were kind enough to read early versions and make valuable comments and suggestions that have improved the final manuscript. We would like to express our special appreciation to Mike Spivak for showing us an early version of his soon to appear "Physics for Mathematicians: Mechanics"; to Michael Nauenberg, for his helpful criticism concerning the early history of Mechanics; to Nelson Beebe and David Eppstein for their help with TeX macros; and to Peter Alfeld, Grady Wright, William Kahan, Gil Strang, Andy Majda, Moe Hirsch, and Carl Wittwer for their helpful contributions.

We would also like to express our deep gratitude to the AMS staff: Sergei Gelfand, Publisher; Stephen Moye, Publications Support Specialist, TeX expert for this project; Mary Medeiros, Electronic Publishing Technician, who formatted the files; and Arlene O'Sean, the production editor. Their patience, flexibility, and expertise were essential to the completion of the project with the highest possible quality.

Introduction

This book is about differential equations—a **very** big subject! It is so extensive, in fact, that we could not hope to cover it completely even in a book many times this size. So we will have to be selective. In the first place, we will restrict our attention almost entirely to *equations of evolution*. That is to say, we will be considering quantities q that depend on a "time" variable t, and we will be considering mainly *initial value problems*. This is the problem of predicting the value of such a quantity q at a time t_1 from its value at some (usually earlier) "initial" time t_0, assuming that we know the "law of evolution" of q. The latter will always be a "differential equation" that tells us how to compute the rate at which q is changing from a knowledge of its current value. While we will concentrate mainly on the easier case of an ordinary differential equation (ODE), where the quantity q depends **only** on the time, we will on occasion consider the partial differential equation (PDE) case, where q depends also on other "spatial variables" x as well as the time t and where the partial derivatives of q with respect to these spatial variables can enter into the law determining its rate of change with respect to time.

Our principal goal will be to help you develop a good intuition for equations of evolution and how they can be used to model a large variety of time-dependent processes—in particular those that arise in the study of classical mechanics. To this end we will stress various metaphors that we hope will encourage you to get started thinking creatively about differential equations and their solutions.

But wait! Just who is this "you" we are addressing? Every textbook author has in mind at least a rough image of some prototypical student for whom he is writing, and since the assumed background and abilities of this model student are sure to have an important influence on how the book gets written, it is only fair that we give you some idea of our own preconceptions about you.

We are assuming that, at a mimimum, the usual reader of this book will have completed the equivalent of two years of undergraduate mathematics in a U.S. college or university and, in particular, will have had a solid introduction to linear algebra and to multi-variable (aka "advanced") calculus. But in all honesty, we have in mind some other hoped-for qualities in our reader, principally that he or she is accustomed to and enjoys seeing mathematics presented conceptually and not as a collection of cookbook methods for solving standard exercises. And finally we hope our readers enjoy working out mathematical details on their own. We will give frequent exercises (usually with liberal hints) that ask the student to fill in some details of a proof or derive a corollary.

A related question is how we expect this book to be used. We would of course be delighted to hear that it has been adopted as the assigned text for many junior and senior level courses in differential equations (and perhaps not surprisingly we would be happy using it ourselves in teaching such a course). But we realize that the book we have written diverges in many ways from the current "standard model" of an ODE text, so it is our real hope and expectation that many students, particularly those of the sort described above, will find it a challenging but helpful source from which to learn about ODEs, either on their own or as a supplement to a more standard assigned text while taking an ODE course.

We should mention here—and explain—a somewhat unusual feature of our exposition. The book consists of two parts that we will refer to as "text" and "appendices". The text is made up of five chapters that together contain about two-thirds of the material, while the appendices consist of ten shorter mini-chapters. Our aim was to make the text relatively easy reading by relegating the more difficult and technical material to the appendices. A reader should be able to get a quick overview of the subject matter of one or more chapters by just reading the text and ignoring the references to material in the

Introduction

appendices. Later, when ready to go deeper or to check an omitted proof, a reading of the relevant appendices should satisfy the reader's hunger for more detail.

Finally we would like to discuss "visual aids"—that is, the various kinds of diagrams and pictures that make it easier for a student to internalize a complicated mathematical concept upon meeting it for the first time. Both of the authors have been very actively involved with the development of software tools for creating such mathematical visualizations and with investigating techniques for using them to enhance the teaching and learning of mathematics, and paradoxically that has made it difficult for us to choose appropriate figures for our text. Indeed, recent advances in technology, in particular the explosive development of the Internet and in particular of the World Wide Web, have not only made it easy to provide visual material online, but moreover the expressiveness possible using the interactive and animated multimedia tools available in the virtual world of the Internet far surpasses that of the classic static diagrams that have traditionally been used in printed texts. As a result we at first considered omitting diagrams entirely from this text, but in the end we decided on a dual approach. We have used traditional diagrams in the text where we felt that they would be useful, and in addition we have placed a much richer assortment of visual material online to accompany the text. Our publisher, the American Mathematical Society, has agreed to set aside a permanent area on its own website to be devoted to this book, and throughout the text you will find references to this area that we will refer to as the "Web Companion".[1] Here, organized by chapter and section, you will find visualizations that go far beyond anything we could hope to put in the pages of a book—static diagrams, certainly, but in addition Flash animations, Java applets, QuickTime movies, Mathematica, Matlab, Maple Notebooks, other interactive learning aids, and also links to other websites that contain material we believe will help and speed your understanding. And not only does this approach allow us to make much more sophisticated visualizations available, but it also will permit us to add new and improved material as it becomes available.

---

[1] Its URL is http://www.ams.org/bookpages/stml-51.

Chapter 1

Differential Equations and Their Solutions

1.1. First-Order ODE: Existence and Uniqueness

What does the following sentence mean, and what image should it cause you to form in your mind?

> Let $V : \mathbf{R}^n \times \mathbf{R} \to \mathbf{R}^n$ be a time-dependent vector field, and let $x(t)$ be a solution of the differential equation $\frac{dx}{dt} = V(x, t)$ satisfying the initial condition $x(t_0) = x_0$.

Let us consider a seemingly very different question. Suppose you know the wind velocity at every point of space and at all instants of time. A puff of smoke drifts by, and at a certain moment you note the precise location of a particular smoke particle. Can you then predict where that particle will be at all future times?

We will see that when this somewhat vague question is translated appropriately into precise mathematical concepts, it leads to the above "differential equation", and that the answer to our prediction question translates to the central existence and uniqueness result in the theory of differential equations. (The answer, by the way, turns out to be a qualified "yes", with several important caveats.)

We interpret "space" to mean the n-dimensional real number space $\mathbf{R}^n$, so a "point of space" is just an n-tuple $x = (x_1, \ldots, x_n)$ of real numbers. If you feel more comfortable thinking $n = 3$, that's fine for the moment, but mathematically it makes no difference, and as we shall see later, even when working with real-world, three-dimensional problems, it is often important to make use of higher-dimensional spaces.

On the other hand, an "instant of time" will always be represented by a single real number t. (There are mathematical situations that do require multi-dimensional time, but we shall not meet them here.) Thus, knowing the wind velocity at every point of space and at all instants of time means that we have a function V that associates to each (x,t) in $\mathbf{R}^n \times \mathbf{R}$ a vector $V(x,t)$ in $\mathbf{R}^n$, the wind velocity at x at time t. We will denote the n components of $V(x,t)$ by $V_1(x,t), \ldots, V_n(x,t)$. (We will always assume that V is at least continuous and usually that it is even continuously differentiable.)

How should we model the path taken by a smoke particle? An ideal smoke particle is characterized by the fact that it "goes with the flow", i.e., it is carried along by the wind. That means that if $x(t) = (x_1(t), \ldots, x_n(t))$ is its location at a time t, then its velocity at time t will be the wind velocity at that point and time, namely $V(x(t),t)$. But the velocity of the particle at time t is $x'(t) = (x_1'(t), \ldots, x_n'(t))$, where primes denote differentiation with respect to t, i.e., $x' = \frac{dx}{dt} = (\frac{dx_1}{dt}, \ldots, \frac{dx_n}{dt})$.

So the path of a smoke particle will be a differentiable curve $x(t)$ in $\mathbf{R}^n$ that satisfies the differential equation $x'(t) = V(x(t),t)$, or $\frac{dx}{dt} = V(x,t)$. If we write this in components, it reads $\frac{dx_i}{dt} = V_i(x_1(t), \ldots, x_n(t), t)$, for $i = 1, \ldots, n$, and for this reason it is often called a system of differential equations. Finally, if at a time t_0 we observe that the smoke particle is at the point x_0 in $\mathbf{R}^n$, then the "initial condition" $x(t_0) = x_0$ is also satisfied.

The page devoted to Chapter 1 in the Web Companion contains a QuickTime movie showing the wind field of a time-dependent two-dimensional system and the path traced out by a "smoke particle". Figure 1.1 shows the direction field and a few such solution curves for an interesting and important one-dimensional ODE called the logistic equation.

1.1. First-Order ODE: Existence and Uniqueness

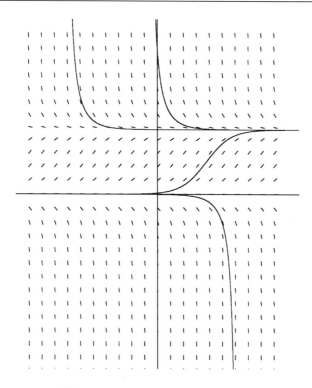

Figure 1.1. The logistic equation.

For the logistic equation, the velocity field is given by $V(x,t) = cx(A-x)$. The vertical x-axis represents the size of some quantity, and the horizontal axis is the time, t. This equation models the growth of x in the presence of environmental constraints. The constant A is called the carrying capacity, and $c(A-x)$ is the "growth rate". Note that the growth rate approaches zero as x approaches the carrying capacity. This equation is discussed in more detail in Section 2.7 on ecological models.

The combination of a differential equation, $\frac{dx}{dt} = V(x,t)$, and an initial condition, $x(t_0) = x_0$, is called an "initial value problem" (IVP), so the above informal prediction question for smoke particles can now be translated into a precise mathematical question: "What can we say about the existence and uniqueness of solutions to such initial value problems?"

We will discuss this central question in detail below, along with important related questions such as how solutions of an IVP change as we vary the initial condition and the vector field. In order not to over-burden the exposition, we will leave many details of proofs to be worked out by the reader in exercises (with liberal hints). Fully detailed proofs can be found in the appendices and various references.

First let us make precise the definition of a solution of the above initial value problem: it is a differentiable map x of some open interval I containing t_0 into $\mathbf{R}^n$ such that $x(t_0) = x_0$ and $x'(t) = V(x(t), t)$ for all t in I.

We first consider uniqueness. The vector field $V : \mathbf{R}^n \times \mathbf{R} \to \mathbf{R}^n$ is called *continuously differentiable* (or C^1) if all of its components $V_i(x_1, \ldots, x_n, t)$ have continuous first partial derivatives with respect to $x_1, \ldots, x_n, t$, and more generally V is called C^k if all partial derivatives of order k or less of its components exist and are continuous.

1.1.1. Uniqueness Theorem. *Let $V : \mathbf{R}^n \times \mathbf{R} \to \mathbf{R}^n$ be a C^1 time-dependent vector field on $\mathbf{R}^n$ and let $x^1(t)$ and $x^2(t)$ be two solutions of the differential equation $\frac{dx}{dt} = V(x,t)$ defined on the same interval $I = (a,b)$ and satisfying the same initial condition, i.e., $x^1(t_0) = x^2(t_0)$ for some $t_0 \in I$. Then in fact $x^1(t) = x^2(t)$ for all $t \in I$.*

▷ **Exercise 1–1.** Show that continuity of V is **not** sufficient to guarantee uniqueness for an IVP. Hint: The classic example (with $n = 1$) is the initial value problem $\frac{dx}{dt} = \sqrt{x}$ and $x(0) = 0$. Show that for each $T > 0$, we get a distinct solution $x_T(t)$ of this IVP by defining $x_T(t) = 0$ for $t < T$ and $x_T(t) = \frac{1}{4}(t-T)^2$ for $t \geq T$.

But what about existence?

1.1.2. Local Existence Theorem. *Let $V : \mathbf{R}^n \times \mathbf{R} \to \mathbf{R}^n$ be a C^1 time-dependent vector field on $\mathbf{R}^n$. Given $p_0 \in \mathbf{R}^n$ and $t_0 \in \mathbf{R}$, there is a neighborhood O of p_0 and an $\epsilon > 0$ such that for every $p*

1.1. First-Order ODE: Existence and Uniqueness

in O there is a solution $x_p : (t_0 - \epsilon, t_0 + \epsilon) \to \mathbf{R}^n$ of the differential equation $\frac{dx}{dt} = V(x, t)$ satisfying the initial condition $x_p(t_0) = p$.

The proofs of existence and uniqueness have been greatly simplified over time, but understanding the details still requires nontrivial effort. Here we will sketch some of the most important ideas and constructs that go into the complete proof, but in order not to interrupt the flow of our exposition, we will defer the details to Appendix B. But even if you choose not to study these proofs now, we urge you to do so at some later time. We think you will find that these proofs are so elegant, and the ideas and constructions that enter into them are of such interest in their own right, that studying them is well worth the time and effort it requires.

We begin with a simple but very important reformulation of the ODE initial value problem $x'(s) = V(x(s), s)$ and $x(t_0) = x^0$. Namely, if we integrate both sides of the first of these equations from t_0 to t, we find that $x(t) = x^0 + \int_{t_0}^{t} V(x(s), s) \, ds$, and we refer to this equation as the integral form of the initial value problem. Note that by substituting $t = t_0$ in the integral form and by differentiating it, we get back the two original equations, so the integral form and the ODE form are equivalent. This suggests that we make the following definition.

1.1.3. Definition. Associated to each time-dependent vector field V on $\mathbf{R}^n$ and $x^0 \in \mathbf{R}^n$, we define a mapping F^{V,x^0} that transforms a continuous function $x : I \to \mathbf{R}^n$ (where I is any interval containing t_0) to another such function $F^{V,x^0}(x) : I \to \mathbf{R}^n$ defined by $F^{V,x^0}(x)(t) = x^0 + \int_{t_0}^{t} V(x(s), s) \, ds$.

▷ **Exercise 1–2.** Show that any y of the form $F^{V,x^0}(x)$ satisfies the initial condition $y(t_0) = x^0$, and moreover y is continuously differentiable with derivative $y'(t) = V(x(t), t)$.

Recall that if f is any mapping, then a point in the domain of f such that $f(p) = p$ is called a *fixed point* of f. Thus we can rephrase the integral form of the initial value problem as follows:

1.1.4. Proposition. *A continuous map $x : I \to \mathbf{R}^n$ is a solution of the initial value problem $x'(t) = V(x(t), t)$, $x(t_0) = x^0$ if and only if x is a fixed point of F^{V,x^0}.*

Now if you have had some experience with fixed-point theorems, that should make your ears perk up a little. Not only are there some very general and powerful results for proving existence and uniqueness of fixed points of maps, but even better, there are nice algorithms for finding fixed points. One such algorithm is the so-called Method of Successive Approximations. (If you are familiar with Newton's Method for finding roots of equations, you will recognize that as a special case of successive approximations.) If we have a set X and a self-mapping $f : X \to X$, then to apply successive approximations, choose some "initial approximation" x_0 in X and then inductively define a sequence $x_{n+1} = f(x_n)$ of "successive approximations".

▷ **Exercise 1–3.** Suppose that X is a metric space, f is continuous, and that the sequence x_n of "successive approximations" converges to a limit p. Show that p is a fixed point of f.

But is there really any hope that we can use successive approximations to find solutions of ODE initial value problems? Let us try a very simple example. Consider the (time-independent) vector field V on $\mathbf{R}^n$ defined by $V(x,t) = x$. It is easy to check that the unique solution with $x(0) = x^0$ is given by $x(t) = e^t x^0$. Let's try using successive approximations to find a fixed point of F^{V,x^0}. For our initial approximation we choose the constant function $x_0(t) = x^0$, and following the general successive approximation prescription, we define x_n inductively by $x_{n+1} = F^{V,x^0}(x_n)$, i.e., $x_{n+1}(t) = x^0 + \int_0^t x_n(s)\,ds$.

▷ **Exercise 1–4.** Show by induction that $x_n(t) = P_n(t)x^0$, where $P_n(t)$ is the polynomial of degree n obtained by truncating the power series for e^t (i.e., $\sum_{j=0}^n \frac{1}{j!}t^j$).

That is certainly a hopeful sign, and while one swallow may not make a spring, it should give us hope that a careful analysis of successive approximations might lead to a proof of the existence and uniqueness theorems for an arbitrary vector field V. This is in fact the case, but

1.1. First-Order ODE: Existence and Uniqueness   11

we will not give further details here. Instead we refer to Appendix B where you will find a complete proof.

1.1.5. Remark. We give a minor technical point. The argument in Appendix B only gives a local uniqueness theorem. That is, it shows that if $x^1 : (a, b) \to \mathbf{R}^n$ and $x^2 : (a, b) \to \mathbf{R}^n$ are two solutions of the same ODE, then if x^1 and x^2 agree at a point, then they also agree in a neighborhood of that point, so that the set of points in (a, b) where they agree is open. But since solutions are by definition continuous, the set of points where x^1 and x^2 agree is also a closed subset of (a, b), and since intervals are connected, it then follows that x^1 and x^2 agree on all of (a, b).

1.1.6. Remark. The existence and uniqueness theorems tell us that for a given initial condition x^0 we can solve our initial value problem (uniquely) for a short time interval. The next question we will take up is for just how long we can "follow a smoke particle". One important thing to notice is the uniformity of the ϵ in the existence theorem—not only do we have a solution for each initial condition, but moreover given any p_0 in $\mathbf{R}^n$, we can find a **fixed** interval $I = (t_0 - \epsilon, t_0 + \epsilon)$ such that a solution with initial condition p exists on the whole interval I for all initial conditions sufficiently close to p_0. Still, this may be less than what you had hoped and expected. You may have thought that for each initial condition p in $\mathbf{R}^n$ we should have a solution $x_p : \mathbf{R} \to \mathbf{R}^n$ of the differential equation with $x_p(t_0) = p$. But such a global existence theorem is too much to expect. For example, taking $n = 1$ again, consider the differential equation $\frac{dx}{dt} = x^2$ with the initial condition $x(0) = x_0$. An easy calculation shows that the unique solution is $x(t) = \frac{x_0}{1 - x_0 t}$. Note that, for each initial condition x_0, this solution "blows up" at time $T = \frac{1}{x_0}$, and by the Uniqueness Theorem, no solution can exist for a time greater than T.

But, you say, a particle of smoke will never go off to infinity in a finite amount of time! Perhaps the smoke metaphor isn't so good after all. The answer is that a real, physical wind field has bounded velocity, and it isn't hard to show that in this case we do indeed have global existence. You will even prove something a lot stronger in a later exercise.

What **can** be said is that for each initial condition, p, there is a unique "maximal" solution of the differential equation with that initial condition. But before discussing this, we are going to make a simplification and restrict our attention to time-**independent** vector fields (which we shall simply call vector fields). That may sound like a tremendous loss of generality, but in fact it is no loss of generality at all!

▷ **Exercise 1–5.** Let $V(x,t) = (V_1(x,t), \ldots, V_n(x,t))$ be a time-dependent vector field in $\mathbf{R}^n$, and define an associated time independent vector field $\tilde{V}$ in $\mathbf{R}^{n+1}$ by $\tilde{V}(y) = (V_1(y), \ldots, V_n(y), 1)$. Show that $y(t) = (x(t), f(t))$ is a solution of the differential equation $\frac{dy}{dt} = \tilde{V}(y)$ if and only if $f(t) = t + c$ and $x(t)$ is a solution of $\frac{dx}{dt} = V(x, t+c)$. Deduce that if $y(t) = (x(t), f(t))$ solves the IVP $\frac{dy}{dt} = \tilde{V}(y)$, $y(t_0) = (x_0, t_0)$, then $x(t)$ is a solution of the IVP $\frac{dx}{dt} = V(x,t)$, $x(t_0) = x_0$.

This may look like a swindle. We don't seem to have done much besides changing the name of the original time variable t to x_{n+1} and considering it a space variable; that is, we switched to space-time notation. But the real change is in making the velocity an $(n+1)$-vector too and setting the last component identically equal to one. In any case this is a true reduction of the time-dependent case to the time-independent case, and as we shall see, that is quite important, since time-independent differential equations have special properties not shared with time-dependent equations that can be used to simplify their study. Time-independent differential equations are usually referred to as *autonomous*, and time-dependent ones as nonautonomous. Here is one of the special properties of autonomous systems.

1.1.7. Proposition. *If $x : (a,b) \to \mathbf{R}^n$ is any solution of the autonomous differentiable equation $\frac{dx}{dt} = V(x)$ and $t_0 \in \mathbf{R}$, then $y : (a+t_0, b+t_0) \to \mathbf{R}^n$ defined by $y(t) = x(t-t_0)$ is also a solution of the same equation.*

1.1. First-Order ODE: Existence and Uniqueness 13

▷ **Exercise 1–6.** Prove the above proposition.

Consequently, when considering the IVP for an autonomous differentiable equation, we can assume that $t_0 = 0$. For if $x(t)$ is a solution with $x(0) = p$, then $x(t-t_0)$ will be a solution with $x(t_0) = p$.

1.1.8. Remark. There is another trick that allows us to reduce the study of higher-order differential equations to the case of first-order equations. We will consider this in detail later, but here is a short preview. Consider the second-order differential equation: $\frac{d^2x}{dt^2} = f(x, \frac{dx}{dt}, t)$. Introduce a new variable v (the velocity) and consider the following related system of first-order equations: $\frac{dx}{dt} = v$ and $\frac{dv}{dt} = f(x, v, t)$. It is pretty obvious there is a close relation between curves $x(t)$ satisfying $x''(t) = f(x(t), x'(t), t)$ and pairs of curves $x(t)$, $v(t)$ satisfying $x'(t) = v(t)$ and $v'(t) = f(x(t), v(t), t)$.

▷ **Exercise 1–7.** Define the notion of an initial value problem for the above second-order differential equation, and write a careful statement of the relation between solutions of this initial value problem and the initial value problem for the related system of first-order differential equations.

We will now look more closely at the uniqueness question for solutions of an initial value problem. The answer is summed up succinctly in the following result.

1.1.9. Maximal Solution Theorem. Let $\frac{dx}{dt} = V(x)$ be an autonomous differential equation in $\mathbf{R}^n$ and p any point of $\mathbf{R}^n$. Among all solutions $x(t)$ of the equation that satisfy the initial condition $x(0) = p$, there is a maximum one, σ_p, in the sense that any solution of this IVP is the restriction of σ_p to some interval containing zero.

▷ **Exercise 1–8.** If you know about connectedness, you should be able to prove this very easily. First, using the local uniqueness theorem, show that any two solutions agree on their overlap, and then define σ_p to be the union of all solutions.

Henceforth whenever we are considering some autonomous differential equation, σ_p will denote this maximal solution curve with initial condition p. The interval on which σ_p is defined will be denoted by $(\alpha(p), \omega(p))$, where of course $\alpha(p)$ is either $-\infty$ or a negative real number and $\omega(p)$ is either ∞ or a positive real number.

As we have seen, a maximal solution σ_p need **not** be defined on all of **R**, and it is important to know just how the solution "blows up" as t approaches a finite endpoint of its interval of definition. A *priori* it might seem that the solution could remain in some bounded region, but it is an important fact that this is impossible—if $\omega(p)$ is finite, then the reason the solution cannot be continued past $\omega(p)$ is simply that it escapes to infinity as t approaches $\omega(p)$.

1.1.10. No Bounded Escape Theorem. *If $\omega(p) < \infty$, then*
$$\lim_{t \to \omega(p)} \|\sigma_p(t)\| = \infty,$$
and similarly, if $\alpha(p) > -\infty$, then
$$\lim_{t \to \alpha(p)} \|\sigma_p(t)\| = \infty.$$

▷ **Exercise 1–9.** Prove the No Bounded Escape Theorem. (Hint: If $\lim_{t \to \omega(p)} \|\sigma(p)\| \neq \infty$, then by Bolzano-Weierstrass there would be a sequence t_k converging to $\omega(p)$ from below, such that $\sigma_p(t_k) \to q$. Then use the local existence theorem around q to show that you could extend the solution beyond $\omega(p)$. Here is where we get to use the fact that there is a neighborhood O of q such that a solution exists with any initial condition q' in O and defined **on the whole interval** $(-\epsilon, \epsilon)$. For k sufficiently large, we will have both $\sigma_p(t_k)$ in O and $t_k > \omega - \epsilon$, which quickly leads to a contradiction.)

Here is another special property of autonomous systems.

▷ **Exercise 1–10.** Show that the images of the σ_p partition $\mathbf{R}^n$ into disjoint smooth curves (the "streamlines" of smoke particles). These curves are referred to as the *orbits* of the ODE. (Hint: If $x(t)$ and $\xi(t)$ are two solutions of the same autonomous ODE and if $x(t_0) = \xi(t_1)$, then show that $x(t_0 + s) = \xi(t_1 + s)$.)

1.1. First-Order ODE: Existence and Uniqueness

1.1.11. Definition. A C^1 vector field $V : \mathbf{R}^n \to \mathbf{R}^n$ (and also the autonomous differential equation $\frac{dx}{dt} = V(x)$) is called *complete* if $\alpha(p) = -\infty$ and $\omega(p) = \infty$ for all p in $\mathbf{R}^n$. In this case, for each $t \in \mathbf{R}$ we define a map $\phi_t : \mathbf{R}^n \to \mathbf{R}^n$ by $\phi_t(p) = \sigma_p(t)$. The mapping $t \mapsto \phi_t$ is called the *flow* generated by the differential equation $\frac{dx}{dt} = V(x)$.

1.1.12. Remark. Using our smoke particle metaphor, the meaning of ϕ_t can be explained as follows: if a puff of smoke occupies a region U at a given time, then t units of time later it will occupy the region $\phi_t(U)$. Note that ϕ_0 is clearly the identity mapping of $\mathbf{R}^n$.

▷ **Exercise 1–11.** Show that the ϕ_t satisfy $\phi_{t_1+t_2} = \phi_{t_1}\phi_{t_2}$, so that in particular $\phi_{-t} = \phi_t^{-1}$. In other words, the flow generated by a complete, autonomous vector field is a homomorphism of the additive group of real numbers into the group of bijective self-mappings of $\mathbf{R}^n$.

In the next section we will see that $(t, p) \mapsto \phi_t(p)$ is jointly continuous, so that the ϕ_t are homeomorphisms of $\mathbf{R}^n$. Later (in Appendix F) we will also see that if the vector field V is C^r, then $(t, p) \mapsto \phi_t(p)$ is also C^r, so that the flow generated by a complete, autonomous, C^r differential equation $\frac{dx}{dt} = V(x)$ is a homomorphism of $\mathbf{R}$ into the group of C^r diffeomorphisms of $\mathbf{R}^n$. The branch of mathematics that studies the properties of flows is called *dynamical systems theory*.

• **Example 1–1. Constant Vector Fields.** The simplest examples of autonomous vector fields in $\mathbf{R}^n$ are the constant vector fields $V(x) = v$, where v is some fixed vector in $\mathbf{R}^n$. The maximal solution curve with initial condition p of $\frac{dx}{dt} = v$ is clearly the linearly parametrized straight line $\sigma_p : \mathbf{R} \to \mathbf{R}^n$ given by $\sigma_p(t) = p + tv$, and it follows that these vector fields are complete. The corresponding flow ϕ_t is given by $\phi_t(p) = p + tv$, so for obvious reasons these are called constant velocity flows. In words, ϕ_t is translation by the vector tv, and indeed these flows are precisely the one-parameter subgroups of the group of translations of $\mathbf{R}^n$.

• **Example 1–2. Exponential Growth.** An important complete vector field in $\mathbf{R}$ is the linear map $V(x) = kx$. The maximal solution curves of $\frac{dx}{dt} = kx$ are again easy to write down explicitly, namely

$\sigma_p(t) = e^{kt}p$; i.e., in this case the flow map ϕ_t is just multiplication by e^{kt}.

• **Example 1–3. Harmonic Oscillator.** If we start from the Harmonic Oscillator Equation, $\frac{d^2x}{dt^2} = -x$, and use the trick above to rewrite this second-order equation as a first-order system, we end up with the linear system in $\mathbf{R}^2$: $\frac{dx}{dt} = -y$, $\frac{dy}{dt} = x$. In this case the maximal solution curve $\sigma_{(x_0,y_0)}(t)$ can again be given explicitly, namely $\sigma_{(x_0,y_0)}(t) = (x_0\cos(t) - y_0\sin(t), x_0\sin(t) + y_0\cos(t))$, so that now ϕ_t is rotation in the plane through an angle t. It is interesting to observe that this can be considered a special case of (a slightly generalized form of) the preceding example. Namely, if we identify $\mathbf{R}^2$ with the complex plane $\mathbf{C}$ in the standard way (i.e., $a+ib := (a,b)$) and write $z = (x,y) = x + iy$, $z_0 = (x_0, y_0) = x_0 + iy_0$, then since $iz = i(x+iy) = -y + ix = (-y, x)$, we can rewrite the above first-order system as $\frac{dz}{dt} = iz$, which has the solution $z(t) = e^{it}z_0$. Of course, multiplication by e^{it} is just rotation through an angle t.

It is very useful to have conditions on a vector field V that will guarantee its completeness.

▷ **Exercise 1–12.** Show that $\|\sigma_p(t) - p\| \leq \int_0^t \|V(\sigma_p(t))\|\, dt$. Use this and the No Bounded Escape Theorem to show that $\frac{dx}{dt} = V(x)$ is complete provided that V is bounded (i.e., $\sup_{x \in \mathbf{R}^n} \|V(x)\| < \infty$).

▷ **Exercise 1–13.** A vector field V may be complete even if it is not bounded, provided that it doesn't "grow too fast". Let $B(r) = \sup_{\|x\|<r} \|V(x)\|$. Show that if $\int_1^\infty \frac{dr}{B(r)} = \infty$, then V is complete. Hint: How long does it take $\sigma_p(t)$ to get outside a ball of radius R?

▷ **Exercise 1–14.** If a vector field is not complete, then given any positive ϵ, there exist points p where either $\alpha(p) > -\epsilon$ or $\omega(p) < \epsilon$.

1.2. Euler's Method

Only a few rather special initial value problems can be solved in closed form using standard elementary functions. For the general case it is

1.2. Euler's Method

necessary to fall back on constructing an approximate solution numerically with the aid of a computer. But what algorithm should we use to program the computer? A natural first guess is successive approximations. But while that is a powerful theoretical tool for studying the general properties of initial value problems (and in particular for proving existence and uniqueness), it does not lead to an efficient algorithm for constructing numerical solutions.

In fact there is no one simple answer to the question of what numerical algorithm to use for solving ODEs, for there is no single method that is "best" in all situations. While there are integration routines (such as the popular fourth-order Runge-Kutta integration) that are fast and accurate when used with many of the equations one meets, there are many situations that require a more sophisticated approach. Indeed, this is still an active area of research, and there are literally dozens of books on the subject. Later, in the chapter on numerical methods, we will introduce you to many of the subtleties of this topic, but here we only want to give you a quick first impression by describing one of the oldest numerical approaches to solving an initial value problem, the so-called "Euler Method". While rarely an optimal choice, it is intuitive, simple, and effective for some purposes. It is also the prototype for the design and analysis of more sophisticated algorithms. This makes it an excellent place to become familiar with the basic concepts that enter into the numerical integration of ODE.

In what follows we will suppose that $\mathbf{f}(t, \mathbf{y})$ is a C^1 time-dependent vector field on $\mathbf{R}^d$, t_o in $\mathbf{R}$ and $\mathbf{y}_o$ in $\mathbf{R}^d$. We will denote by $\sigma(\mathbf{f}, \mathbf{y}_o, t_o, t)$ the solution operator taking this data to the values $\mathbf{y}(t)$ of the maximal solution of the associated initial value problem. By definition, $\mathbf{y}(t)$ is the function defined on a maximal interval $I = [t_o, t_o+T_*)$, with $0 < T_* \leq \infty$, satisfying the differential equation $\frac{d\mathbf{y}}{dt} = \mathbf{f}(t, \mathbf{y})$ and the initial condition $\mathbf{y}(t_o) = \mathbf{y}_o$. The goal in the numerical integration of ODE is to devise effective methods for approximating such a solution $\mathbf{y}(t)$ on an interval $I = [t_o, t_o + T]$ for $T < T_*$. The strategy that many methods use is to discretize the interval I using $N + 1$ equally spaced gridpoints $t_n := t_o + nh, n = 0, \ldots, N$ with $h = \frac{T}{N}$ so that $t_0 = t_o$ and $t_N = t_o + T$ and then use some algorithm to define values $\mathbf{y}_0, \ldots, \mathbf{y}_N$ in $\mathbf{R}^d$, in such a way that when N is large, each $\mathbf{y}_n$ is close to the corresponding $\mathbf{y}(t_n)$. The quantity $\max_{0 \leq n \leq N} \|\mathbf{y}(t_n) - \mathbf{y}_n\|$ is called the *global error* of the algorithm on the interval. If the global error converges to zero as N tends to infinity (for every choice of $\mathbf{f}$ satisfying some Lipschitz condition, t_o, $\mathbf{y}_o$, and $T < T_*$), then we say that we have a *convergent algorithm*. Euler's Method is a convergent algorithm of this sort.

One common way to construct the algorithm that produces the values $y_1, \ldots, y_N$ uses a recursion based on a so-called (one-step) "stepping procedure". This is a discrete approximate solution operator, $\Sigma(\mathbf{f}, \mathbf{y}_n, t_n, h)$, having as inputs

1) a time-dependent vector field $\mathbf{f}$ on $\mathbf{R}^d$,
2) a time t_n in $\mathbf{R}$,
3) a value $\mathbf{y}_n$ in $\mathbf{R}^d$ corresponding to the initial time, and
4) a "time-step" h in $\mathbf{R}$

and as output a point of $\mathbf{R}^d$ that approximates the solution of the initial value problem $\mathbf{y}' = f(t, \mathbf{y})$, $\mathbf{y}(t_i) = \mathbf{y}_i$ at $t_i + h$ well when h is small. (More precisely, the so-called "local truncation error", $\|\sigma(\mathbf{f}, \mathbf{y}(t_n), t_n, t_n + h) - \Sigma(\mathbf{f}, \mathbf{y}(t_n), t_n, h)\|$, should approach zero at least superlinearly in the time-step h.) Given such a stepping procedure, the approximations $\mathbf{y}_n$ of the $\mathbf{y}(t_n)$ are defined recursively by $\mathbf{y}_{n+1} = \Sigma(\mathbf{f}, \mathbf{y}_n, t_n, h)$. Numerical integration methods that use discrete approximations of derivatives defining the vector field $\mathbf{f}$ to obtain the operator Σ are referred to as finite difference methods.

1.2.1. Remark. Notice that there will be two sources that contribute to the global error, $\|\mathbf{y}(t_n) - \mathbf{y}_n\|$. First, at each stage of the recursion there will be an additional local truncation error added to what has already accumulated up to that point. Moreover, because the recursion uses $\mathbf{y}_n$ rather than $\mathbf{y}(t_n)$, after the first step there will be an additional error that includes accumulated local truncation errors, in addition to amplification or attenuation of these errors by the method. (In practice there is a third source of error, namely machine round-off error from using floating-point arithmetic. Since these are

1.3. Stationary Points and Closed Orbits

amplified or attenuated in the same manner as truncation errors, we will often consolidate them and pretend that our computers do precise real number arithmetic, but there are situations where it is important to take it into consideration.)

For Euler's Method the stepping procedure is particularly simple and natural. It is defined by $\Sigma_E(\mathbf{f}, \mathbf{y}_n, t_n, h) := \mathbf{y}_n + h\,\mathbf{f}(t_n, \mathbf{y}_n)$. It is easy to see why this is a good choice. If as above we denote $\sigma(\mathbf{f}, \mathbf{y}_n, t_n, t)$ by $\mathbf{y}(t)$, then by Taylor's Theorem,

$$\begin{aligned}\mathbf{y}(t_n + h) &= \mathbf{y}(t_n) + h\,\mathbf{y}'(t_n) + O(h^2) \\ &= \mathbf{y}_n + h\,\mathbf{f}(t_n, \mathbf{y}_n) + O(h^2) \\ &= \Sigma_E(\mathbf{f}, \mathbf{y}_n, t_n, h) + O(h^2),\end{aligned}$$

so that $\|\sigma(\mathbf{f}, \mathbf{y}_n, t_n, t_n + h) - \Sigma_E(\mathbf{f}, \mathbf{y}_n, t_n, h)\|$, the local truncation error for Euler's Method, does go to zero quadratically in h. When we partition $[t_o, t_o + T]$ into N equal parts, $h = \frac{T}{N}$, each step in the recursion for computing $\mathbf{y}_n$ will contribute a local truncation error that is $O(h^2) = O(\frac{1}{N^2})$. Since there are N steps in the recursion and at each step we add $O(\frac{1}{N^2})$ to the error, this suggests that the global error will be $O(\frac{1}{N})$ and hence will go to zero as N tends to infinity. However, because of the potential amplification of prior errors, this is not a complete proof that Euler's Method is convergent, and we will put off the details of the rigorous argument until the chapter on numerical methods.

▷ **Exercise 1–15.** Show that Euler's Method applied to the initial value problem $\frac{dy}{dt} = y$ with $y(0) = 1$ gives $\lim_{N \to \infty}(1 + \frac{T}{N})^N = e^T$. For $T = 1$ and $N = 2$, show that the global error is indeed greater than the sum of the two local truncation errors.

1.3. Stationary Points and Closed Orbits

We next describe certain special types of solutions of a differential equation that play an important role in the description and analysis of the global behavior of its flow. For generality we will also consider the case of time-dependent vector fields, but these solutions are really most important in the study of autonomous equations.

If a constant map $\sigma: I \to \mathbf{R}^n$, $\sigma(t) = p$ for all $t \in I$, is a solution of the equation $\frac{dx}{dt} = V(x,t)$, then $V(p,t) = \sigma'(t) = 0$ for all t, and conversely this implies $\sigma(t) \equiv p$ is a solution. In particular, in the autonomous case, the maximal solution σ_p is a constant map if and only if $V(p) = 0$. Such points p are of course called zeros of the time-independent vector field V, but because of their great importance they have also been given many more aliases, including critical point, singularity, stationary point, rest point, equilibrium point, and fixed point.

A related but more interesting type of solution of $\frac{dx}{dt} = V(x,t)$ is a so-called *closed orbit*, also referred to as a *periodic solution*. To define these, we start with an arbitrary solution σ defined on the whole real line. A real number T is called a *period* of σ if $\sigma(t) = \sigma(t+T)$ for all $t \in \mathbf{R}$, and we will denote by $\mathbf{Per}(\sigma)$ the set of all periods of σ. Of course 0 is always a period of σ, and one possibility is that it is the only period, in which case σ is called a nonperiodic orbit. At the other extreme, σ is a constant solution if and only if every real number is a period of σ.

What other possibilities are there for $\mathbf{Per}(\sigma)$? To answer that, let us look at some obvious properties of the set of periods. First, $\mathbf{Per}(\sigma)$ is clearly a closed subset of $\mathbf{R}$—this follows from the continuity of σ. Secondly, if T_1 and T_2 are both periods of σ, then $\sigma(t + (T_1 - T_2)) = \sigma((t - T_2) + T_1) = \sigma(t - T_2) = \sigma(t - T_2 + T_2) = \sigma(t)$, so we see that the difference of any two periods is another period. Thus $\mathbf{Per}(\sigma)$ is a closed subgroup of the group of real numbers under addition. But the structure of such groups is well known.

1.3.1. Proposition. *If Γ is a closed subgroup of $\mathbf{R}$, then either $\Gamma = \mathbf{R}$, or $\Gamma = \{0\}$, or else there is a smallest positive element γ in Γ and Γ consists of all integer multiples of γ.*

▷ **Exercise 1–16.** Prove this proposition. (Hint: If Γ is nontrivial, then the set of positive elements of Γ is nonempty and hence has a greatest lower bound γ which is in Γ since Γ is closed. If $\gamma = 0$, show that Γ is dense in $\mathbf{R}$ and hence it is all of $\mathbf{R}$. If $\gamma \neq 0$, it is the smallest positive element of Γ. In this case, if $n \in \Gamma$, then dividing n

1.3. Stationary Points and Closed Orbits

by γ gives $n = q\gamma + r$ with $0 \leq r < \gamma$. Show that the remainder, r, must be zero.)

A solution σ is called periodic if it is nonconstant and has a nontrivial period, so that by the proposition all its periods are multiples of a smallest positive period γ, called the *prime period* of σ.

A real number T is called a period of the time-dependent vector field V if $V(x, t) = V(x, t + T)$ for all $t \in T$ and $x \in \mathbf{R}$. A repeat of the arguments above show that the set **Per**(V) of all periods of V is again a closed subgroup of $\mathbf{R}$, so again there are three cases: 1) **Per**$(V) = \mathbf{R}$, i.e., V is time-independent, 2) **Per**$(V) = \{0\}$, i.e., V is nonperiodic, or 3) there is a smallest positive element T_0 of **Per**(V) (the prime period of V) and **Per**(V) consists of all integer multiples of this prime period.

▷ **Exercise 1–17.** Show that if T is a period of the time-dependent vector field V and σ is a solution of $\frac{dx}{dt} = V(x, t)$, then T is also a period of σ provided there exists a real number t_1 such that $\sigma(t_1) = \sigma(t_1 + T)$. (Hint: Use the uniqueness theorem.)

Note the following corollary: in the autonomous case, if an orbit σ "comes back and meets itself", i.e., if there are two distinct times t_1 and t_2 such that $\sigma(t_1) = \sigma(t_2)$, then σ is a periodic orbit and $t_2 - t_1$ is a period. For this reason, periodic solutions of autonomous ODEs are also referred to as closed orbits. Another way of stating this same fact is as follows:

1.3.2. Proposition. *Let ϕ_t be the flow generated by a complete, autonomous ODE, $\frac{dx}{dt} = V(x)$. A necessary and sufficient condition for the maximum solution curve σ_p with initial condition p to be periodic with period T is that p be a fixed point of ϕ_T.*

• **Example 1–4.** For the harmonic oscillator system in $\mathbf{R}^2$: $\frac{dx}{dt} = -y$, $\frac{dy}{dt} = x$, we have seen that the solution with initial condition (x_0, y_0) is $x(t) = x_0 \cos(t) - y_0 \sin(t)$, $y(t) = x_0 \sin(t) + y_0 \cos(t)$. Clearly the origin is a stationary point, and every other solution is periodic with the same prime period 2π.

1.3.3. Remark. The ODEs modeling many physical systems have periodic orbits, and each such orbit defines a physical "clock" whose natural unit is the prime period of the orbit. We simply choose a configuration of the system that lies on this periodic orbit and tick off the successive recurrences of that configuration to "tell time". The resolution to which before and after can be distinguished with such a clock is limited to approximately the prime period of the orbit. There seems to be no limit to the benefits of ever more precise chronometry—each time a clock has been constructed with a significantly shorter period, it has opened up new technological possibilities. Humankind has always had a 24-hour period clock provided by the rotation of the earth on its axis, but it was only about four hundred years ago that reasonably accurate clocks were developed with a period in the 1-second range. In recent decades the resolution of clocks has increased dramatically. For example, the fundamental clock period for the computer on which we are writing this text is about 0.4×10^{-9} seconds. The highest resolution (and most accurate) of current clocks is the cesium vapor atomic clocks used by international standards agencies. These have a period of about 10^{-11} seconds (with a drift error of about 1 second in 300,000 years!). This means that if two events occur only one hundred billionth of a second apart, one of these clocks can in principle tell which came first.

1.4. Continuity with Respect to Initial Conditions

We consider next how the maximal solutions σ_p of a first-order ODE $\frac{dx}{dt} = V(x)$ depends on the initial condition p. Eventually we will see that this dependence is as smooth as the vector field V, but as a first step we will content ourselves with proving just continuity. The argument rests on a simple but important general principle called Gronwall's Inequality.

1.4.1. Gronwall's Inequality. *Let $u : [0, T) \to [0, \infty)$ be a continuous, nonnegative, real-valued function and assume that $u(t) \le U(t) := C + K \int_0^t u(s)\, ds$ for certain constants $C \ge 0$ and $K > 0$. Then $u(t) \le Ce^{Kt}$.*

1.4. Continuity with Respect to Initial Conditions

▷ **Exercise 1–18.** Prove Gronwall's Inequality.
Hint: Since $u \leq U$, it is enough to show that $U(t) \leq Ce^{Kt}$, or equivalently that $e^{-Kt}U(t) \leq C$, and since $U(0) = C$, it will suffice to show that $e^{-Kt}U(t)$ is nonincreasing, i.e., that $(e^{-Kt}U(t))' \leq 0$. Since $(e^{-Kt}U(t))' = e^{-Kt}(U'(t) - KU)$ and $U' = Ku$, this just says that $Ke^{-Kt}(u - U) \leq 0$.

1.4.2. Theorem on Continuity w.r.t. Initial Conditions.
Let V be a C^1 vector field on $\mathbf{R}^n$ and let $\sigma_p(t)$ denote the maximal solution curve of $\frac{dx}{dt} = V(x)$ with initial condition p. Then as q tends to p, $\sigma_q(t)$ approaches $\sigma_p(t)$, and the convergence is uniform for t in any bounded interval I on which σ_p is defined.

Proof. We have seen that $\sigma_p(t) = p + \int_0^t V(\sigma_p(s), s)\, ds$, and it follows that $\|\sigma_p(t) - \sigma_q(t)\| \leq \|p - q\| + \int_0^t \|V(\sigma_p(s), s) - V(\sigma_q(s), s)\|\, ds$. On the other hand, it is proved in Appendix A that on any bounded set (and in particular on a bounded neighborhood of $\sigma_p(I) \times I$) V satisfies a Lipschitz condition $\|V(x,t) - V(y,t)\| \leq K\|x - y\|$, so it follows that $\|\sigma_p(t) - \sigma_q(t)\| \leq \|p - q\| + K \int_{t_0}^t \|\sigma_p(s) - \sigma_q(s)\|\, ds$. It now follows from Gronwall's Inequality that $\|\sigma_p(t) - \sigma_q(t)\| \leq \|p - q\| e^{Kt}$. ∎

1.4.3. Remark.
For the differential equation $\frac{dx}{dt} = kx$, the maximal solution is $\sigma_p(t) = e^{kt} p$, so $\|\sigma_p(t) - \sigma_q(t)\| = e^{kt} \|p - q\|$. Thus if k is positive, then any two solutions diverge from each other exponentially fast, while if k is negative, all solutions approach the origin (and hence each other) exponentially fast.

But continuity with respect to initial conditions is not the whole story.

1.4.4. Theorem on Smoothness w.r.t. Initial Conditions.
Let V be a C^r vector field on $\mathbf{R}^n$, $r \geq 1$, and let $\sigma_p(t)$ denote the maximal solution curve of $\frac{dx}{dt} = V(x)$ with initial condition p. Then the map $(p, t) \mapsto \sigma_p(t)$ is C^r.

The proof of this theorem is one of the most difficult in elementary ODE theory, and we have deferred it to Appendix F.

Let $V : \mathbf{R}^n \times \mathbf{R}^k \to \mathbf{R}^n$ be a smooth function. Then to each α in $\mathbf{R}^k$ we can associate a vector field $V(\cdot, \alpha)$ on $\mathbf{R}^k$, defined by $x \mapsto V(x, \alpha)$. For this reason it is customary to consider V as a "vector field on $\mathbf{R}^n$ depending on a parameter α in $\mathbf{R}^{k}$". It is often important to know how solutions of $\frac{dx}{dt} = V(x, \alpha)$ depend on the parameter α, and this is answered by the following theorem.

1.4.5. Theorem on Smoothness w.r.t. Parameters. Let $V : \mathbf{R}^n \times \mathbf{R}^k \to \mathbf{R}^n$ be a C^r map, $r > 1$, and let σ_p^α denote the maximum solution curve of $\frac{dx}{dt} = V(x, \alpha)$ with initial condition p. Then the map $(p, \alpha, t) \mapsto \sigma_p^\alpha(t)$ is C^r.

▷ **Exercise 1–19.** Deduce this from the Theorem on Smoothness w.r.t. Initial Conditions. Hint: This is another one of those cute reduction arguments that this subject is full of. The idea is to consider the vector field $\tilde{V}$ on $\mathbf{R}^n \times \mathbf{R}^k$ defined by $\tilde{V}(x, \alpha) = (V(x, \alpha), 0)$ and to note that its maximal solution with initial condition (p, α) is $t \mapsto (\sigma_p^\alpha(t), \alpha)$.

You may have noticed an ambiguity inherent in our use of σ_p to denote the maximal solution curve with initial condition p of a vector field V. After all, this maximal solution clearly depends on V as well as on p, so let us now be more careful and denote it by σ_p^V. Of course, this immediately raises the question of just how σ_p^V depends on V. If V changes just a little, does it follow that σ_p^V also does not change by much? If we return to our smoke particle in the wind metaphor, then this seems reasonable; if we make a tiny perturbation of the direction and speed of the wind at every point, it seems that the path of a smoke particle should not be grossly different. This intuition is correct, and all that is required to prove it is another tricky application of Gronwall's Inequality.

1.4.6. Theorem on the Continuity of σ_p^V w.r.t. V. Let V be a C^1 time-dependent vector field on $\mathbf{R}^n$ and let K be a Lipschitz constant for V, in the sense that $\|V(x,t) - V(y,t)\| \le K\|x - y\|$ for all x, y, and t. If W is another C^1 time-dependent vector field on $\mathbf{R}^n$ such that $\|V(x,t) - W(x,t)\| \le \epsilon$ for all x and t, then $\|\sigma_p^V(t) - \sigma_p^W(t)\| \le \frac{\epsilon}{K}\left(e^{Kt} - 1\right)$.

1.5. Chaos—Or a Butterfly Spoils Laplace's Dream

▷ **Exercise 1–20.** Prove the above theorem. Hint: If we define $u(t) = \left\|\sigma_p^V(t) - \sigma_p^W(t)\right\| + \frac{\epsilon}{K}$, then the conclusion may be written as $u(t) \leq \frac{\epsilon}{K} e^{Kt}$, which follows from Gronwall's Inequality provided we can prove $u(t) \leq \frac{\epsilon}{K} + K \int_0^t u(s)\,ds$. To show that, start from $u(t) - \frac{\epsilon}{K} = \left\|\sigma_p^V(t) - \sigma_p^W(t)\right\| \leq \int_0^t \left\|V(\sigma_p^V(s)) - W(\sigma_p^W(s))\right\| ds$ and use

$$\left\|V(\sigma_p^V(s)) - W(\sigma_p^W(s))\right\| \leq \left\|V(\sigma_p^V(s)) - V(\sigma_p^W(s))\right\| \\ + \left\|V(\sigma_p^W(s)) - W(\sigma_p^W(s))\right\| \\ \leq (Ku(s) - \epsilon) + \epsilon = Ku(s).$$

1.5. Chaos—Or a Butterfly Spoils Laplace's Dream

L'état présent du système de la Nature est évidemment une suite de ce qu'elle était au moment précédent et, si nous concevons une intelligence qui, pour un instant donné, embrasse tous les rapports des êtres de cet Univers, elle pourra déterminer pour un temps quelconque pris dans le passé ou dans l'avenir la position respective, les motions et généralement toutes les affections de ces êtres...

—Pierre Simon de Laplace, 1773[1]

The so-called "scientific method" is a loosely defined iterative process of experimentation, induction, and deduction with the goal of deriving general "laws" for describing various aspects of reality. Prediction plays a central role in this enterprise. During the period of discovery and research, comparing experiments against predictions helps eliminate erroneous preliminary versions of a theory and conversely can provide confirming evidence when a theory is correct. And when a theory finally has been validated, its predictive power can lead to valuable new technologies. In the physical sciences, the laws frequently take the form of differential equations (of just the sort we have been

---

[1] The current state of Nature is evidently a consequence of what it was in the preceding moment, and if we conceive of an intelligence that at a given moment knows the relations of all things of this Universe, it could then tell the positions, motions and effects of all of these entities at any past or future time...

considering) that model the time-evolution of various real-world processes. So it should not be surprising that the sort of issues that we have just been discussing have important practical and philosophical ramifications when it comes to evaluating and interpreting the predictive power of such laws, and indeed some of the above theorems were developed for just such reasons.

At first glance, it might appear that theory supports Laplace's ringing deterministic manifesto quoted above. But if we examine matters with more care, it becomes evident that, while making dependable predictions might be possible for a god who could calculate with infinite precision and who knew the laws with perfect accuracy, for any lesser beings there are severe problems not only in practice but even in principle.

First let us look at the positive side of things. In order to make reliable predictions based on a differential equation $\frac{dx}{dt} = V(x)$, at least the following two conditions must be satisfied:

1) There should be a unique solution for each initial condition, and it should be defined for all $t \in \mathbf{R}$.

2) This solution should depend continuously on the initial condition and also on the vector field V.

Initial value problems that satisfy these two conditions are often referred to as "well-posed" problems.

The importance of the first condition is obvious, and we will not say more about it. The second is perhaps less obvious, but nevertheless equally important. The point is that even if we know the initial conditions with perfect accuracy (which we usually do not), the finite precision of machine representation of numbers as well as round-off and truncation errors in computer algorithms would introduce small errors. So if arbitrarily small differences in initial conditions resulted in wildly different solutions, then prediction would be impossible.

Similarly we do not in practice ever know the vector field V perfectly. For example, in the problem of predicting the motions of the planets, it is not just their mutual positions that determine the force law V, but also the positions of all their moons and of the great multitude of asteroids and comets that inhabit the solar system. If the

1.5. Chaos—Or a Butterfly Spoils Laplace's Dream

tiny force on Jupiter caused by a small asteroid had a significant effect on its motion, then predicting the planetary orbits would be an impossible task.

In the preceding section we saw that complete, C^1 vector fields do give rise to a well-posed initial value problem, so Laplace seems to be on solid ground. Nevertheless, even though the initial value problems that arise in real-world applications may be technically well-posed in the above sense, they often behave as if they were ill-posed. For a class of examples that turns up frequently—the so-called chaotic systems—predictability is only an unachievable theoretical ideal. While their short-term behavior is predictable, on longer time-scales prediction becomes, for practical purposes, impossible. This may seem paradoxical at first; if we have an algorithm for predicting accurately for ten seconds, then should not repeating it with that first prediction as a new initial condition provide an accurate prediction for twenty seconds? Unfortunately, a hallmark feature of chaotic systems, called "sensitive dependence on initial conditions", defeats this strategy.

Let us consider an initial value problem $\frac{dx}{dt} = V(x)$, $x(0) = p_0$ and see how things go wrong for a chaotic system when we try to compute $\sigma_{p_0}(t)$ for large t. Suppose that p_1 is very close to p_0, say $\|p_0 - p_1\| < \delta$, and let us compare $\sigma_{p_1}(t)$ and $\sigma_{p_0}(t)$. Continuity with respect to initial conditions tells us that for δ small enough $\sigma_{p_1}(t)$ at least initially will not diverge too far from $\sigma_{p_1}(t)$. In fact, for a chaotic system, a typical behavior—when p_0 is near a so-called "strange attractor"—is for $\sigma_{p_1}(t)$ to at first "track" $\sigma_{p_0}(t)$ in the sense that $\|\sigma_{p_0}(t) - \sigma_{p_1}(t)\|$ initially stays nearly constant or even decreases—so in particular the motions of $\sigma_{p_0}(t)$ and $\sigma_{p_1}(t)$ are highly correlated. But then, suddenly, there will be a period during which $\sigma_{p_1}(t)$ starts to veer off in a different direction, following which $\|\sigma_{p_0}(t) - \sigma_{p_1}(t)\|$ will grow exponentially fast for a while. Soon they will be far apart, and although their distance remains bounded, from that time forward their motions become completely uncorrelated. If we make δ smaller, then we can guarantee that $\sigma_{p_1}(t)$ will track $\sigma_{p_0}(t)$ for a longer period, but (and this is the essence of sensitive dependence on initial conditions) **no matter how small we make δ, the veering away and loss of correlation will always occur**. The reason this is relevant is that when we try to compute $\sigma_{p_0}(t)$, there will always be some tiny error in the initial condition, and in addition there will be systematic rounding, discretization, and truncation errors in our numerical integration process, so we are always in essence computing $\sigma_{p_1}(t)$ for some p_1 near p_0 rather than computing $\sigma_{p_0}(t)$ itself. The important thing to remember is that even the most miniscule of deviations will get enormously amplified after the loss of correlation occurs.

While there is no mathematical proof of the fact, it is generally believed that the fluid mechanics equations that govern the evolution of weather are chaotic. The betting is that accurate weather predictions more than two weeks in advance will never be feasible, no matter how much computing power we throw at the problem. As the meteorologist Edward Lorenz once put it, "... the flap of a butterfly's wings in Brazil can set off a tornado in Texas." This metaphor has caught on, and you will often hear sensitive dependence on initial conditions referred to as the "butterfly effect".

In Figure 1.2 we show a representation of the so-called "Lorenz attractor". This shows up in an ODE that Lorenz was studying as a highly over-simplified meteorological model . The Web Companion has a QuickTime Movie made with 3D-XplorMath that shows the Lorenz attractor being generated in real time. What is visible from the movie (and not in the static figure) is how two points of the orbit that are initially very close will moments later be far apart, on different "wings" of the attractor. (By the way, the fact that the Lorenz attractor resembles a butterfly is totally serendipitous!)

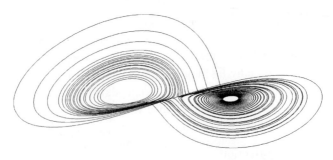

Figure 1.2. The Lorenz attractor.

1.5. Chaos—Or a Butterfly Spoils Laplace's Dream

Another strange attractor, shown in Figure 1.3, appears in an ODE called the Rikitake Two-Disk Dynamo. Like the Lorenz system, the Rikitake ODE was invented to model an important real-world phenomenon, namely the Earth's geomagnetic field. The flipping back and forth between attractor "wings" in this case corresponds to the flipping of the Earth's North and South Magnetic Poles that has long been known from the geologic record.

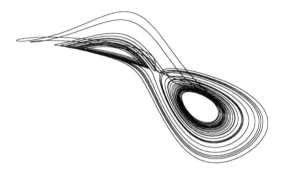

Figure 1.3. The Rikiatke attractor.

Fortunately, even though systems that exhibit sensitive dependence on initial conditions do not permit long-time a priori prediction, it does not follow that such systems cannot be used to **control** processes that go on over long time periods. For example, when NASA sends a space-probe to a distant planet, the procedure is to look at all initial conditions and times that end up at the appropriate point on the given planet and then among these optimize for some variable (such as the transit time or payload weight). Of course they are using the prediction that with this choice of time and initial condition the probe will end up on the planet, but they realize that this prediction is only a first approximation. After lift-off, the current position and velocity of the probe is measured at intervals small enough to assure only small deviation from the previous predicted values. Then, these actual position and velocity are compared with the desired values and a "mid-course correction" is programmed that will bring the actual values back in line with the desired values. The equations governing a space probe are not actually chaotic, but this same sort of controllability has also been proved rigorously for certain chaotic systems.

Experiment. Balance a broomstick vertically as best you can and let it fall. Repeat this many times, each time measuring the angle it makes with a fixed direction. You will see that the angles are randomly distributed around a circle, suggesting sensitive dependence on initial conditions (even though this system is not technically chaotic). Now place the broomstick on your fingertip and try to control it in a nearly upright position by making rapid slight finger motions—most people know almost instinctively how to do this. It is also instructive to note that you can make small rapid back-and-forth motions with your finger in a pre-planned direction, adding small perturbations as required to maintain the broomstick in approximate balance. (It is a fact that this actually serves to stabilize the control problem.)

We hope you have asked yourself an obvious question. If the weather is too chaotic to predict, can we perhaps nevertheless control it? After all, if a tiny butterfly can really perturb things enough to cause a storm a week later, it should not be beyond the power of humans to sense the effects of this perturbation while it is still small enough to counteract. (Of course this is not an entirely new idea—people have been seeding clouds to produce rain for decades. But the real challenge is to learn enough about how large weather systems evolve to be able to guide their development effectively with available amounts of energy.)

▷ **Exercise 1–21.** Learn how to control the weather. Hint: It could easily take you a lifetime to complete this exercise, but if you succeed, it will have been a life well spent.

1.5.1. Further Notes on Chaos. The study of chaotic systems is a relatively new field of mathematics, and even the "correct" definition of chaos is still a matter of some debate. In fact chaos should probably be thought of more as a "syndrome"—a related collection of symptoms—than as a precisely defined concept. We have concentrated here on one particular symptom of chaotic systems, their sensitive dependence on initial conditions, but there are others that

1.6. Analytic ODE and Their Solutions

are closely related and equally as important, such as having a positive "Lyapounov Exponent", the existence of so-called "strange attractors", "homoclinic tangles", and "horseshoe maps". These latter concepts are quite technical, and we will not attempt to define or describe them here (but see the references below).

In recent years chaos theory and the related areas of dynamical systems and nonlinear science, have been the focus of enormous excitement and enthusiasm, giving rise to a large and still rapidly growing literature consisting of literally hundreds of books, some technical and specialized and others directed at the lay public. Two of the best nontechnical expositions are David Ruelle's "Chance and Chaos" and James Gleick's "Chaos: Making a New Science". For an excellent introduction at a more mathematically sophisticated level see the collection of articles in "Chaos and Fractals: The Mathematics Behind the Computer Graphics", edited by Robert Devaney and Linda Keen. Other technical treatment we can recommend are Steven Strogatz' "Nonlinear Dynamics and Chaos", Hubbard and West's "Differential Equations: A Dynamical Systems Approach", Robert Devaney's "A First Course in Chaotic Dynamical Systems", and Tom Mullin's "The Nature of Chaos".

1.6. Analytic ODE and Their Solutions

Until now we have worked entirely in the real domain, but we can equally well consider complex-valued differential equations. Of course we should be precise about how to interpret this concept, and in fact there are several different interpretations with different levels of interest and sophistication. Using the most superficial generalization, it seems as if there is nothing really new—since we can identify $\mathbf{C}$ with $\mathbf{R}^2$, a smooth vector field on $\mathbf{C}^n$ is just a smooth vector field on $\mathbf{R}^{2n}$. But even here there are some advantages in using a complex approach. Recall the important two-dimensional real linear system $\frac{dx}{dt} = -y$, $\frac{dy}{dt} = x$, mentioned earlier, that arises when we reduce the harmonic oscillator equation $\frac{d^2x}{dt^2} = -x$ to a first-order system. We saw that if we regard $\mathbf{R}^2$ as $\mathbf{C}$ and write $z = x + iy$ as usual, then our system becomes $\frac{dz}{dt} = iz$, so the solution with initial condition

z_0 is evidently $z(t) = e^{it}z_0$, and we recover the usual solution of the harmonic oscillator by taking the real part of this complex solution.

But if you have had a standard course on complex function theory, then you can probably guess what the really important generalization should be. First of all, we should replace the time t by a complex variable τ, demand that the vector field V that occurs on the right-hand side of our equation $\frac{dz}{d\tau} = V(z)$ is an analytic function of z, and look for analytic solutions $z(\tau)$.

To simplify the notation, we will consider the case of a single equation, but everything works equally well for a system of equations $\frac{dz_i}{d\tau} = V_i(z_1, \ldots, z_n)$. We shall also assume that V is an entire function (i.e., defined and analytic on all of $\mathbf{C}$), but the generalization to the case that V is only defined in some simply connected region $\Omega \subset \mathbf{C}$ presents little extra difficulty.

Let us write $H(B_r, \mathbf{C})$ for the space of continuous, complex-valued functions defined on B_r (the closed disk of radius r in $\mathbf{C}$) that are analytic in the interior. Just as in the real case, we can define the map $F = F^{V,z_0}$ of $H(B_r, \mathbf{C})$ into itself by $F(\zeta)(\tau) = z_0 + \int_0^\tau V(\zeta(\sigma))\,d\sigma$. Note that by Cauchy's Theorem the integral is well-defined, independent of the path joining 0 to τ, and since the indefinite integral of an analytic function is again analytic, F does indeed map $H(B_r, \mathbf{C})$ to itself. Clearly $F(\zeta)(0) = z_0$ and $\frac{d}{d\tau}F(\zeta)(\tau) = V(\zeta(\tau))$, so $\zeta \in H(B_r, \mathbf{C})$ satisfies the initial value problem $\frac{dz}{d\tau} = V(z)$, $z(0) = z_0$ if and only if it is a fixed point of F^{V,z_0}. The fact that a uniform limit of a sequence of analytic functions is again analytic implies that $H(B_r, \mathbf{C})$ is a complete metric space in the metric $\rho(\zeta_1, \zeta_2) = \|\zeta_1 - \zeta_2\|_\infty$ given by the "sup" norm, $\|\zeta\|_\infty = \sup_{\tau \in B_r} |\zeta(\tau)|$. We now have all the ingredients required to extend to this new setting the same Banach Contraction Principle argument used in Appendix B to prove the existence and uniqueness theorem in the real case. It follows that given $z \in \mathbf{C}$, there is a neighborhood O of z and a positive ϵ such that for each $z_0 \in O$ there is a unique $\zeta_{z_0} \in H(B_\epsilon, \mathbf{C})$ that solves the initial value problem $\frac{dz}{d\tau} = V(z)$, $z(0) = z_0$. And the proof in Appendix F that solutions vary smoothly with the initial condition generalizes to show that ζ_{z_0} is holomorphic in the initial condition z_0.

1.7. Invariance Properties of Flows

Now let us consider the case of a real ODE, $\frac{dx}{dt} = V(x)$, but assume that the vector field $V : \mathbf{R}^n \to \mathbf{R}^n$ is analytic. This means simply that each component $V_i(x_1, \ldots, x_n)$ is given by a convergent power series. Then these same power series extend the definition of V to an analytic map of $\mathbf{C}^n$ to itself, and we are back to the situation above. (In fact, this is just the special case of what we considered above when the coefficients of the power series are all real.) Of course, if we consider only the solutions of this "complexified" ODE whose initial conditions z_0 are real and also restrict the time parameter τ to real values, then we get the solutions of the original real equation. So what we learn from this excursion to $\mathbf{C}^n$ and back is that when the right-hand side of the ODE $\frac{dx}{dt} = V(x)$ is an analytic function of x, then the solutions are also analytic functions of the time and the initial conditions.

This complexification trick is already useful in the simple case that the vector field V is linear, i.e., when $V_i(x) = \sum_i^n A_{ij} x_j$ for some $n \times n$ real matrix A. The reason is that the characteristic polynomial of A, $P(\lambda) = \det(A - \lambda I)$, always factors into linear factors over $\mathbf{C}$, but not necessarily over $\mathbf{R}$. In particular, if P has distinct roots, then it is diagonalizable over $\mathbf{C}$ and it is trivial to write down the solutions of the IVP in an eigenbasis. We will explore this in detail in Chapter 2 on linear ODEs.

1.7. Invariance Properties of Flows

In this section we suppose that V is some complete vector field on $\mathbf{R}^n$ and that ϕ_t is the flow on $\mathbf{R}^n$ that it generates. For many purposes it is important to know what things are "preserved" (i.e., left fixed or "invariant") under a flow.

For example, the function $F : \mathbf{R}^n \to \mathbf{R}$ is said to be invariant under the flow (or to be a "constant of the motion") if $F \circ \phi_t = F$ for all t. Note that this just means that each solution curve σ_p lies on the level surface $F = F(p)$ of the function F. (In particular, in case $n = 2$, where the level "surfaces" are level curves, the solution curves will in general be entire connected components of these curves.)

▷ **Exercise 1–22.** Show that a differentiable function F is a constant of the motion if and only if its directional derivative at any point x in the direction $V(x)$ is zero, i.e., $\sum_k \frac{\partial F(x)}{\partial x_k} V_k(x) = 0$.

The flow is called *isometric* (or distance preserving) if for all points p, q in $\mathbf{R}^n$ and all times t, $\|\phi_t(p) - \phi_t(q)\| = \|p - q\|$, and it is called *volume preserving* if for all open sets O of $\mathbf{R}^n$, the volume of $\phi_t(O)$ equals the volume of O.

Given a linear map $B : \mathbf{R}^n \to \mathbf{R}^n$, we get a bilinear map $\hat{B} : \mathbf{R}^n \times \mathbf{R}^n \to \mathbf{R}$ by $\hat{B}(u, v) = \langle Bu, v \rangle$, where $\langle Bu, v \rangle$ is just the inner product (or dot product) of Bu and v. We say that the flow preserves the bilinear form $\hat{B}$ if $\hat{B}((D\phi_t)_x(u), (D\phi_t)_x(v)) = \hat{B}(u, v)$ for all u, v in $\mathbf{R}^n$ and all x in $\mathbf{R}^n$.

Here, the linear map $D(\phi_t)_x : \mathbf{R}^n \to \mathbf{R}^n$ is the differential of ϕ_t at x; i.e., if the components of $\phi_t(x)$ are $\Phi_i(x, t)$, then the matrix of $D(\phi_t)_x$ is just the Jacobian matrix $\frac{\partial \Phi_i(x,t)}{\partial x_j}$. We will denote the determinant of this latter matrix (the Jacobian determinant) by $J(x, t)$. We note that because $\phi_0(x) = x$, $\frac{\partial \Phi_i(x,0)}{\partial x_j}$ is the identity matrix, and it follows that $J(x, 0) = 1$.

▷ **Exercise 1–23.** Since, by definition, $t \mapsto \phi_t(x)$ is a solution of $\frac{dx}{dt} = V(x)$, $\frac{\partial \Phi_i(x,t)}{\partial t} = V_i(\phi_t(x))$. Using this, deduce that $\frac{\partial}{\partial t} \frac{\partial \Phi_i(x,t)}{\partial x_j} = \sum_k \frac{\partial V_i(\phi_t(x))}{\partial x_k} \frac{\partial \Phi_k(x,t)}{\partial x_j}$ and in particular that $\left(\frac{\partial}{\partial t}\right)_{t=0} \frac{\partial \Phi_i(x,t)}{\partial x_j} = \frac{\partial V_i(x)}{\partial x_j}$.

▷ **Exercise 1–24.** We define a scalar function $\mathbf{div}(V)$, the divergence of V, by $\mathbf{div}(V) := \sum_i \frac{\partial V_i}{\partial x_i}$. Using the formula for the derivative of a determinant, show that $\left(\frac{\partial}{\partial t}\right)_{t=0} J(x, t) = \mathbf{div}(V)(x)$.

▷ **Exercise 1–25.** Now, using the "change of variable formula" for an n-dimensional integral, you should be able to show that the flow generated by V is volume preserving if and only if $\mathbf{div}(V)$ is identically zero. Hint: You will need to use the group property, $\phi_{t+s} = \phi_t \circ \phi_s$.

▷ **Exercise 1–26.** Let B_{ij} denote the matrix of the linear map B. Show that a flow preserves $\hat{B}$ if and only if $\sum_k \left(B_{ik} \frac{\partial V_k}{\partial x_j} + \frac{\partial V_k}{\partial x_i} B_{kj} \right) = 0$. Show that the flow is isometric if and only if it preserves $\hat{I}$ (i.e.,

1.7. Invariance Properties of Flows

the inner product) and hence if and only if the matrix $\frac{\partial V_i}{\partial x_j}$ is everywhere skew-symmetric. Show that isometric flows are also measure preserving.

▷ **Exercise 1–27.** Show that the translation flows generated by constant vector fields are isometric and also that the flow generated by a linear vector field $V(x) = Ax$ is isometric if and only if A is skew-adjoint. Conversely show that if $V(x)$ is a vector field generating a one-parameter group of isometries of $\mathbf{R}^n$, then $V(x) = v + Ax$, where v is a point of $\mathbf{R}^n$ and A is a skew-adjoint linear map of $\mathbf{R}^n$. Hint: Show that $\frac{\partial^2 V_i}{\partial x_j \partial x_k}$ vanishes identically.

Chapter 2

Linear Differential Equations

2.1. First-Order Linear ODE

Two differential equations that students usually meet very early in their mathematical careers are the first-order "equation of exponential growth", $\frac{dx}{dt} = ax$, with the explicit solution $x(t) = x(0)e^{at}$, and the second-order "equation of simple harmonic motion", $\frac{d^2x}{dt^2} = -\omega^2 x$, whose solution can also be written down explicitly: $x(t) = x(0)\cos(\omega t) + \frac{x'(0)}{\omega}\sin(\omega t)$. The interest in these two equations goes well beyond the fact that they have simple and explicit solutions. Much more important is the fact that they can be used to model successfully many real-world situations. Indeed, they are so important in both pure and applied mathematics that we will devote this and the next several sections to studying various generalizations of these equations and their applications to building models of real-world phenomena. Let us start by looking at (and behind) the property that gives these two equations their special character.

One of the most obvious features common to both of these equations is that their right-hand sides are linear functions. Now, in many real-world situations the response of a system to an influence is well approximated by a linear function of that influence, so granting that the dynamics of such problems can be described by an ODE, it should be no surprise that the dynamical equations for such systems are linear. In particular, if x measures the deviation of some system from an equilibrium configuration, then there will usually be a restoring force driving the system back towards equilibrium, the magnitude of which is linear in x—this is the general formulation of Hooke's Law that "stress is proportional to strain". From a mathematical point of view, there is nothing mysterious about this—the restoring force is actually only approximately linear, with the approximation getting better as we approach equilibrium. If we assume only that the restoring force is a differentiable function of the deviation from equilibrium, then, since it vanishes at the equilibrium, we see that the approximate linearity of the force near equilibrium is just a manifestation of Taylor's Theorem with Remainder. This observation points to a further reason for why linear equations play such a central role. Suppose we have a nonlinear differential equation $\frac{dx}{dt} = V(x)$. At an "equilibrium point" p, i.e., a point where $V(p) = 0$, define A to be the differential of V at p. Then for small x, Ax is a good approximation of $V(p+x)$, so we can hope to approximate solutions of the nonlinear equation near p with solutions of the linear equation $\frac{dx}{dt} = Ax$ near 0. In fact this technique of "linearization" is one of the most powerful tools for analyzing nonlinear differential equations and one that we shall return to repeatedly.

The most natural generalization of the equation of exponential growth to an n-dimensional system is an equation of the form $\frac{dx}{dt} = Ax$, where now x represents a point of $\mathbf{R}^n$ and $A : \mathbf{R}^n \to \mathbf{R}^n$ is a linear operator, or equivalently an $n \times n$ matrix. Such an equation is called an autonomous, first-order, linear ordinary differential equation.

▷ **Exercise 2–1. The Principle of Superposition.** Show that any linear combination of solutions of such a system is again a solution, so that if as usual σ_p denotes the solution of the initial value problem with initial condition p, then $\sigma_{p_1+p_2} = \sigma_{p_1} + \sigma_{p_2}$.

When $n = 1$, A is just a scalar, and we know that $\sigma_p(t) = e^{tA}p$, or in other words, the flow ϕ_t generated by the differential equation is just multiplication by e^{tA}. What we shall see below is that for $n > 1$ we can still make good sense out of e^{tA}, and this same formula still gives the flow.

2.1. First-Order Linear ODE

We saw very early that in one-dimensional space successive approximations worked particularly well for the linear case, so we will begin by attempting to repeat that success in higher dimensions.

Denote by $C(\mathbf{R}, \mathbf{R}^n)$ the continuous maps of $\mathbf{R}$ into $\mathbf{R}^n$, and as earlier let $F = F^{A,x_0}$ be the map of $C(\mathbf{R}, \mathbf{R}^n)$ to itself defined by $F(x)(t) := x_0 + \int_0^t A(x(s))\,ds$. Since A is linear, this can also be written as $F(x)(t) := x_0 + A\int_0^t x(s)\,ds$. We know that the solution of the IVP with initial value x_0 is just the unique fixed point of F, so let's try to find it by successive approximations starting from the constant path $x^0(t) = x_0$. If we recall that the sequence of successive approximations, x^n, is defined recursively by $x^{n+1} = F(x^n)$, then an elementary induction gives $x^n(t) = \sum_{k=0}^n \frac{1}{k!}(tA)^k x_0$, suggesting that the solution to the initial value problem should be given by the limit of this sequence, namely the infinite series $\sum_{k=0}^\infty \frac{1}{k!}(tA)^k x_0$. Now (for obvious reasons) given a linear operator T acting on $\mathbf{R}^n$, the limit of the infinite series of operators $\sum_{k=0}^\infty \frac{1}{k!}T^k$ is denoted by e^T or $\exp(T)$, so we can also say that the solution to our IVP should be $e^{tA}x_0$.

The convergence properties of the series for $e^T x$ follow easily from the Weierstrass M-test. If we define $M_k = \frac{1}{k!}\|T\|^k r$, then $\sum M_k$ converges to $e^{\|T\|}r$, and since $\left\|\frac{1}{k!}T^k x\right\| < M_k$ when $\|x\| < r$, it follows that $\sum_{k=0}^\infty \frac{1}{k!}T^k x$ converges absolutely and uniformly to a limit, $e^T x$, on any bounded subset of $\mathbf{R}^n$.

▷ **Exercise 2–2.** Provide the details for the last statement. (Hint: Since the sequence of partial sums $\sum_{k=0}^n M_k$ converges, it is Cauchy; i.e., given $\epsilon > 0$, we can choose N large enough that $\sum_m^{m+k} M_k < \epsilon$ provided $m > N$. Now if $\|x\| < r$, $\left\|\sum_{k=0}^{m+k} \frac{1}{k!}T^k x - \sum_{k=0}^m \frac{1}{k!}T^k x\right\| < \sum_m^{m+k} M_k < \epsilon$, proving that the infinite series defining $e^T x$ is uniformly Cauchy and hence uniformly convergent in $\|x\| < r$.)

Since the partial sums of the series for $e^T x$ are linear in x, so is their limit, so e^T is indeed a linear operator on $\mathbf{R}^n$.

Next observe that since a power series in t can be differentiated term by term, it follows that $\frac{d}{dt}e^{tA}x_0 = Ae^{tA}x_0$; i.e., $x(t) = e^{tA}x_0$ is a solution of the ODE $\frac{dx}{dt} = Ax$. Finally, substituting zero for t in the power series gives $e^{0A}x_0 = x_0$. This completes the proof of the following proposition.

2.1.1. Proposition. *If A is a linear operator on $\mathbf{R}^n$, then the solution of the linear differential equation $\frac{dx}{dt} = Ax$ with initial condition x_0 is $x(t) = e^{tA}x_0$.*

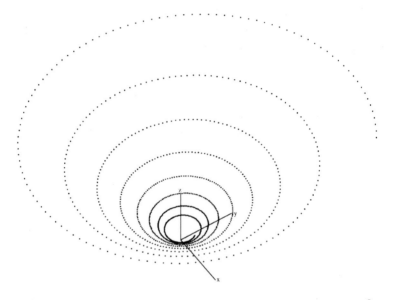

Figure 2.1. A typical solution of a first-order linear ODE in $\mathbf{R}^3$.
Note: The dots are placed along the solution at fixed time
intervals. This gives a visual clue to the speed
at which the solution is traversed.

As a by-product of the above discussion we see that a linear ODE $\frac{dx}{dt} = Ax$ is complete, and the associated flow ϕ_t is just e^{tA}. By a general fact about flows it follows that $e^{(s+t)A} = e^{sA}e^{tA}$ and $e^{-A} = (e^A)^{-1}$, so $\exp : A \mapsto e^A$ is a map of the vector space $\mathbf{L}(\mathbf{R}^n)$ of all linear maps of $\mathbf{R}^n$ into the group $\mathbf{GL}(\mathbf{R}^n)$ of invertible elements of $\mathbf{L}(\mathbf{R}^n)$ and for each $A \in \mathbf{L}(\mathbf{R}^n)$, $t \mapsto e^{tA}$ is a homomorphism of the additive group of real numbers into $\mathbf{GL}(\mathbf{R}^n)$.

2.1. First-Order Linear ODE

▷ **Exercise 2–3.** Show more generally that if A and B are commuting linear operators on $\mathbf{R}^n$, then $e^{A+B} = e^A e^B$. (Hint: Since A and B commute, the Binomial Theorem is valid for $(A+B)^k$, and since the series defining e^{A+B} is absolutely convergent, it is permissible to rearrange terms in the infinite sum. For a different proof, show that $e^{tA}e^{tB}x_0$ satisfies the initial value problem $\frac{dx}{dt} = (A+B)x$, $x(0) = x_0$, and use the Uniqueness Theorem.)

At first glance it might seem hopeless to attempt to solve the linear ODE $\frac{dx}{dt} = Ax$ by computing the power series for e^{tA}—if A is a 10×10 matrix, then computing just the first dozen powers of A will already be pretty time consuming. However, suppose that v is an eigenvector of A belonging to the eigenvalue λ, i.e., $Av = \lambda v$. Then $A^n v = \lambda^n v$, so that in this case $e^{tA}v = e^{t\lambda}v$! If we combine this fact with the Principle of Superposition, then we see that we are in good shape whenever the operator A is diagonalizable. Recall that this just means that there is a basis of $\mathbf{R}^n$, $e_1, \ldots, e_n$, consisting of eigenvectors of A, so that $Ae_i = \lambda_i e_i$. We can expand an arbitrary initial condition $x_0 \in \mathbf{R}^n$ in this basis, i.e., $x_0 = \sum_i a_i e_i$, and then $e^{tA}x_0 = \sum_i a_i e^{t\lambda_i} e_i$ is the explicit solution of the initial value problem (a fact we could have easily verified without introducing the concept of the exponential of a matrix).

Nothing in this section has depended on the fact that we were dealing with real rather than complex vectors and matrices. If $A : \mathbf{C}^n \to \mathbf{C}^n$ is a complex linear map (or a complex $n \times n$ matrix), then the same argument as above shows that the power series for $e^{tA}z$ converges absolutely for all z in $\mathbf{C}^n$ (and for all t in $\mathbf{C}$).

If A is initially given as an operator on $\mathbf{R}^n$, it can be useful to "extend" it to an operator on $\mathbf{C}^n$ by a process called complexification. The inclusion of $\mathbf{R}$ in $\mathbf{C}$ identifies $\mathbf{R}^n$ as a real subspace of $\mathbf{C}^n$, and $\mathbf{C}^n$ is the direct sum (as a real vector space) $\mathbf{C}^n = \mathbf{R}^n \oplus i\mathbf{R}^n$. If $z = (z_1, \ldots, z_n) \in \mathbf{C}^n$, then we project on these subspaces by taking the real and imaginary parts of z (i.e., the real vectors x and y whose components x_i and y_i are the real and imaginary parts of z_i). This is clearly the unique decomposition of z in the form $z = x + iy$ with both x and y in $\mathbf{R}^n$. We extend A to $\mathbf{C}^n$ by defining $Az = Ax + iAy$, and it is easy to see that this extended map is complex linear. (Hint: It is enough to check that $Aiz = iAz$.)

▷ **Exercise 2–4.** Show that if we complexify an operator A on $\mathbf{R}^n$ as above and if a curve $z(t)$ in $\mathbf{C}^n$ is a solution of $\frac{dz}{dt} = Az$, then its real and imaginary parts are also solutions of this equation.

What is the advantage of complexification? As the following example shows, a nondiagonalizable operator A on $\mathbf{R}^n$ may become diagonalizable after complexification, allowing us to solve $\frac{dz}{dt} = Az$ easily in $\mathbf{C}^n$. Moreover, we can then apply the preceding exercise to solve the initial value problem in $\mathbf{R}^n$ from the solution in $\mathbf{C}^n$.

• **Example 2–1.** We can write the system $\frac{dx_1}{dt} = x_2$, $\frac{dx_2}{dt} = -x_1$ as $\frac{dx}{dt} = Ax$, where A is the linear operator on $\mathbf{R}^2$ that is defined by $A(x_1, x_2) = (x_2, -x_1)$. Since A^2 is minus the identity, A has no real eigenvalues and so is not diagonalizable. But, if we complexify A, then the vectors $e_1 = (1, i)$ and $e_2 = (1, -i)$ in $\mathbf{C}^2$ satisfy $Ae_1 = ie_1$ and $Ae_2 = -ie_2$, so they are an eigenbasis for the complexification of A, and we have diagonalized A in $\mathbf{C}^2$. The solution of $\frac{dz}{dt} = Az$ with initial value $e_1 = (1, i)$ is $e^{it}e_1 = (e^{it}, ie^{it})$. Taking real parts, we find that the solution of the initial value problem for $\frac{dx}{dt} = Ax$ with initial condition $(1, 0)$ is $(\cos(t), -\sin(t))$, while taking imaginary parts, we see that the solution with initial condition $(0, 1)$ is $(\sin(t), \cos(t))$. By the Principle of Superposition the solution $\sigma_{(a,b)}(t)$ with initial condition (a, b) is $(a\cos(t) + b\sin(t), -a\sin(t) + b\cos(t))$.

Next we will analyze in more detail the properties of the flow e^{tA} on $\mathbf{C}^n$ generated by a linear differential equation $\frac{dz}{dt} = Az$. We have seen that this flow is transparent for the case that A is diagonalizable, but we want to treat the general case, so we will not assume this. Our approach is based on the following elementary consequence of the Principle of Superposition.

2.1.2. Reduction Principle. Let $\mathbf{C}^n$ be the direct sum of subspaces V_i, each of which is mapped into itself by the operator A, and let $v \in \mathbf{C}^n$ and $v = v_1 + \cdots + v_k$, with $v_i \in V_i$. If σ_p denotes the solution of $\frac{dz}{dt} = Az$ with initial condition p, then $\sigma_{v_i}(t) \in V_i$ for all t and $\sigma_v(t) = \sigma_{v_1}(t) + \cdots + \sigma_{v_k}(t)$.

2.1. First-Order Linear ODE

▷ **Exercise 2–5.** Use the Uniqueness Theorem for solutions of ODE to show that if σ is a solution of $\frac{dz}{dt} = Az$ and if V is a subspace of $\mathbf{C}^n$ that is mapped into itself by A, then if $\sigma(t_0) \in V$ at one instant t_0, it follows that $\sigma(t) \in V$ for all t.

Let $\lambda_1, \ldots, \lambda_k$ denote the eigenvalues of A— the complex numbers λ for which there exists a nonzero vector v in $\mathbf{C}^n$ with $Av = \lambda v$. Because λ is an eigenvalue of A if and only if $(A - \lambda I)v = 0$ has a nontrivial solution, the eigenvalues of A are just the roots of the so-called characteristic polynomial of A, $\chi_A(\lambda) := \det(A - \lambda I)$. Since we are working over the complex numbers, polynomials factor into products of powers of their distinct linear factors, so $\chi_A(\lambda) = \Pi_{j=1}^{k}(\lambda - \lambda_j)^{m_j}$. The m_j are called the multiplicities of the eigenvalues λ_j, and their sum is clearly n, the degree of χ_A.

A vector v is called a generalized eigenvector of A belonging to the eigenvalue λ if $(A - \lambda)^j v = 0$ for some positive integer j. The set $E(A, \lambda)$ of all such vectors is called the generalized eigenspace of A for the eigenvalue λ.

▷ **Exercise 2–6.** Let λ be an eigenvalue of A. Show that

(a) $E(A, \lambda)$ is a linear subspace of $\mathbf{C}^n$,

(b) A maps $E(A, \lambda)$ into itself, and

(c) N, the restriction of $A - \lambda I$ to $E(A, \lambda)$, is nilpotent; i.e., some power of N is zero.

(Hints: Part (a) is easy, and so is part (b) in the special case that $\lambda = 0$. Since $E(A, \lambda) = E(A - \lambda I, 0)$, it follows from this special case that $E(A, \lambda)$ is mapped into itself by $A - \lambda I$, and (b) follows easily. Finally, for (c), choose a basis $e_1, \ldots, e_r$ for $E(A, \lambda)$ and let $N^{r_j} e_j = 0$. Show that $N^r = 0$, where r is the maximum of the r_j.)

The analysis of the operator A and of its associated differential equation $\frac{dz}{dt} = Az$ is greatly simplified by the following important and basic theorem of linear algebra.

2.1.3. Primary Decomposition Theorem. *If A is a linear operator on a finite-dimensional complex vector space V, then V is the direct sum of the generalized eigenspaces of A.*

The proof of the Primary Decomposition Theorem can be found in Theorem 1 (Section 1, Chapter 6) on page 331 of [HS].

This result is important for our current project—trying to understand better the IVP for $\frac{dz}{dt} = Az$—because in combination with the Reduction Principle it shows that to compute the solution of the IVP for an arbitrary initial condition, it suffices to find the solution when the initial conditions belongs to a generalized eigenspace of A. Thus without loss of generality we will (temporarily) assume that A has a single eigenvalue λ, so that the generalized eigenspace of A belonging to the eigenvalue λ is all of $\mathbf{C}^n$.

By an exercise above, the operator $N := A - \lambda I$ is nilpotent, say $N^r = 0$. Since all higher powers of N then also vanish, the power series for e^{tN} truncates to a polynomial of degree $r - 1$: $e^{tN} = \sum_{j=0}^{r-1} \frac{t^j}{j!} N^j$. On the other hand, since the operator λI is diagonal with all eigenvalues equal to λ, $e^{t\lambda I}$ is just multiplication by $e^{\lambda t}$. Finally since $A = N + \lambda I$ and since N and λI commute, $e^{tA} = e^{\lambda t} e^{tN} = \sum_{j=0}^{r-1} (e^{t\lambda} t^j) \frac{1}{j!} N^j$, and this is the explicit formula we have been after for the flow generated by the differential equation $\frac{dz}{dt} = Az$ on the generalized eigenspace $E(A, \lambda)$. Note that when A is diagonalizable, $N = 0$, or equivalently $r = 0$, so we get back the fact that in this case e^{tA} is just multiplication by $e^{t\lambda}$.

The above formula can be made even more explicit if we choose a basis that puts the nilpotent operator N into *Jordan canonical form*. We recall what this means from an operator theoretic perspective. An s-dimensional linear subspace B is called a Jordan block for the nilpotent operator N if it has a basis $e_1, \ldots, e_s$, satisfying $e_j = Ne_{j-1}$ for $j < s$, and $Ne_s = 0$, and such a basis is called a Jordan basis for B (with respect to N). The Jordan Canonical Form Theorem (see Theorem 1 (Section 3, Chapter 6) on page 334 of [HS]) says that if N is any nilpotent operator on a finite-dimensional complex vector space V, then V can be written as the direct sum of Jordan blocks,

2.1. First-Order Linear ODE

and putting N into Jordan canonical form means choosing such blocks and a Jordan basis for each of them.

▷ **Exercise 2–7.** Let B be a Jordan block for N and $e_1, \ldots, e_s$ a Jordan basis for B. If $A = N + \lambda I$, write the solution of the IVP for $\frac{dz}{dt} = Az$ with initial condition e_i as a linear combination of the e_j with explicit coefficients.

There are simple algorithms for putting a matrix in Jordan canonical form, and together with the above exercise this permits us to give a completely explicit solution of the IVP for any first-order autonomous linear ODE on $\mathbf{C}^n$. If you are ever called upon to actually carry this out, in practice, there are routines that are built in to programs such as Matlab to simplify the process for you. Basically, you only need to input a matrix and an initial condition and the computer will take it from there.

The formula above for e^{tA} on generalized eigenspaces of A leads to very important asymptotic estimates for the growth of solutions as t tends to $\pm\infty$. These depend on the following well-known fact.

▷ **Exercise 2–8.** Let $\lambda = \mu + i\nu$, with μ and ν real, and let k be a nonnegative integer.

(a) If $\mu < \mu_0 < 0$, then $\lim_{t \to \infty} e^{\lambda t} t^k = 0$. In fact $|e^{\lambda t} t^k| = o(e^{\mu_0 t})$ as $t \to \infty$.

(b) If $\mu > \mu_0 > 0$, then $\lim_{t \to -\infty} e^{\lambda t} t^k = 0$. In fact $|e^{\lambda t} t^k| = o(e^{\mu_0 t})$ as $t \to -\infty$.

(Hint: $|e^{\lambda t} t^k| = e^{\mu_0 t} \frac{t^k}{e^{(\mu_0 - \mu)t}}$, and by L'Hôpital's Rule, applied recursively k times, $\frac{t^k}{e^{(\mu_0 - \mu)t}} \to 0$.)

For an operator A on $\mathbf{C}^n$, we will denote by $E^U(A)$ the direct sum of the generalized eigenspaces $E(A, \lambda)$ for eigenvalues λ having positive real part and by $E^S(A)$ the direct sum of the generalized eigenspaces for eigenvalues having negative real part. These are called, respectively, the unstable subspace and stable subspace of A. If there are no purely imaginary eigenvalues, so that $\mathbf{C}^n$ is the direct sum of the stable and unstable subspaces, then A is called a *hyperbolic operator*. As usual, σ_p denotes the maximal solution of $\frac{dz}{dt} = Az$.

2.1.4. Asymptotics Theorem for Linear ODE. Let A be a linear operator on $\mathbf{C}^n$. If $p \in E^S(A)$ and $q \in E^U(A)$ are nonzero vectors, then
$$\lim_{t \to \infty} \|\sigma_p(t)\| = 0,$$
$$\lim_{t \to -\infty} \|\sigma_p(t)\| = \infty,$$
$$\lim_{t \to -\infty} \|\sigma_q(t)\| = 0,$$
and
$$\lim_{t \to \infty} \|\sigma_q(t)\| = \infty.$$
More precisely, if we choose $\nu_0 > 0$ smaller than the real parts of all eigenvalues of A having positive real part and choose $\mu_0 < 0$ larger than the real parts of all eigenvalues having negative real part, then there are positive constants C_i such that $\|\sigma_p(t)\| < C_1 e^{\mu_0 t}$ for $t \to \infty$, $\|\sigma_p(t)\| > C_2 e^{\mu_0 t}$ for $t \to -\infty$, $\|\sigma_q(t)\| < C_3 e^{\nu_0 t}$ for $t \to -\infty$, and $\|\sigma_q(t)\| > C_4 e^{\nu_0 t}$ for $t \to \infty$.

Proof. Replacing A by $-A$ interchanges $E^S(A)$ and $E^U(A)$, so it is enough to prove the inequalities for p, and by the triangle inequality, we can even assume that p is in an eigenspace $E(A, \lambda)$, with $\lambda = \mu + i\nu$ and $\mu < \mu_0 < 0$. Then borrowing notation from above, $\|e^{tA}p\| \leq \sum_{j=0}^{r-1} \frac{1}{j!} |e^{t\lambda} t^j| \|N\|^j \|p\|$, and by part (a) of the above exercise this is $o(e^{\mu_0 t})$ for $t \to \infty$, proving the first inequality. For the second, let s be the highest power of N such that $N^s p \neq 0$, so that $\|e^{tA}p\| = |e^{\lambda t} t^s| \left\| \frac{1}{s!} N^s p + \frac{1}{t} \frac{1}{(s-1)!} N^{s-1} p + \cdots + \frac{1}{t^s} p \right\|$. For $t \to -\infty$, $|e^{\lambda t} t^s| > e^{\mu_0 t}$, and $\left\| \frac{1}{s!} N^s p + \frac{1}{t} \frac{1}{(s-1)!} N^{s-1} p + \cdots + \frac{1}{t^s} p \right\| > \frac{1}{2s!} \|N^s p\|$, so the second inequality follows also. ∎

▷ **Exercise 2–9.** Verify the following fact (used implicitly in the above proof). If $\mathbf{C}^n$ is the direct sum of linear subspaces $V_1, \ldots, V_k$, then there is a positive constant C such that if $v_i \in V_i$ and $v = \sum_i v_i$, then $\|v\| \geq C \sum_i \|v_i\|$. (Hint: The linear map T of $\bigoplus_i V_i$ to $\mathbf{C}^n$ defined by $(v_1, \ldots, v_k) \mapsto v_1 + \cdots + v_k$ is invertible, and $\|T^{-1} v\| \leq \|T^{-1}\| \|v\|$.)

2.1. First-Order Linear ODE

2.1.5. Corollary. If the operator A is hyperbolic, then $E^S(A)$ can be characterized purely topologically, as the set of all z in $\mathbf{C}^n$ for which $\sigma_z(t)$ converges to 0 as $t \to \infty$, and similarly $E^U(A)$ can be characterized as the set of all z for which $\sigma_z(t)$ converges to 0 as $t \to -\infty$.

Proof. Since A is hyperbolic, we can write z uniquely in the form $z = p+q$ with $p \in E^S(A)$ and $q \in E^U(A)$. Then $\sigma_z(t) = \sigma_p(t) + \sigma_q(t)$, so by the theorem, $\sigma_z(t) \to 0$ as $t \to \infty$ if and only if $q = 0$, and similarly $\sigma_z(t) \to 0$ as $t \to -\infty$ if and only if $p = 0$. ∎

2.1.6. Definition. Let p be an equilibrium point of an autonomous first-order (not necessarily linear) differential equation $\frac{dx}{dt} = V(x)$. We call p a *stable equilibrium* of this equation if given any neighborhood O of p, there is a smaller neighborhood U of p such that for any q in U, $\sigma_q(t) \in O$ for all positive t. If in addition we can choose U so that $\sigma_q(t) \to p$ as $t \to \infty$ for all $q \in U$, then we say that p is *asymptotically stable*.

2.1.7. Asymptotic Stability Theorem for Linear ODE. The origin is an asymptotically stable equilibrium of the linear ODE $\frac{dz}{dt} = Az$ if and only if all of the eigenvalues of A have negative real parts. In fact, if $\alpha > 0$ and the real part of every eigenvalue of A is less than $-\alpha$, then there is a positive constant C such that $\left\|e^{tA}\right\| < Ce^{-\alpha t}$ for all $t \geq 0$.

Proof. If A has an eigenvalue with positive real part, then it follows from the Asymptotics Theorem that the origin cannot even be stable. Similarly, if A has a purely imaginary eigenvalue $i\mu$ and if v is a corresponding eigenvector, then $\left\|e^{tA}v\right\| = \left\|e^{i\mu t}v\right\| = \|v\|$ for all t, so again the origin is not asymptotically stable. Conversely, assume all eigenvalues have real part negative, say less than $-\alpha$. It follows from the Asymptotics Theorem that for t sufficiently large, say greater than T, $\left\|e^{tA}\right\| < e^{-\alpha t}$, so we can take C to be anything larger than 1 and the maximum of $e^{\alpha t}\left\|e^{tA}\right\|$ for $0 \leq t \leq T$. ∎

▷ **Exercise 2–10.** Determine under what conditions the origin is stable but not asymptotically stable.

2.2. Nonautonomous First-Order Linear ODE

The first-order linear ODEs studied in the preceding section were autonomous, but there are also situations in which one must deal with time-dependent first-order linear ODEs. Perhaps the simplest example of such an equation is one of the form $\frac{dx}{dt} = a(t)x$, where a is a continuous real-valued function on **R**, and for an equation of this form, the solution to the initial value problem is clearly $x(t) = x(t_0) \exp(\int_{t_0}^{t} a(s)\,ds)$. In this section we will consider the natural generalization of this equation and its solution to $\mathbf{R}^n$. (And in fact everything we will say works equally well in $\mathbf{C}^n$.)

A continuous map $A : \mathbf{R} \to \mathbf{L}(\mathbf{R}^n)$ of **R** into the space of linear operators on $\mathbf{R}^n$ defines a continuous, time-dependent vector field V^A on $\mathbf{R}^n$: $V^A(x,t) = A(t)x$, and so a nonautonomous linear ODE $\frac{dx}{dt} = A(t)x$ on $\mathbf{R}^n$. You may be worried that since V^A is only continuous and not C^1, our existence and uniqueness results do not apply to this ODE, but we will now see that the linearity in x more than compensates for the lack of smoothness in t. The first thing to notice is that V^A satisfies a Lipschitz condition with respect to its x variable: $\|V^A(x_1,t) - V^A(x_2,t)\| \leq \|A(t)\| \|x_1 - x_2\|$. So if we define $K = \sup_{|t| \leq M} \|A(t)\|$, we have the following uniform Lipschitz estimate:

2.2.1. Proposition. $\|V^A(x_1,t) - V^A(x_2,t)\| \leq K \|x_1 - x_2\|$ for all $x_1, x_2 \in \mathbf{R}^n$ and all $t \in \mathbf{R}$ with $|t| \leq M$.

We shall next see that this estimate is just what we need to prove that the sequence of successive approximations, defined recursively by $x^0(t) = x^0$ and $x^{k+1}(t) = x^0 + \int_{t_0}^{t} A(s) x^k(s)\,ds$, converges globally to a solution of the initial value problem.

▷ **Exercise 2–11.** Prove $\|x^{k+1}(t) - x^k(t)\| \leq \frac{(K|t|)^k}{k!}$ for $|t| \leq M$. (Hint: Use induction and the proposition.)

Then, by the triangle inequality, $\|x^{N+m}(t) - x^m(t)\| \leq \sum_{k=m}^{N+m} \frac{(KM)^k}{k!}$ for all $t \in [-M, M]$. Now the series $\sum_{k=0}^{\infty} \frac{(KM)^k}{k!}$ converges to e^{KM},

2.2. Nonautonomous First-Order Linear ODE

so its sequence of partial sums is Cauchy, proving that the sequence x^k is uniformly Cauchy and hence converges to a limit x, uniformly on every finite interval. But uniform convergence is just what we need to pass to the limit in the recursive definition of the sequence x^k, so we end up with the fixed point formula $x(t) = x^0 + \int_{t_0}^{t} A(s)x(s)\,ds$, which of course implies that x solves the initial value problem. The Lipschitz condition gives uniqueness too: if $\tilde{x}$ is a second solution, then $\tilde{x}(t) = x^0 + \int_{t_0}^{t} A(s)\tilde{x}(s)\,ds$, so subtracting, taking norms, and using the Lipschitz condition gives $\|x(t) - \tilde{x}(t)\| \le \int_{t_0}^{t} K\,\|x(s) - \tilde{x}(s)\|\,ds$, and Gronwall's Inequality implies $\|x(t) - \tilde{x}(t)\| = 0$. This proves

2.2.2. Theorem. *Let A be a continuous map of $\mathbf{R}$ into the space $\mathbf{L}(\mathbf{R}^n)$ of linear operators on $\mathbf{R}^n$. For each t_0 in $\mathbf{R}$ and p in $\mathbf{R}^n$ there is a unique solution $x : \mathbf{R} \to \mathbf{R}^n$ of the time-dependent ODE $\frac{dx}{dt} = A(t)x$ satisfying the initial condition $x(t_0) = p$. This solution is linear in p, so that there is a uniquely determined map Σ^A of $\mathbf{R} \times \mathbf{R}$ into $\mathbf{L}(\mathbf{R}^n)$ such that $x(t) = \Sigma^A(t, t_0)p$.*

2.2.3. Definition. The map $\Sigma^A : \mathbf{R} \times \mathbf{R} \to \mathbf{L}(\mathbf{R}^n)$ is called the *propagator* for the ODE $\frac{dx}{dt} = A(t)x$.

2.2.4. Remark. Nothing in the above requires that $A(t)$ be defined for all $t \in \mathbf{R}$. If A is only defined on some interval I, then the same considerations apply, but of course the propagator function is then only defined on $I \times I$.

2.2.5. Properties of Propagators. The following properties of the propagator function Σ^A follow directly from its definition (plus Gronwall's Inequality for the fifth property).

1) $\Sigma^A(t, t) = I$.
2) $\frac{d}{dt}\Sigma^A(t, t_0) = A(t)\Sigma^A(t, t_0)$.
3) $\Sigma^A(t_2, t_1)\Sigma^A(t_1, t_0) = \Sigma^A(t_2, t_0)$.
4) $\Sigma^A(t_1, t_2) = \Sigma^A(t_2, t_1)^{-1}$.
5) $\|\Sigma^A(t_1, t_2)\| \le e^{K|t_1 - t_2|}$ if $\|A(s)\| \le K$ for s between t_1 and t_2.

▷ **Exercise 2–12.** Check that these properties always hold.

2.2.6. Remark. Let's use Euler's Method (Section 1.2) to approximate $\Sigma(t,t_0)x^0$. We take N equal time-steps of length $\Delta t = \frac{(t-t_0)}{N}$ and write $t_k = t_0 + k\Delta t$. The Euler approximation is

$$(I + A(t_N)\Delta t)(I + A(t_{N-1})\Delta t)\ldots(I + A(t_0)\Delta t)x^0.$$

Since in general the operators $A(t)$ at different times do not commute, it is important to take the product of operators in the "time-ordered" sense, i.e., the factors $(I + A(t_k))$ must be ordered from right to left in order of increasing k. Letting N tend to infinity, we get $\Sigma(t) = \lim_{N\to\infty} \Pi_{k=0}^{N}(I + A(t_k)\Delta t)$. Notice that this looks a lot like an integral, with "Riemann products" replacing Riemann sums, and for this reason $\Sigma(t)$ is often referred to as a "time-ordered product integral". In the special case that $A(t)$ is a constant matrix A, we get a familiar looking formula, $e^{tA} = \lim_{N\to\infty}(I + \frac{t}{N}A)^N$.

2.3. Coupled and Uncoupled Harmonic Operators

The natural generalization of the equation of simple harmonic motion, $\frac{d^2x}{dt^2} = -\omega^2 x$, to higher dimensions is the equation for n coupled harmonic oscillators, namely an equation of the form $\frac{d^2x}{dt^2} = -Kx$ (or, in coordinates, $\frac{d^2x_i}{dt^2} = -\sum_{j=1}^{n} K_{ij}x_j$) where now x denotes a point of $\mathbf{R}^n$ and K is a positive definite self-adjoint operator on $\mathbf{R}^n$. Recall that self-adjointness means that $\langle Kx, y\rangle = \langle x, Ky\rangle$, or equivalently that the matrix K_{ij} is symmetric. As we shall see in Appendix G, it follows that K has an orthonormal basis $e_1,\ldots,e_n$ of eigenvectors, with corresponding eigenvalues $\lambda_1,\ldots,\lambda_n$, and saying that K is positive just means that each eigenvalue λ_i is a positive real number ω_i^2.

▷ **Exercise 2–13.** Show that K has a unique positive, self-adjoint square root Ω and that Ω is characterized by $\Omega e_i = \omega_i e_i$.

Thus, we can rewrite the coupled harmonic oscillator equation as $\frac{d^2x}{dt^2} = -\Omega^2 x$, and if we use orthogonal coordinates $y_1,\ldots,y_n$ given by these eigenvectors e_i, then since the matrix for Ω is diagonal in this basis, the coupled harmonic oscillator equation reduces to an uncoupled system of n simple harmonic motion equations $\frac{d^2y_i}{dt^2} = -\omega_i^2 y_i$. We call these equations "uncoupled" because now the right-hand side of the

2.3. Coupled and Uncoupled Harmonic Operators

equation for $\frac{d^2 y_i}{dt^2}$ involves only y_i and not the other y_j, so that solving the system reduces to solving each one individually without reference to the others, and we have really reduced the problem of solving a system of n coupled harmonic oscillators to the already solved initial value problem for simple harmonic motion. We can summarize this observation as follows.

2.3.1. Proposition. *The solution of the coupled harmonic oscillator equation $\frac{d^2 x}{dt^2} = -\Omega^2 x$ with initial conditions $x(0) = x^0$ and $x'(0) = u^0$ is $x(t) = \cos(t\Omega)x^0 + \sin(t\Omega)(\Omega^{-1} u^0)$. (See Figure 2.2.)*

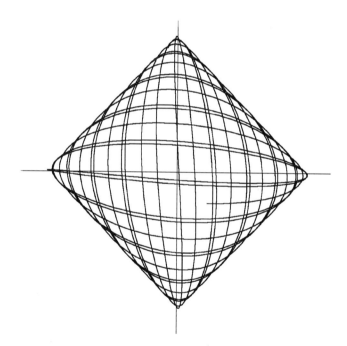

Figure 2.2. Two coupled harmonic oscillators.

▷ **Exercise 2–14.** Rederive this proposition by combining the exponential formula for the flow generated by a linear first-order equation with the general approach to reducing second-order equations to first order. Hint: Denote elements of $\mathbf{R}^n \times \mathbf{R}^n$ by a column vector $\binom{x}{v}$ of two elements of $\mathbf{R}^n$, and rewrite the harmonic oscillator system on $\mathbf{R}^n$ as the equivalent first-order linear system on $\mathbf{R}^n \times \mathbf{R}^n$ $\frac{d}{dt}\binom{x}{v} = \mathcal{A}\binom{x}{v}$ where $\mathcal{A}$ is the linear operator on $\mathbf{R}^n \times \mathbf{R}^n$ defined by $\mathcal{A}\binom{x}{v} = \binom{v}{-\Omega^2 x}$. Show by induction that

$$\mathcal{A}^{2n}\binom{x}{v} = \binom{(-1)^n \Omega^{2n} x}{(-1)^n \Omega^{2n+1} \Omega^{-1} v}$$

and

$$\mathcal{A}^{2n+1}\binom{x}{v} = \binom{(-1)^n \Omega^{2n+1} \Omega^{-1} v}{(-1)^{n+1} \Omega^{2n+2} x}.$$

You can now easily evaluate $\exp(t\mathcal{A})$, and if you recall the power series for sin and cos, you should have no problem completing the exercise.

We will return several times to the consideration of the coupled harmonic oscillator equation. It will reappear both as an approximation for any conservative Newtonian system close to equilibrium and also as the prototype for a so-called completely integrable Hamiltonian system.

2.4. Inhomogeneous Linear Differential Equations

We can also use the explicit solution of the initial value problem for linear differential equations to solve explicitly an important class of nonlinear time-dependent equations referred to as inhomogeneous linear ordinary differential equations. These are equations of the form

$$\frac{dx}{dt} = Ax + g(t),$$

where as above $A: \mathbf{R}^n \to \mathbf{R}^n$ is a linear operator and $g: \mathbf{R} \to \mathbf{R}^n$ is an arbitrary smooth function of time.

▷ **Exercise 2–15.** For $n = 1$, an inhomogeneous linear differential equation has the form $\frac{dx}{dt} = ax + g(t)$, where a is a real number and $g: \mathbf{R} \to \mathbf{R}$ is a smooth real-valued function. In this case, show by direct verification that the solution with initial value x^0 is given by $x(t) = e^{at}(x^0 + \int_0^t e^{-as} g(s)\, ds)$.

It does not require much imagination to guess the correct generalization.

2.5. Asymptotic Stability of Nonlinear ODE

2.4.1. Variation of Parameters Formula. The solution, $x(t)$, of the inhomogeneous linear differential equation $\frac{dx}{dt} = Ax + g(t)$ with initial condition $x(0) = x^0$ is given by

$$x(t) = e^{tA}x^0 + \int_0^t e^{(t-s)A} g(s)\, ds.$$

▷ **Exercise 2–16.** Verify the validity of this formula (and see if you can find the reason behind its strange name).

▷ **Exercise 2–17.** Assume that the linear operator A is asymptotically stable, i.e., all of its eigenvalues have negative real part, and assume that the forcing term $g(t)$ is periodic with period $T > 0$. Show that there is a unique point $p \in \mathbf{R}^n$ for which the solution $x(t)$ with initial value $x(0) = p$ is periodic with period T. (Hint: x is given by the above Variation of Parameters Formula, so the condition that it be periodic of period T is that $p = e^{TA}p + \int_0^T e^{(T-s)A} g(s)\, ds$, or $p = (I - e^{TA})^{-1} \int_0^T e^{(T-s)A} g(s)\, ds$. Why is the operator $(I - e^{tA})$ invertible?)

▷ **Exercise 2–18.** If we start with a nonautonomous linear equation $\frac{dx}{dt} = A(t)x$ with propagator $\Sigma^A(t_1, t_0)$ (see Section 2.2), we can again add a time-dependent "forcing" term, $g(t)$, to get the inhomogeneous version $\frac{dx}{dt} = A(t)x + g(t)$. Verify the following generalization of the Variation of Parameters Formula for this case:

$$x(t) = \Sigma^A(t, t_0) \left(x^0 + \int_{t_0}^t \Sigma^A(t_0, s) g(s)\, ds \right).$$

2.5. Asymptotic Stability of Nonlinear ODE

In addition to the straightforward use of the Variation of Parameters Formula to solve $\frac{dx}{dt} = Ax + g(t)$ explicitly when the function g is known, there is another clever application of the formula to obtain qualitative information about solutions of autonomous nonlinear systems. This works as follows. Suppose we are given a smooth vector field $V : \mathbf{R}^n \to \mathbf{R}^n$ and we have a decomposition of V as $V = A + r$, where A is a linear map and r is a "remainder". If $x(t)$ is a solution of $\frac{dx}{dt} = V(x)$ with $x(0) = x^0$, then $x'(t) = A(x(t)) + g(t)$, where $g(t) := r(x(t))$. It follows from the Variation of Parameters Formula that $x(t) = e^{tA}x^0 + \int_0^t e^{(t-s)A}r(x(s))\,ds$, and taking norms, $\|x(t)\| \le \|e^{tA}\|\,\|x^0\| + \int_0^t \|e^{(t-s)A}\|\,\|r(x(s))\|\,ds$.

At first sight this formula seems to lead nowhere, since the unknown function $x(t)$ also occurs on the right-hand side of the equation. But as we shall now see, it allows us to use Gronwall's Inequality to bootstrap from partial information about a solution x, together with qualitative information about A and r, to derive more precise information concerning x.

Consider a C^2 vector field $V : \mathbf{R}^n \to \mathbf{R}^n$ that has an equilibrium point (i.e., vanishes) at the origin, 0. (If V has an equilibrium at some other point p, then we could translate p to 0, so this is no loss of generality.) We will investigate the behavior of solutions $x(t)$ with initial condition $x(0) = x^0$ near 0. If $A = DV_0$ is the differential of V at 0, then by Taylor's Theorem, $V(x) = Ax + r(x)$, where $r(x) = \|x\|R(x)$, with $R(x)$ a function that is smooth except at 0 and $\lim_{x \to 0} \|R(x)\| = 0$. Now suppose that 0 is an asymptotically stable equilibrium of the linearized equation $\frac{dx}{dt} = Ax$. By the Asymptotic Stability Theorem for Linear ODE, this just means that all the eigenvalues of A have negative real part. Moreover, if we choose $\alpha > 0$ such that the real part of every eigenvalue of A is less than $-\alpha$, then we saw that there is a positive constant C such that $\|e^{tA}\| < Ce^{-\alpha t}$ for $t \ge 0$. The norm estimate above now implies that $\|x(t)\| \le Ce^{-\alpha t}\|x^0\| + Ce^{-\alpha t}\int_0^t e^{\alpha s}\|x(s)\|\|R(x(s))\|\,ds$. If we define $u(t) := e^{\alpha t}\|x(t)\|$, then we can write this as $u(t) \le C\|x^0\| + C\int_0^t u(s)\|R(x(s))\|\,ds$. Can you see where we are headed?

Choose a positive number K less than α, so that $\kappa = K - \alpha$ is negative. Then, since $\lim_{x \to 0} \|R(x)\| = 0$, we can choose an $\epsilon > 0$ so that $\|R(y)\| < \frac{K}{C}$ if $\|y\| < \epsilon$. Thus if our solution $x(t)$ satisfies $\|x(t)\| < \epsilon$ for all t in some interval $I = [0, T]$, we will have $u(t) \le C\|x^0\| + K\int_0^t u(s)\,ds$ for $t \in I$, and so by Gronwall's Inequality, $u(t) \le C\|x^0\|e^{Kt}$, i.e., $\|x(t)\| \le C\|x^0\|e^{\kappa t}$ for all t in I.

2.6. Forced Harmonic Oscillators

Note that taking $t = 0$ gives $C \geq 1$, so if we choose $\delta < \frac{\epsilon}{C}$, then $\delta < \epsilon$. We now assert that if $\|x^0\| < \delta$, then the solution $x(t)$ with initial condition $x(0) = x^0$ satisfies $\|x(t)\| < \epsilon$ for all $t > 0$, from which it then follows that $\|x(t)\| \leq C\|x^0\|e^{\kappa t}$ for all $t > 0$ and hence that $x(t)$ converges to 0 exponentially, and in particular it follows that 0 is an asymptotically stable equilibrium point of V. It is easy to prove the assertion. If it were false, then there would be a first time T that $\|x(t)\| \geq \epsilon$, so that $\|x(t)\| \leq \epsilon$ for all t in $I = [0, T]$. Thus by the above computation, $\|x(T)\| \leq C\|x^0\|e^{\kappa T}$. But $C\|x^0\|e^{\kappa T} < C\delta e^{\kappa T} < \epsilon e^{\kappa T} < \epsilon$ since κ is negative. This implies that $\|x(T)\| < \epsilon$, a contradiction, and the assertion is proved. We have now proved the following important theorem:

2.5.1. Asymptotic Stability Theorem for Nonlinear ODE. *Let V be a C^2 vector field in $\mathbf{R}^n$. Assume that $V(p) = 0$, and let $A = DV_p$. Then p is an asymptotically stable equilibrium point of the nonlinear equation $\frac{dx}{dt} = V(x)$ provided that the origin is an asymptotically stable equilibrium point of the linearized ODE $\frac{dx}{dt} = Ax$ (i.e., if all the eigenvalues of A have negative real parts).*

2.6. Forced Harmonic Oscillators

Next let's look at the important inhomogeneous version of the coupled harmonic oscillator equation: $\frac{d^2x}{dt^2} = -\Omega^2 x + g(t)$. The function $g : \mathbf{R} \to \mathbf{R}^n$ has the interpretation of a time-dependent (but spatially constant) force acting on the oscillators, and the equation is usually referred to as the forced harmonic oscillator equation.

▷ **Exercise 2–19.** Show that the solution of the initial value problem for the forced harmonic oscillator with initial position x^0 and initial velocity u^0 is

$$x(t) = \xi(t) + \Omega^{-1}\int_0^t \sin((t-s)\Omega)g(s)\,ds,$$

where $\xi(t) = \cos(t\Omega)x^0 + \sin(t\Omega)(\Omega^{-1}u^0)$ is the solution of the unforced harmonic oscillator with the same initial conditions.

We can rewrite the solution of the forced harmonic oscillator as $x(t) = \xi(t) + \int_0^t G(t-s)g(s)\,ds$, where $G(t) = \Omega^{-1}\sin(t\Omega)$ (called the Green's operator). We can think of $\Delta(t) = \int_0^t G(t-s)g(s)\,ds$ as the deviation from the unforced solution, and it has two interesting physical interpretations. First it is the solution of the forced oscillator as seen by an observer moving with the unforced oscillator, and second, it is the solution of the forced oscillator corresponding to zero initial position and velocity.

▷ **Exercise 2–20.** (Resonance) Consider a one-dimensional forced harmonic oscillator equation with periodic forcing: $\frac{d^2x}{dt^2} = -\omega^2 x + \cos(\omega_0 t)$. Solve this explicitly, treating separately the "resonant" case, when the forcing frequency ω_0 is an integer multiple of the natural frequency ω. Note that in this case the maximum amplitude of the oscillations goes to infinity essentially linearly in time, while in the nonresonant case the motion remains bounded. (Think about pushing a swing.)

2.7. Exponential Growth and Ecological Models

The linear differential equation $\frac{dq}{dt} = kq$ and its integrated form $q(t) = q(t_0)e^{k(t-t_0)}$ turn up with great regularity as equations purporting to model the time evolution of real-world systems. While such models can be very useful within a limited domain, it is not uncommon to see them extended far beyond the limits of their validity, leading to nonsensical or paradoxical predictions of extreme exponential growth. It is important to keep in mind when developing models of evolving systems that, while death and decay may unfortunately be inevitable, birth and growth usually are dependent on the availability of limited resources.

Let us start by recalling briefly the standard reasoning that leads to models of exponential growth (or decay). Usually $q(t)$ measures the quantity of some "substance" present at time t, and the dynamic assumptions are that there are two processes going on simultaneously: a "birth" or "growth" process that in a small time interval Δt adds to an amount Q of the substance an additional amount $aQ\Delta t$ and a "death" or "decay" process that in this same time interval subtracts

2.7. Exponential Growth and Ecological Models

an amount $bQ\Delta t$. If we imagine the substance as consisting of many discrete units or "atoms", then a measures the probability that an atom will give birth to another atom in unit time and b similarly measures the probability that it will die in unit time. If we define $k = a - b$, then the net effect of both processes is to change Q by the amount $\Delta Q = kQ\Delta t$ in the time interval Δt, so that $\frac{\Delta Q}{\Delta t} = kQ$. Letting Δt tend to zero, we arrive at the above differential equation.

Since it is an autonomous equation, we will as usual choose $t_0 = 0$. Changing the units by which we measure the quantity of the substance will replace q by some positive multiple Cq, and while this does not change the equation, it replaces $q(0)$ by $Cq(0)$, so when units are not important, it is natural to choose them to make $q(0) = 1$. Similarly, changing units of time replaces t by a multiple mt, which has the effect of replacing k by mk, so again if the unit of time is not important, we could choose it so that $k = 1$. However we will not use either of these simplifying normalizations below.

▷ **Exercise 2–21.** Let us define an *exponential growth law* more abstractly as a continuous function $f : \mathbf{R} \to \mathbf{R}^+ := (0, \infty)$ having the property that on any interval $[t, t+\delta]$ it changes by a factor that depends only on the length, δ, of the interval and not on its location. If we denote this "growth factor" by $g(\delta)$, then in symbols we can write this defining property as the functional equation $f(t+\delta) = f(t)g(\delta)$.

1) Check that $t \mapsto Ce^{kt}$ is an exponential growth law. What is g in this case?

2) Given that f is an exponential growth law, suppose that $f(3) = 5$ and $f(7) = 20$. Find $f(5)$, $f(11)$, $f(1)$, $f(0)$.

▷ **Exercise 2–22.** Let f be an exponential growth law as in the preceding exercise, with growth factor g.

1) Show that $g(\delta) = f(\delta)/f(0)$.

2) Deduce that $g(\delta_1 + \delta_2) = g(\delta_1)g(\delta_2)$ and hence that g must have the form e^{kt} so $f(t) = f(0)e^{kt}$.

There is nothing sacrosanct about using e as a base for exponentiation in an exponential growth law; any other positive number B would do as well, since $e^{kt} = e^{\log_e B(k/\log_e B)t} = (e^{\log_e B})^{(k/\log_e B)t} = B^{\kappa t}$, where $\kappa = \frac{k}{\log_e B}$. Since most people have a better feeling for powers of 2 or powers of 10 than for powers of e, there is often merit in using 2 or 10 as a base, and so it is worth remembering that (approximately) $\log_e 2 = 0.6931$ and $\log_e 10 = 2.3026$.

Let us now return to our growth law in its standard form, $q(t) = q(0)e^{kt}$. In each time interval of length $\tau = \frac{\log_e 2}{|k|}$, q will be doubled if k is positive or halved if k is negative, and for this reason τ is called the doubling-time or half-life, respectively. Let us (somewhat arbitrarily) define the "critical interval" to be the interval $[-T, T]$, where T is 30τ, or approximately $\frac{20.8}{|k|}$. Since 2^{30} is approximately a billion, what we see is that if $q(0)$ represents a "practical" amount of the substance, then at one end of the critical interval $q(t)$ will be invisibly small, while at the other end of the critical interval it will be enormous.

One typical application of an exponential growth law is to model the decay of a radioactive isotope. It is not difficult to measure the decay constant $k = -b$ (and hence the half-life) with considerable precision and to verify that the long term behavior is as predicted. Another common application is to model biological growth, and it is here that one must proceed with some care. Just as the decay rates of radioactive atoms can be measured using a Geiger counter or ionization chamber, so the growth rates of yeast, bacteria, algae, etc., can be measured using test tubes and Petri dishes, and doubling-times are often on the order of hours or days. Similarly, the growth rate for the population of some animal species in a more or less limited geographic area can be estimated by counting and statistical sampling, and here doubling-times are on the order of months or years.

Now, if these exponential growth models for biological systems were rigorously true, our test tubes and Petri dishes would quickly overflow and our laboratory would soon be knee-deep in a mixture of slime molds, pond scum, and other unpleasant substances. It is not hard to see where the assumptions for the exponential growth model have gone wrong. The birth and death rates for, say, yeast are

2.7. Exponential Growth and Ecological Models

not in fact constants, but they depend on environmental variables such as the temperature and the composition of the medium in which they grow. If a culture is started off in a fresh medium with ample nutrients, the growth rate will be optimal and the death rate low. But if nutrients are not replenished and fermentation products such as alcohol not removed, then the rate of cell division will gradually lessen and older cells will starve or be poisoned, so the growth rate k will gradually decrease. A better equation for modeling the growth of such systems, one that attempts to take such effects into account, is the so-called *logistic equation*, $\frac{dq}{dt} = c(Q-q)q$. Here Q is a constant (called the carrying capacity) and the growth rate $c(Q-q)$ decreases to zero as q approaches this limiting size. The convex exponential curve is replaced by an "S-curve" that starts out approximately exponential (when q is small and the growth rate is approximately cQ) but then has a point of inflection and becomes concave, approaching Q asymptotically as its limiting value.

Species in nature do not exist in isolation, but rather as components of inter-related families of species called eco-systems, and the birth and death rates of any one species in an eco-system will usually depend on the population sizes of one or more other species of the system. If there are n species, with sizes $q_1(t), \ldots, q_n(t)$, the dynamics of the eco-system will be governed by a system of n equations, $\frac{dq_i}{dt} = k_i q_i$, with $k_i = a_i - b_i$, the a_i and b_i being, respectively, the growth and death rates of the ith species. For a closed system (i.e., one where every species on which a given species depends is part of the system) the growth and death rates will not depend explicitly on time, so the governing system of ODEs is autonomous. **But** since the rates k_i will be functions of the q_i, the equations more explicitly are $\frac{dq_i}{dt} = k_i(q_1, \ldots, q_n) q_i$, and we see that this system of ODEs is no longer linear.

A classic example of an eco-system model is the Volterra-Lotka equations, also known as the predator-prey model. One of its major virtues is its simplicity; while it is more of a toy model than a realistic description of biological competition, it is fairly easy to analyze and yet exhibits many salient features of more detailed models that while more realistic are also considerably harder to analyze. In this model

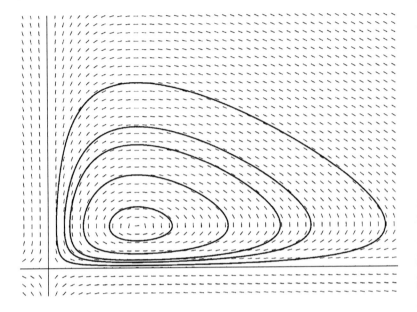

Figure 2.3. The Volterra-Lotka equation.

we assume a closed system consisting of two species, a predator species (e.g., foxes or cats) with population $q_1(t)$ and a prey species (e.g., rabbits or mice) with population $q_2(t)$. We assume that the prey have an adequate food supply, so that their birth rate a_2 is a constant γ, and we assume that the predators are at the top of the food chain so that they have a constant death rate $b_1 = \beta$. On the other hand we assume that the prey death rate is proportional to the number of predators, so $b_2 = \delta q_1$, and that the predator birth rate is proportional to the food (prey) available to them, so $a_1 = \alpha q_2$. Thus $k_1 = a_1 - b_1 = (\alpha q_2 - \beta)$ and $k_2 = a_2 - b_2 = (\gamma - \delta q_1)$ and the Volterra-Lotka equations are

$$\frac{dq_1}{dt} = (\alpha q_2 - \beta) q_1,$$
$$\frac{dq_2}{dt} = (\gamma - \delta q_1) q_2.$$

A particularly clear and careful analysis of the Volterra-Lotka system and its solutions can be found in [HS]. Here we will only give an abbreviated qualitative description. Solutions in the first quadrant (the

2.7. Exponential Growth and Ecological Models

only meaningful region for the model) are all periodic. Starting from the point on an orbit nearest the origin (where both predator and prey populations are small), first the prey start increasing approximately exponentially (since there are few predators eating them), while the population of predators remains nearly constant since they have little food (prey). Eventually, when the prey population gets large enough, the predators start growing rapidly in numbers since they now have adequate food, and as this happens, since the prey are now getting eaten in larger numbers, their growth rate slows and eventually becomes negative. With their food now decreasing, the growth rate of the predators also first slows and then becomes negative, and soon both populations are back to where they started. Figure 2.3 is a picture of the Volterra-Lotka direction field and several typical orbits.

You perhaps have begun to wonder where *Homo sapiens* fits into this picture. We are at the top of the food chain—there are no longer any saber-toothed tigers out there to eat us. Moreover, the Green Revolution and modern medical technology have greatly lessened the toll taken by famine and pestilence. And while serious local conflicts are as common as ever, we even seem to be making progress in lessening the risk of the cataclysmic kind of global war that in earlier times was responsible for wholesale death. As a result, the day is in sight when the average death rate for the global human population will approximate the constant value that is the lower limit set by our biology. On the other side of the ledger—births—the picture is far more hazy and only very poorly understood. What is involved is a highly complex combination of social factors and human psychology that go into the choices people make that in turn determine total fertility, i.e., the average number of children a woman will have in her lifetime. This number varies considerably from one geographic region to another at a given instant and also can change quite rapidly in time in a given region. Since the growth rate of Earth's human population depends very sensitively on this number, its significant potential variability and the uncertainty in forecasting it beyond a few years make almost any long-term prediction demographers come up with something to be treated with considerable caution.

That being said, when both sides of the ledger are added up and averaged over the whole planet, the human population growth rate has been fairly constant over the past seventy-five years. It grew slowly until it reached a maximum of about two percent in the mid 1960s (corresponding to a doubling-time of about thirty-five years) and then dropped somewhat and in recent decades has been hovering around 1.6 percent. In that period the estimated total world population has gone from 2 billion in 1930 to 4 billion in 1975 and 6 billion in 2000. It is expected to reach 8 billion by 2020—suggesting a doubling-time of approximately forty-five years.

No one who has considered the matter seriously believes that it is other than dangerous hubris to think that we humans can avoid the imperatives of unlimited exponential growth indefinitely, even with our ability to make continuing technological advances. Order of magnitude estimates of the basic resources that would be required, such as fossil fuel and other energy sources, food, water, and sunlight, suggest that it is highly unlikely that Earth could support even close to fifty billion people at anything approaching the current level of consumption in the developed countries.[1] But, at the present growth rate, that is the number of humans who will inhabit our planet in less than two hundred years. The Four Horsemen are waiting in the wings, and if we do not ourselves find a more benign way to limit our population, then well before that time has passed, they will return to help us do the job.

---

[1] For a detailed, illuminating, and very well-written discussion of this matter see J. E. Cohen's "How Many People Can the Earth Support?" (W. W. Norton, 1995), especially Chapter 13 and Appendix 3. (But don't expect easy answers!)

Chapter 3

Second-Order ODE and the Calculus of Variations

3.1. Tangent Vectors and the Tangent Bundle

Let $\sigma : I \to \mathbf{R}^n$ be a C^1 curve in $\mathbf{R}^n$ and suppose that $\sigma(t_0) = p$ and $\sigma'(t_0) = v$. Up until this point we have referred to v as either the velocity or the tangent vector to σ at time t_0. From now on we will make a small but important distinction between these two concepts. While the distinction is not critical in dealing with first-order ODE, it will simplify our discussion of second-order ODE. Henceforth we will refer to v as the velocity of σ at time t_0 and define its tangent vector at time t_0 to be the ordered pair $\dot{\sigma}(t_0) := (p, v)$. (If you are familiar with the distinction that is sometimes made between "free" and "based" vectors, you will recognize this as a special case of that.)

The set of all tangent vectors (p, v) we get in this way is called the *tangent bundle* of $\mathbf{R}^n$, and we will denote it by $\mathbf{TR}^n$. Clearly $\mathbf{TR}^n = \mathbf{R}^n \times \mathbf{R}^n$, so you might think that it is a useless complication to introduce this new symbol, but in fact there is a good reason for using this kind of redundant notation. As mathematical constructs get more complex, it is important to have notation that gives visual clues about what symbols mean. So, in particular, when we want

3. Second-Order ODE and the Calculus of Variations

to emphasize that a pair (p,v) is referring to a tangent vector, it is better to write $(p,v) \in \mathbf{TR}^n$ rather than $(p,v) \in \mathbf{R}^n \times \mathbf{R}^n$.

The projection of $\mathbf{TR}^n$ onto its first factor, taking (p,v) to its base point p, will be denoted by $\Pi : \mathbf{TR}^n \to \mathbf{R}^n$, and it is called the tangent bundle projection. (There is of course another projection, onto the second factor, but it will play no role in what follows and we will not give it a name.) The set of all tangent vectors that project onto a fixed base point p will be denoted by $\mathbf{T}_p\mathbf{R}^n$. It is called the tangent space to $\mathbf{R}^n$ at the point p, and its elements are called tangent vectors at p. It is clearly an n-dimensional real vector space, and its dual space, the space of linear maps of $\mathbf{T}_p\mathbf{R}^n$ into $\mathbf{R}$, is called the cotangent space of $\mathbf{R}^n$ at p and is denoted by $\mathbf{T}_p^*\mathbf{R}^n$.

Let f be a smooth real-valued function defined in an open set O of $\mathbf{R}^n$. If $p \in O$ and $V = (p,v) \in \mathbf{T}_p\mathbf{R}^n$, recall that Vf, the directional derivative of f at p in the direction v, is defined as $Vf := \sum_{i=1}^n v_i \frac{\partial f}{\partial x_i}(p)$ (see Appendix C). If as above $\sigma : I \to \mathbf{R}^n$ is a smooth curve and $V = \dot\sigma(t_0)$ is its tangent vector at time t_0, then by the chain rule $\left(\frac{d}{dt}\right)_{t=t_0} f(\sigma(t)) = Vf$. We will follow the customary practice of using the symbolic differentiation operator $\sum_{i=1}^n v_i(\frac{\partial}{\partial x_i})_p$ as an alternative notation for the tangent vector V. If we fix f, then $V \mapsto Vf$ is a linear functional, df_p, on $\mathbf{T}_p\mathbf{R}^n$ (i.e., an element of $\mathbf{T}_p^*\mathbf{R}^n$) called the differential of f at p.

Note that the function f defined on O gives rise to two associated functions in $\Pi^{-1}(O)$. The first is just $f \circ \Pi$, $(p,v) \mapsto f(p)$, and the second is df, $(p,v) \mapsto df_p(v)$. This last remark is the basis of a very important construction that "promotes" a system of local coordinates $(x_1, \ldots, x_n)$ for $\mathbf{R}^n$ in O (see Appendix D) to a system of local coordinates $(q_1, \ldots, q_n, \dot q_1, \ldots, \dot q_n)$ for $\mathbf{TR}^n$ in $\Pi^{-1}(O)$ called the associated *canonical coordinates* for the tangent bundle. Namely, we define $q_i := x_i \circ \Pi$ and $\dot q_i := dx_i$. Suppose $V_1 = (p_1, v_1)$ and $V_2 = (p_2, v_2)$ are in $\Pi^{-1}(O)$ and that $q_i(V_1) = q_i(V_2)$ and $\dot q_i(V_1) = \dot q_i(V_2)$ for all $i = 1, \ldots, n$. Since the x_i are local coordinates in O and $x_i(p_1) = q_i(V_1) = q_i(V_2) = x_i(p_2)$, it follows that $p_1 = p_2 = p$. It also follows from the fact that x_i is a local coordinate system that the $(dx_i)_p$ are a basis for $\mathbf{T}_p\mathbf{R}^n$; hence $(dx_i)_p(v_1) = \dot q_i(V_1) = \dot q_i(V_2) = (dx_i)_p(v_2)$ implies $v_1 = v_2$. Thus $V_1 = V_2$, so $(q_1, \ldots, q_n, \dot q_1, \ldots, \dot q_n)$ really are

3.1. Tangent Vectors and the Tangent Bundle

local coordinates for $\mathbf{TR}^n$ in $\Pi^{-1}(O)$. Note that for each p in O, $(\dot{q}_1, \ldots, \dot{q}_n)$ are actually Cartesian coordinates on $\mathbf{T}_p\mathbf{R}^n$. We will be using canonical coordinates almost continuously from now on.

▷ **Exercise 3–1.** Check that if $(x_1, \ldots, x_n)$ are the standard coordinates for $\mathbf{R}^n$, then $(q_1, \ldots, q_n, \dot{q}_1, \ldots, \dot{q}_n)$ are the standard coordinates for $\mathbf{TR}^n = \mathbf{R}^{2n}$.

In Appendix C we define a certain natural vector field R on $\mathbf{R}^n$—called the radial or Euler vector field—by $R(x) := x$, and we point out that it has the same expression in every Cartesian coordinate system $(x_1, \ldots, x_n)$, namely $R = \sum_{i=1}^n x_i \frac{\partial}{\partial x_i}$. We also recall there Euler's famous theorem on homogeneous functions, which states that if $f : \mathbf{R}^n \to \mathbf{R}$ is C^1 and is positively homogeneous of degree k (meaning $f(tx) = t^k f(x)$ for all $t > 0$ and $x \neq 0$), then $Rf = \sum_{i=1}^n x_i \frac{\partial f}{\partial x_i} = kf$. We now define a vector field R on $\mathbf{TR}^n$, also called the radial or Euler vector field, by defining it on each tangent space $\mathbf{T}_p\mathbf{R}^n$ to be the radial vector field on $\mathbf{T}_p\mathbf{R}^n$. Explicitly, $R(p, v) = (0, v)$. Recalling that $(\dot{q}_1, \ldots, \dot{q}_n)$ are Cartesian coordinates on $\mathbf{T}_p\mathbf{R}^n$ we have

3.1.1. Proposition. *If $(q_1, \ldots, q_n, \dot{q}_1, \ldots, \dot{q}_n)$ are canonical coordinates for $\mathbf{TR}^n$ in $\Pi^{-1}(O)$, then the radial vector field R on $\mathbf{TR}^n$ has the expression*

$$R = \sum_{i=1}^n \dot{q}_i \frac{\partial}{\partial \dot{q}_i}$$

in $\Pi^{-1}(O)$. Hence if $F : \mathbf{TR}^n \to \mathbf{R}$ is a C^1 real-valued function that is positively homogeneous of degree k on each tangent space $\mathbf{T}_p\mathbf{R}^n$, then $RF = \sum_{i=1}^n \dot{q}_i \frac{\partial F}{\partial \dot{q}_i} = kF$.

Suppose that $\sigma : I \to \mathbf{R}^n$ is a C^1 curve. A path $\tilde{\sigma} : I \to \mathbf{TR}^n$ is called a *lifting* of σ if it projects onto σ under Π, i.e., if it is of the form $\tilde{\sigma}(t) = (\sigma(t), \lambda(t))$ for some C^1 map $\lambda : I \to \mathbf{R}^n$. You should think of a lifting of σ as being a vector field defined along σ. There are many different possible liftings of σ. For example, the lifting $t \mapsto (\sigma(t), 0)$ is the zero vector field along σ, and $t \mapsto (\sigma(t), \sigma''(t))$ is the acceleration vector field of σ. The tangent vector field $\dot{\sigma}$, $t \mapsto (\sigma(t), \sigma'(t))$ plays an especially important role, and we shall also refer to it as the *canonical lifting* of σ to the tangent bundle. We note that it takes C^k curves in $\mathbf{R}^n$ ($k \geq 1$) to C^{k-1} curves in $\mathbf{TR}^n$.

Let $x^1, \ldots, x^n$ be local coordinates in O, and assume σ maps I into O. In these coordinates the curve σ is described by its so-called parametric representation, $x_i(t) := x^i(\sigma(t))$. Let's see what the parametric representation is for the canonical lifting, $\dot\sigma(t)$, with respect to the canonical coordinates $(q_1, \ldots, q_n, \dot q_1, \ldots, \dot q_n)$ defined by the x^i. Since $\Pi \circ \dot\sigma(t) = \sigma(t)$, it follows from the definition of the q_i that
$$q_i(\dot\sigma(t)) = x^i(\Pi(\dot\sigma(t))) = x_i(t).$$
On the other hand,
$$\dot q_i(\dot\sigma(t)) = dx^i(\sigma'(t)) = \frac{d}{dt} x^i(\sigma(t)) = \frac{dx_i(t)}{dt}.$$

3.2. Second-Order Differential Equations

We have seen that solving a first-order differential equation in $\mathbf{R}^n$ involves finding a path $x(t)$ in $\mathbf{R}^n$ given

1) its initial position and

2) its velocity as a function of its position and the time.

Similarly, solving a second-order differential equation in $\mathbf{R}^n$ involves finding a path $x(t)$ in $\mathbf{R}^n$ given

1) its initial position **and** its initial velocity and

2) its acceleration as a function of its position, its velocity, and the time.

To make this precise, suppose $A : \mathbf{R}^n \times \mathbf{R}^n \times \mathbf{R} \to \mathbf{R}^n$ is a C^1 function. A C^2 curve $x(t)$ in $\mathbf{R}^n$ is said to be a solution of the second-order differential equation in $\mathbf{R}^n$, $\frac{d^2 x}{dt^2} = A\left(x, \frac{dx}{dt}, t\right)$, if $x''(t) = A(x(t), x'(t), t)$ holds for all t in the domain of x. Given such a second-order differential equation on $\mathbf{R}^n$, the associated initial value problem (or IVP) is to find a solution $x(t)$ for which both the position $x(t_0)$ and the velocity $x'(t_0)$ have some specified values at a particular time t_0.

Fortunately, we do not have to start all over from scratch to develop theory and intuition concerning second-order equations. There

3.2. Second-Order Differential Equations

is an easy trick that we already remarked on in Chapter 1 that effectively reduces the consideration of a second-order differential equation in $\mathbf{R}^n$ to the consideration of a first-order equation in $\mathbf{TR}^n = \mathbf{R}^n \times \mathbf{R}^n$. Namely, given $A : \mathbf{R}^n \times \mathbf{R}^n \times \mathbf{R} \to \mathbf{R}^n$ as above, define a time-dependent vector field V on $\mathbf{TR}^n$ by $V(p, v, t) = (v, A(p, v, t))$. Suppose first that $(x(t), v(t))$ is a C^1 path in $\mathbf{TR}^n$. Note that its derivative is just $(x'(t), v'(t))$, so this path is a solution of the first-order differential equation $\frac{d(x,v)}{dt} = V(x, v, t)$ if and only if $(x'(t), v'(t)) = V(x(t), v(t), t) = (v(t), A(x(t), v(t), t))$, which of course means that $x'(t) = v(t)$ while $v'(t) = A(x(t), v(t), t)$. But then $x''(t) = v'(t) = A(x(t), v(t), t)$, so $x(t)$ is a solution of the second-order equation $\frac{d^2x}{dt^2} = A\left(x, \frac{dx}{dt}, t\right)$. Conversely, if $x(t)$ is a solution of $\frac{d^2x}{dt^2} = A\left(x, \frac{dx}{dt}, t\right)$ and we define $v(t) = x'(t)$, then $(x(t), v(t))$ is the canonical lifting of $x(t)$ and is a solution of $\frac{d(x,v)}{dt} = V(x, v, t)$. What we have shown is

3.2.1. Reduction Theorem for Second-Order ODE. *The canonical lifting of C^2 curves in $\mathbf{R}^n$ to C^1 curves on $\mathbf{TR}^n$ sets up a bijective correspondence between solutions of the second-order differential equation $\frac{d^2x}{dt^2} = A\left(x, \frac{dx}{dt}, t\right)$ on $\mathbf{R}^n$ and solutions of the first-order equation $\frac{d(x,v)}{dt} = V(x, v, t)$ on $\mathbf{TR}^n$, where $V(x(t), v(t), t) = (v(t), A(x(t), v(t), t))$.*

▷ **Exercise 3–2.** Use this correspondence to formulate and prove existence, uniqueness, and smoothness theorems for second-order differential equations from the corresponding theorems for first-order differential equations. Extend this to higher-order differential equations.

According to the Reduction Theorem, a second-order ODE in $\mathbf{R}^n$ is described by some vector field on $\mathbf{TR}^n$. But be careful, not every vector field V on $\mathbf{TR}^n$ arises in this way.

▷ **Exercise 3–3.** Show that a time-dependent vector field $V(p, v, t)$ on $\mathbf{TR}^n$ arises as above from some second-order ODE on $\mathbf{R}^n$, $\frac{d^2x}{dt^2} = A\left(x, \frac{dx}{dt}, t\right)$, if and only if $\Pi(V(p, v, t)) = v$ for all $(p, v, t) \in \mathbf{TR}^n \times \mathbf{R}$.

▷ **Exercise 3–4.** Let $x^1, \ldots, x^n$ be local coordinates in some open set O of $\mathbf{R}^n$ and let $q_1, \ldots, q_n, \dot{q}_1, \ldots, \dot{q}_n$ be the associated canonical coordinates in $\mathbf{TR}^n$. Suppose that $\sigma : I \to O$ is a smooth path and $x_i(t) := x^i(\sigma(t))$ its parametric representation, and let $q_i(t) := q_i(\dot{\sigma}(t))$, $\dot{q}_i(t) := \dot{q}_i(\dot{\sigma}(t))$ be the parametric representation of the canonical lifting $\dot{\sigma}$. Show that σ is a solution of the second-order ODE $\frac{d^2x}{dt^2} = A\left(x, \frac{dx}{dt}, t\right)$ if and only if for all $t \in I$, $\frac{d\dot{q}_i(t)}{dt} = A_i(q_1(t), \ldots, q_n(t), \dot{q}_1(t), \ldots, \dot{q}_n(t), t)$. (Hint: Recall that $q_i(t) := x_i(t)$ and $\dot{q}_i(t) := \frac{dx_i(t)}{dt}$.)

3.2.2. Definition. Let $\frac{d^2x}{dt^2} = A\left(x, \frac{dx}{dt}\right)$ be a time-independent second-order ODE on $\mathbf{R}^n$. A function $f : \mathbf{TR}^n \to \mathbf{R}$ is called a *constant of the motion* (or a *conserved quantity* or a *first integral*) of this ODE if f is constant along the canonical lifting, $\dot{\sigma}$, of every solution curve σ (equivalently, if whenever $x(t)$ satisfies the ODE, then $f(x(t), x'(t))$ is a constant).

▷ **Exercise 3–5.** Let V be a vector field in $\mathbf{TR}^n$ defined by $V(p, v) := (v, A(p, v))$. Show that $f : \mathbf{TR}^n \to \mathbf{R}$ is a constant of the motion of $\frac{d^2x}{dt^2} = A\left(x, \frac{dx}{dt}\right)$ if and only if the directional derivative, Vf, of f in the direction V, is identically zero.

3.3. The Calculus of Variations

Where do second-order ordinary differential equations come from—or to phrase this question somewhat differently, what sort of things get represented mathematically as solutions of second-order ODEs?

Perhaps the first answer that will spring to mind for many people is Newton's Second Law of Motion, "$F = ma$", which without doubt inspired much of the early work on second-order ODE. But as we shall soon see, important as Newton's Equations of Motion are, they are best seen mathematically as a special case of a much more general class of second-order ODE, called Euler-Lagrange Equations, and a more satisfying answer to our question will grow out of an understanding of this family of equations.

Euler-Lagrange Equations arise as a necessary condition for a particular curve to be a maximum (or minimum) for certain real-valued

3.3. The Calculus of Variations

functions on spaces of curves. We will introduce these equations formally in the following section. Here, to help set the stage, we look at the simplest of all second-order ODE—one that happens to be of Euler-Lagrange type—namely $\frac{d^2x}{dt^2} = 0$. The solution $x(t)$ with the initial condition $x(0) = p$ and $x'(0) = v$ is clearly $x(t) = p + tv$, so in particular all of its solutions are straight lines parametrized proportionally to arclength. Of course, this is just the special case of Newton's Second Law when the force is zero. (Newton considered this special case to be so important that he called it the First Law of Motion—every body remains in a state of rest or of uniform motion in a straight line unless compelled to change that state by forces acting on it.) But from a geometers point of view, what characterizes a straight line is that it has minimal length among all paths joining two points (see Appendix E). Thus the simple equation $\frac{d^2x}{dt^2} = 0$ succinctly encodes the answer to an important optimization question. This is a typical example of a Calculus of Variations Problem and of its solution and we will now look at the general theory behind such problems.

First a little warning: many people find the traditional notation used in the Calculus of Variations somewhat confusing (and even bizarre) at first. Once you get used to it, you will probably come to appreciate its succinctness and expressiveness, but to try to get you over this initial hurdle, we will start with a careful explanation of this notation.

To fix the notation, we choose local coordinates $x^1, \ldots, x^n$ in $\mathbf{R}^n$ and let $q_1, \ldots, q_n, \dot{q}_1, \ldots, \dot{q}_n$ denote the associated canonical coordinates for $\mathbf{TR}^n$. If F is a function on $\mathbf{TR}^n$, we will adopt the usual abuse of notation and write $F(q_1, \ldots, q_n, \dot{q}_1, \ldots, \dot{q}_n)$ for its coordinate representation with respect to these coordinates, i.e., $F(V) = F(q_1(V), \ldots, q_n(V), \dot{q}_1(V), \ldots, \dot{q}_n(V))$. Moreover, to simplify the notation further, we will usually abbreviate $F(q_1, \ldots, q_n, \dot{q}_1, \ldots, \dot{q}_n)$ to $F(q_i, \dot{q}_i)$. Let $\sigma(t)$ be a path in O and let $x_i(t) := x^i(\sigma(t))$ be its parametric representation in these coordinates. We recall that the parametric representation of the canonical lifting, $\dot{\sigma}(t)$, with respect to these canonical coordinates is given by $q_i(t) = x_i(t)$ and $\dot{q}_i(t) = \frac{dx_i(t)}{dt}$,

3. Second-Order ODE and the Calculus of Variations

so that
$$F(\dot\sigma(t)) = F\left(x_i(t), \frac{dx_i(t)}{dt}\right).$$

A Calculus of Variations Problem starts from a smooth real-valued function $\mathcal{L}$ on $\mathbf{TR}^n$, usually referred to as the *Lagrangian function*. The important thing to understand about the Lagrangian is how it gets used. Namely, given any smooth path, $\sigma : [a,b] \to \mathbf{R}^n$ as above, we define a smooth real-valued function on $[a,b]$, $\mathcal{L}(\dot\sigma) : [a,b] \to \mathbf{R}$, by composing $\mathcal{L}$ with the canonical lifting $\dot\sigma$ of σ, i.e., $t \mapsto \mathcal{L}(\dot\sigma(t)) = \mathcal{L}(q_i(t), \dot q_i(t))$. The next step is to associate a single real number, $F_\mathcal{L}(\sigma)$, to the Lagrangian $\mathcal{L}$ and the path σ by integrating $\mathcal{L}(\dot\sigma)$ from a to b:

$$F_\mathcal{L}(\sigma) := \int_a^b \mathcal{L}(\dot\sigma(t))\,dt = \int_a^b \mathcal{L}(q_i(t), \dot q_i(t))\,dt.$$

This mapping $F_\mathcal{L}$ from paths σ to real numbers $F_\mathcal{L}(\sigma)$ is called the *associated functional* defined by the Lagrangian $\mathcal{L}$. (And it is these functionals that we were referring to in the introduction to this section when we spoke of "certain real-valued functions on spaces of curves".)

The main goal of the Calculus of Variations is to find the maxima, the minima, and the other "extrema" of such functionals, $F_\mathcal{L}$, or rather of the restriction of $F_\mathcal{L}$ to certain special spaces of curves. The precise definition of extrema will be given below, but roughly speaking, they include in addition to maxima and minima, analogues of points of inflection for functions defined on $\mathbf{R}$ and saddle-points for functions defined on $\mathbf{R}^2$.

3.4. The Euler-Lagrange Equations

The so-called Euler-Lagrange Equations associated to a Lagrangian $\mathcal{L}$ is a system of n second-order ordinary differential equations on $\mathbf{R}^n$, written symbolically as

$$\frac{\partial \mathcal{L}}{\partial q_i} - \frac{d}{dt}\frac{\partial \mathcal{L}}{\partial \dot q_i} = 0.$$

At first sight this may look puzzling. In the first place, they look more like a system of partial differential equations than ordinary differential equations. How should one interpret these equations as ordinary

3.4. The Euler-Lagrange Equations

differential equations, and what does it mean to say that a curve

$$\sigma(t) = (x_1(t), \ldots, x_n(t))$$

in $\mathbf{R}^n$ satisfies this set of equations?

Since $\mathcal{L}$ is a function on $\mathbf{TR}^n$, it has coordinate expressions $\mathcal{L}(q_i, \dot{q}_i)$ with respect to a system of canonical coordinates, and in the Euler-Lagrange Equations, $\frac{\partial \mathcal{L}}{\partial q_i}$ and $\frac{\partial \mathcal{L}}{\partial \dot{q}_i}$ refer to the partial derivatives of these functions of $(q_i, \dot{q}_i)$ **evaluated along the canonical lifting** $\dot{\sigma}$ of σ; i.e., after taking the partial derivatives, we substitute $q_i(t)$ and $\dot{q}_i(t)$ for q_i and $\dot{q}_i$, where $q_i(t) := q_i(\dot{\sigma}(t)) = x(t)$ and $\dot{q}_i(t) := \dot{q}_i(\dot{\sigma}(t)) = \frac{dx_1(t)}{dt}$. The proper interpretation of $\frac{d}{dt}\frac{\partial \mathcal{L}}{\partial \dot{q}_i}$ is of course a simple consequence of this interpretation of $\frac{\partial \mathcal{L}}{\partial \dot{q}_i}$, and if we apply the chain rule, the Euler-Lagrange Equations become

$$\sum_{j=1}^{n}\left(\frac{\partial^2 \mathcal{L}}{\partial \dot{q}_i \partial \dot{q}_j}\right)\frac{d^2 x_j}{dt^2} + \sum_{j=1}^{n}\left(\frac{\partial^2 \mathcal{L}}{\partial \dot{q}_i \partial q_j}\right)\frac{dx_j}{dt} = \frac{\partial \mathcal{L}}{\partial q_i},$$

where of course $\frac{\partial^2 \mathcal{L}}{\partial \dot{q}_i \partial \dot{q}_j}$, $\frac{\partial^2 \mathcal{L}}{\partial \dot{q}_i \partial q_j}$, and $\frac{\partial \mathcal{L}}{\partial q_i}$ are all to be evaluated at

$$(q_i(t), \dot{q}_i(t)) = \left(x_i(t), \frac{dx_i(t)}{dt}\right).$$

This now looks more like a second-order ODE for the functions $x_i(t)$ defining the path σ. However it is unlike the second-order systems that we have dealt with up until now, in that the second derivatives $\frac{d^2 x_j}{dt^2}$ are **not** given explicitly but rather implicitly, and it is not immediately clear that we can solve the Euler-Lagrange Equations uniquely for these second derivatives.

3.4.1. Definition. A Lagrangian $\mathcal{L}(q, \dot{q})$ is called *regular* or *nondegenerate* if the matrix $\frac{\partial^2 \mathcal{L}}{\partial \dot{q}_i \partial \dot{q}_j}$ is invertible for all points $(q, \dot{q})$ in $\mathbf{TR}^n$. In this case we denote by M_{ij} the matrix-valued function on $\mathbf{TR}^n$ which at each point $(q_i, \dot{q}_i)$ is the inverse matrix of $\frac{\partial^2 \mathcal{L}}{\partial \dot{q}_i \partial \dot{q}_j}$ at that point.

We will assume that our Lagrangians are nondegenerate in what follows—this is usually the case for Lagrangians that come up in applications. With this assumption, we can now rewrite the Euler-Lagrange Equations in the standard form for a system of n second-order ODE for the n functions $x_k(t)$ that define the path σ:

$$\frac{d^2 x_k}{dt^2} = \sum_{i=1}^{n} M_{ki} \left(\frac{\partial \mathcal{L}}{\partial q_i} - \sum_{j=1}^{n} \left(\frac{\partial^2 \mathcal{L}}{\partial \dot{q}_i \partial q_j} \right) \frac{dx_j}{dt} \right).$$

The matrix elements M_{ki}, as well as the partial derivatives of $\mathcal{L}$, are of course all evaluated at

$$(q_i(t), \dot{q}_i(t)) = \left(x(t), \frac{dx_1(t)}{dt} \right),$$

so this latter equation has exactly the form of a system of n second-order ODE for the n functions $x_i(t)$ that we have considered previously.

3.4.2. Definition. A C^2 curve $\sigma(t) = (x_1(t), \ldots, x_n(t))$ in $\mathbf{R}^n$ is called an *extremal* for the functional $F_\mathcal{L}$ if the $x_i(t)$ satisfy the Euler-Lagrange Equations.

As we will see a little later, if $\sigma : [a, b] \to \mathbf{R}^n$ is a C^2 curve such that $F_\mathcal{L}$ assumes its maximum (or a minimum) at σ among all C^2 curves with the same parameter interval $[a, b]$ and with the same endpoints $\sigma(a)$ and $\sigma(b)$, then σ will be an extremal of $F_\mathcal{L}$. Thus the Euler-Lagrange Equations play the role in the Calculus of Variations analogous to that played by the equation $f'(x) = 0$ in ordinary calculus when one looks for the maxima and minima of a real-valued function $f(x)$. But they are **only** necessary conditions for a maximum or minimum. Just as in the case of ordinary calculus (where one can have inflection points as well as maxima and minima) although a maximum or minimum of $F_\mathcal{L}$ will satisfy the Euler-Lagrange Equations, there can also be other paths that satisfy the Euler-Lagrange Equations but are neither maxima nor minima of $F_\mathcal{L}$.

By the Reduction Theorem for Second-Order ODE, we can replace the above system of n second-order equations for the n functions $x_i(t)$ by a system of $2n$ first-order equations for $2n$ functions

3.5. Conservation Laws for Euler-Lagrange Equations

$q_i(t)$ and $\dot{q}_i(t)$:

$$\frac{dq_k}{dt} = \dot{q}_k,$$

$$\frac{d\dot{q}_k}{dt} = \sum_{i=1}^{n} M_{ki} \left(\frac{\partial \mathcal{L}}{\partial q_i} - \sum_{j=1}^{n} \left(\frac{\partial^2 \mathcal{L}}{\partial \dot{q}_i \partial q_j} \right) \dot{q}_j \right).$$

Given a solution $(q_i(t), \dot{q}_i(t))$ to this first-order system, it follows that $x_i(t) = q_i(t)$ will solve the Euler-Lagrange Equations, and so $\sigma(t) = (x_1(t), \ldots, x_n(t))$ will be a extremal of $F_\mathcal{L}$.

3.4.3. Proposition. Assume that $\mathcal{L} : \mathbf{TR}^n \to \mathbf{R}$ is regular and of class C^{r+2}, $r \geq 0$. Then every extremal of $F_\mathcal{L}$ is of class at least C^r.

▷ **Exercise 3–6.** Prove this. (Hint: The argument is an almost trivial induction. Can you see why it is referred to as "bootstrapping"?)

3.5. Conservation Laws for Euler-Lagrange Equations

While there are some advantages to transforming the Euler-Lagrange Equations as above into a more standard looking system of first- or second-order ODE, as we shall now see, there are also advantages to working with these equations in their original form, $\frac{\partial \mathcal{L}}{\partial q_i} - \frac{d}{dt}\frac{\partial \mathcal{L}}{\partial \dot{q}_i} = 0$, or equivalently, $\frac{d}{dt}\frac{\partial \mathcal{L}}{\partial \dot{q}_i} = \frac{\partial \mathcal{L}}{\partial q_i}$.

3.5.1. Definition. If $(q_i, \dot{q}_i)$ are canonical coordinates in $\mathbf{TR}^n$ and $\mathcal{L} : \mathbf{TR}^n \to \mathbf{R}$ is a Lagrangian function, then for each $k = 1, \ldots, n$ we define a function p_k on $\mathbf{TR}^n$ called the *conjugate momentum* of q_k by $p_k := \frac{\partial \mathcal{L}}{\partial \dot{q}_i}$. We call q_k an *ignorable coordinate* with respect to the Lagrangian function $\mathcal{L}$ if $\mathcal{L}(q_i, \dot{q}_i)$ is independent of q_k, i.e., if $\frac{\partial \mathcal{L}}{\partial q_k}$ is identically zero, or equivalently if $\mathcal{L}(q_i, \dot{q}_i)$ is unchanged by the substitution $q_k \mapsto q_k + c$ for every constant c.

It follows directly from this definition that

3.5.2. Proposition. Let $\mathcal{L} : \mathbf{TR}^n \to \mathbf{R}$ be a Lagrangian function and let $(q_i, \dot{q}_i)$ be canonical coordinates in $\mathbf{TR}^n$. If q_k is an ignorable coordinate with respect to the Lagrangian $\mathcal{L}$, then the conjugate

74   3. Second-Order ODE and the Calculus of Variations

momentum $p_k = \frac{\partial \mathcal{L}}{\partial \dot{q}_i}$ is a constant of the motion for solutions of the Euler-Lagrange Equations $\frac{d}{dt}\frac{\partial \mathcal{L}}{\partial \dot{q}_i} = \frac{\partial \mathcal{L}}{\partial q_i}$.

Each Lagrangian function, $\mathcal{L} : \mathbf{TR}^n \to \mathbf{R}$, has associated to it an important related function, called the *total energy function* for $\mathcal{L}$.

3.5.3. Definition. If $\mathcal{L} : \mathbf{TR}^n \to \mathbf{R}$ is a Lagrangian function, we define the associated total energy function $E_\mathcal{L}$ on $\mathbf{TR}^n$ by $E_\mathcal{L} := R\mathcal{L} - \mathcal{L}$, where $R\mathcal{L}$ denotes the radial derivative of $\mathcal{L}$. Equivalently, in any canonical coordinate system we can define $E_\mathcal{L} := \sum_{i=1}^n \dot{q}_i \frac{\partial \mathcal{L}}{\partial \dot{q}_i} - \mathcal{L} = \sum_{i=1}^n p_i \dot{q}_i - \mathcal{L}$, where p_i is the conjugate momentum of q_i.

3.5.4. Remark. By Euler's theorem on homogeneous functions, if $\mathcal{L}$ is positively homogeneous of degree k on each tangent space $\mathbf{T}_p\mathbf{R}^n$, then $E_\mathcal{L} = (k-1)\mathcal{L}$. There is an important class of examples from geometry that we will study later (Riemannian metrics) for which $\mathcal{L}$ is actually a quadratic polynomial on each $\mathbf{T}_p\mathbf{R}^n$, so that in this case $E_\mathcal{L}$ is just $\mathcal{L}$ itself.

What makes $E_\mathcal{L}$ important is that it is **always** a constant of the motion for the Euler-Lagrange Equations defined by $\mathcal{L}$.

3.5.5. Conservation of Total Energy Theorem. If $\mathcal{L}$ is any Lagrangian function on $\mathbf{TR}^n$, then its associated total energy function, $E_\mathcal{L}$, is a constant of the motion for solutions of the Euler-Lagrange Equations, $\frac{\partial \mathcal{L}}{\partial q_i} - \frac{d}{dt}\frac{\partial \mathcal{L}}{\partial \dot{q}_i} = 0$, defined by $\mathcal{L}$.

▷ **Exercise 3–7.** Prove the Conservation of Total Energy Theorem. (Hint: Compute

$$\frac{d}{dt}\left(\sum_{i=1}^n \frac{\partial \mathcal{L}}{\partial \dot{q}_i}\dot{q}_i - \mathcal{L}\right)$$

using the chain rule and make the appropriate substitutions from the Euler-Lagrange Equations.)

3.5.6. Remark. The subject of conserved quantities (or constants of the motion) plays a very important role in the Calculus of Variations, and we will return to it frequently. In particular we will see later that there is an intimate connection between conserved quantities and symmetry properties of the Lagrangians.

3.6. Two Classic Examples

Before demonstrating that the Euler-Lagrange Equations are in fact necessary conditions for minima and maxima of $F_\mathcal{L}$, we will look in detail at two example Lagrangians: the *Speed*, $S(p,v) := \|v\|$, and the *Energy Density*, $\boldsymbol{E}(p,v) := \frac{1}{2}\|v\|^2$. Working a little with these examples should give you some feeling for how all the above formalism works out in practice, and moreover both happen to be of considerable interest in their own right.

Notice that $F_S(\sigma) := \int_a^b \|\sigma'(t)\|\, dt$ is just (by definition) the length of the curve σ, and we will also denote it by $L(\sigma)$. Recall that it is unchanged by reparametrization. $F_{\boldsymbol{E}}(\sigma) := \frac{1}{2}\int_a^b \|\sigma'(t)\|^2\, dt$ is usually called the *energy* of the curve σ and we will denote it by $\mathcal{E}(\sigma)$. These two Lagrangians are clearly very closely related—indeed, $S = \sqrt{2\boldsymbol{E}}$, and not surprisingly this leads to important relations between $L(\sigma)$ and $\mathcal{E}(\sigma)$.

▷ **Exercise 3–8.** Show that if $\sigma : [a,b] \to \mathbf{R}^n$ is any C^1 path, then $L(\sigma) \leq \sqrt{2(b-a)}\sqrt{\mathcal{E}(\sigma)}$, and show that this inequality is an equality if and only if σ is parametrized proportionally to arclength. (Hint: Apply the Schwarz inequality to the functions $\|\sigma'(t)\|$ and the constant function 1 on the interval $[a,b]$.)

▷ **Exercise 3–9.** Show that of all C^1 paths joining two points p and q in $\mathbf{R}^n$ and parametrized by the interval $[a,b]$, the unique one of minimal energy is the straight line parametrized proportionally to arclength: $\sigma(t) = p + \frac{t-a}{b-a}(q-p)$, which has energy $\mathcal{E}(\sigma) = \frac{\|p-q\|^2}{2(b-a)}$. (Hint: Use the preceding exercise together with the fact that this σ minimizes the length of all such paths.)

We next look for the coordinate expressions $S(q_i, \dot q_i)$ and $\boldsymbol{E}(q_i, \dot q_i)$ associated to these Lagrangians. First consider the case of Cartesian coordinates $x^1, \ldots, x^n$ associated to an orthonormal basis $e_1, \ldots, e_n$ for $\mathbf{R}^n$. In this case the canonical coordinates $(q_i, \dot q_i)$ of a point $V = (p,v) \in \mathbf{TR}^n$ are defined by $p = \sum_{i=1}^n q_i e_i$ and $v = \sum_{i=1}^n \dot q_i e_i$. Since the e_i are orthonormal, $\boldsymbol{E}(q_i, \dot q_i) = \boldsymbol{E}(V) = \frac{1}{2}\|v\|^2 = \frac{1}{2}\sum_{i=1}^n \dot q_i^2$, and it follows that $S(q_i, \dot q_i) = \left(\sum_{i=1}^n \dot q_i^2\right)^{\frac{1}{2}}$. Then $\frac{\partial \boldsymbol{E}}{\partial q_i} = 0$, while $\frac{\partial \boldsymbol{E}}{\partial \dot q_i} = \dot q_i$.

3. Second-Order ODE and the Calculus of Variations

▷ **Exercise 3–10.** Use the fact that E is homogeneous of degree 2 to conclude that the total energy function E_E is just E itself, and check that conservation of total energy in this case just says that solutions of the Euler-Lagrange Equations are parametrized proportionally to arclength. On the other hand, use the fact that S is homogeneous of degree 1 to deduce that the total energy function E_S is identically zero (so that conservation of total energy for solutions of the Euler-Lagrange Equations associated to the speed provides no useful information).

Now that we have an explicit formula for $\boldsymbol{E}(q_i, \dot{q}_i)$, it is easy to write down the system of Euler-Lagrange Equations for the Lagrangian $\mathcal{L} = \boldsymbol{E}$, and we see that it is just the trivial system $\frac{d^2 x_i}{dt^2} = 0$ we discussed earlier, and its solutions as expected are straight lines parametrized proportionally to arclength. Another way to integrate these Euler-Lagrange Equations is to note that since all the q_i are clearly ignorable coordinates, all the conjugate momenta $p_i = \frac{\partial E}{\partial \dot{q}_i} = \dot{q}_i$ are constants of the motion, say $\dot{q}_i = v_i$, so $\frac{dq_i}{dt} = \dot{q}_i = v_i$ and hence $x_i(t) = q_i(t) = q_i(0) + t v_i$. While this is a typical example of how to employ ignorable coordinates, their true usefulness is masked here because the Euler-Lagrange Equations are so simple anyway. In more complex situations they can greatly simplify the integration problem.

In fact, the Euler-Lagrange Equations for the Lagrangian $\mathcal{L} = S$ is a good example of this. The system of second-order ODE now looks fairly complicated, but again all the q_i are ignorable, and now their conjugate momenta are $p_i = \frac{\partial S}{\partial \dot{q}_i} = \dot{q}_i (\sum_{i=1}^{n} \dot{q}_i^2)^{-\frac{1}{2}}$. Since the $\dot{q}_i(t)$ are just the components of $\sigma'(t)$, clearly the p_i are the components of the unit vector with the same direction as $\sigma'(t)$, i.e., the normalized tangent vector $\frac{\sigma'(t)}{\|\sigma'(t)\|}$, so the fact that these are constants of the motion just says that the normalized tangent of the solution $\sigma(t)$ is constant—i.e., that the solution curve σ points in a fixed direction and hence is a straight line.

Now let's look at the corresponding expressions for polar coordinates (r, θ) in $\mathbf{R}^2$, related as usual to the Cartesian coordinates (x, y) by $x = r \cos(\theta)$ and $y = r \sin(\theta)$. It is traditional to denote the canonical coordinates associated to these polar coordinates by

3.6. Two Classic Examples

$r, \theta, \dot{r}, \dot{\theta}$, rather than the generic $q_1, q_2, \dot{q}_1, \dot{q}_2$, and we will follow this custom. Suppose a path $\sigma(t)$ is given parametrically in the polar coordinates by $r = r(t)$ and $\theta = \theta(t)$. Then its parametric expression in Cartesian coordinates is $x(t) = r(t)\cos(\theta(t))$ and $y(t) = r(t)\sin(\theta(t))$. Now the calculation above gives $\boldsymbol{E}(\dot{\sigma}(t)) = \frac{1}{2}(x'(t)^2 + y'(t)^2)$, and an easy calculation then gives $\boldsymbol{E}(\dot{\sigma}(t)) = r'(t)^2 + r(t)^2\theta'(t)^2$. On the other hand, by definition of canonical coordinates, $\dot{r}(t) = r'(t)$ and $\dot{\theta}(t) = \theta'(t)$, and it follows that the coordinate expression for the energy density in the canonical coordinates associated to polar coordinates is $\boldsymbol{E}(r, \theta, \dot{r}, \dot{\theta}) = \frac{1}{2}(\dot{r}^2 + r^2\dot{\theta}^2)$. It follows of course that $S(r, \theta, \dot{r}, \dot{\theta}) = \sqrt{\dot{r}^2 + r^2\dot{\theta}^2}$, a formula that you should be familiar with from elementary calculus. The Euler-Lagrange Equations for the energy $\boldsymbol{E}$ in polar coordinates are $\frac{d}{dt}\frac{\partial \boldsymbol{E}}{\partial \dot{r}} = \frac{\partial \boldsymbol{E}}{\partial r}$ and $\frac{d}{dt}\frac{\partial \boldsymbol{E}}{\partial \dot{\theta}} = \frac{\partial \boldsymbol{E}}{\partial \theta}$, i.e., $\frac{d}{dt}\dot{r} = r\dot{\theta}^2$ and $\frac{d}{dt}(r^2\dot{\theta}) = 0$.

The second of these equations just expresses the fact that θ is an ignorable coordinate, so that its conjugate momentum $p_\theta = r^2\dot{\theta}$ is a constant to the motion. (Since θ is an angular variable, p_θ is called angular momentum.) We already know that conservation of total energy implies that solution curves are parametrized proportionally to arclength. Let's try to show that they are straight lines. A straight line is a curve satisfying $ax + by = 1$, or $r(a\cos\theta + b\sin\theta) = 1$ for some constants a and b, so if we define $u = \frac{1}{r}$, then a straight line is characterized by $u = a\cos\theta + b\sin\theta$. But these are just the solutions of the harmonic oscillator equation $\frac{d^2 u}{d\theta^2} = -u$.

▷ **Exercise 3–11.** Show that the Euler-Lagrange Equations $\frac{d}{dt}\dot{r} = r\dot{\theta}^2$ and $\frac{d}{dt}(r^2\dot{\theta}) = 0$ imply $\frac{d^2 u}{d\theta^2} = -u$. (Hint: The first equation gives $\dot{\theta} = \frac{c}{r^2} = cu^2$, so the chain rule says that $\frac{d}{dt} = \frac{d\theta}{dt}\frac{d}{d\theta} = \dot{\theta}\frac{d}{d\theta} = cu^2 \frac{d}{d\theta}$, and in particular $\dot{r} = \frac{d}{dt}r = \frac{d}{dt}\frac{1}{u} = -\frac{1}{u^2}\frac{du}{dt} = -\frac{1}{u^2}(cu^2\frac{du}{d\theta}) = -c\frac{du}{d\theta}$. Then on the one hand $\frac{d}{dt}\dot{r} = cu^2 \frac{d}{d\theta}\dot{r} = -c^2 u^2 \frac{d^2 u}{d\theta^2}$, while on the other hand the second Euler-Lagrange Equation now becomes $\frac{d}{dt}\dot{r} = r\dot{\theta}^2 = \frac{1}{u}(cu^2)^2 = c^2 u^3$, and equating these two expressions for $\frac{d}{dt}\dot{r}$ gives the desired result.)

▷ **Exercise 3–12.** Another Lagrangian we will consider in detail later plays a central role in particle dynamics, namely $\mathcal{L}(q, \dot{q}) =$

$K(\dot{q}) - U(q)$, where $K(\dot{q}) = \frac{1}{2}\sum_{i=1}^{n} m_i \dot{q}_i^2$ is called the kinetic energy and U is a smooth function on $\mathbf{R}^n$ called the potential energy function. The $m_1, \ldots, m_n$ are positive real numbers, called the particle masses. For this Lagrangian, $\frac{\partial \mathcal{L}}{\partial q_i} = -\frac{\partial U}{\partial q_i}$, while $\frac{\partial \mathcal{L}}{\partial \dot{q}_i} = m_i \dot{q}_i$, so the matrix $\frac{\partial^2 \mathcal{L}}{\partial \dot{q}_i \partial \dot{q}_j}$ is diagonal with the constants $m_1, \ldots, m_n$ on the diagonal. The Euler-Lagrange Equations reduce to Newton's Equations $m_i \frac{d^2 x_i}{dt^2} = F_i$, provided we identify the force with the negative gradient of the potential: $F_i = -\frac{\partial U}{\partial q_i}$. Show that for this Lagrangian the total energy function is the sum of the kinetic energy and the potential energy: $E_\mathcal{L}(q, \dot{q}) = K(\dot{q}) + U(q)$. (Hint: The kinetic energy is homogeneous of degree 2.)

3.7. Derivation of the Euler-Lagrange Equations

We fix a Lagrangian $\mathcal{L}$ on $\mathbf{TR}^n$ and will write simply F for the functional $F_\mathcal{L}$. We next specify the domain of F; namely we define it to be the vector space $C^2(I, \mathbf{R}^n)$ of all paths $\sigma : I \to \mathbf{R}^n$, where I is a closed interval $[a, b]$ and σ has two continuous derivatives on I. Given σ and h in $C^2(I, \mathbf{R}^n)$, we call $s \mapsto \sigma + sh$ the straight line in $C^2(I, \mathbf{R}^n)$ through σ in the direction h, and we call $DF(\sigma, h) = (\frac{d}{ds})_{s=0} F(\sigma + sh)$ the directional derivative of F at σ in the direction h. Our next goal is to find a formula for this. If we write $q_i(t, s) = \sigma_i(t) + sh_i(t)$ and $\dot{q}_i(t, s) = \sigma_i'(t) + sh_i'(t)$, then

$$F(\sigma + sh) = \int_a^b \mathcal{L}(q_1(t,s), \ldots, q_n(t,s), \dot{q}_1(t,s), \ldots, \dot{q}_n(t,s))\, dt,$$

and so differentiating under the integral sign and using the chain rule gives

$$DF(\sigma, h) = \int_a^b \sum_i \left(\frac{\partial \mathcal{L}}{\partial q_i} h_i(t) + \sum_i \frac{\partial \mathcal{L}}{\partial \dot{q}_i} h_i'(t) \right) dt,$$

and integrating each term in the second sum by parts gives

3.7.1. The Variational Formula.

$$DF(\sigma, h) = \int_a^b \sum_i \left(\frac{\partial \mathcal{L}}{\partial q_i} - \frac{d}{dt}\left(\frac{\partial \mathcal{L}}{\partial \dot{q}_i}\right)\right) h_i(t)\, dt + \sum_i \left[\frac{\partial \mathcal{L}}{\partial \dot{q}_i} h_i\right]_b^a.$$

3.7. Derivation of the Euler-Lagrange Equations

The following is the key fact in proving E. Noether's Theorem on the relation between symmetries of a Lagrangian and constants of the motion for the Euler-Lagrange Equations.

3.7.2. Corollary. *If σ is an extremal of the functional F, then*

$$DF(\sigma, h) = \sum_i \left[\frac{\partial \mathcal{L}}{\partial \dot{q}_i} h_i \right]_b^a.$$

Proof. By definition of an extremal, the Euler-Lagrange Equations are satisfied; hence the first summand in the variational formula vanishes. ∎

3.7.3. Corollary. *If $DF(\sigma, h) = 0$ for all variations h vanishing at a and b, then for $i = 1, \ldots, n$,*

$$\int_a^b \left(\frac{\partial \mathcal{L}}{\partial q_i} - \frac{d}{dt}\left(\frac{\partial \mathcal{L}}{\partial \dot{q}_i} \right) \right) f(t)\, dt = 0$$

for all C^2 real-valued functions f vanishing at a and b.

Proof. Apply the variational formula to variations h with $h_i = f$ and $h_j = 0$ for $j \neq i$. ∎

▷ **Exercise 3–13.** If $a \le c < d \le b$, check that the function g on $[a, b]$ that is zero for $a < c$ and that is defined for $x \ge c$ by $g(x) = \int_c^x (t-c)^2 (\frac{c+d}{2} - t)^2\, dt$ is C^2, is positive on $(c, b]$, and has its first two derivatives zero at c and $\frac{c+d}{2}$. By "reflecting" this function in $\frac{c+d}{2}$ and multiplying the resulting function by g, show that you get a C^2 function on $[a, b]$ that is positive on (c, d) and zero elsewhere.

3.7.4. Fundamental Lemma of the Calculus of Variations. *Let $F : [a, b] \to \mathbf{R}$ be continuous and assume that $\int_a^b F(x) f(x)\, dx = 0$ for all C^2 functions $f : [a, b] \to \mathbf{R}$ that vanish at a and b. Then F is identically zero.*

80   3. Second-Order ODE and the Calculus of Variations

▷ **Exercise 3–14.**   Prove this. Hint: If not, then $F(x) \neq 0$ for some x in $[a, b]$, and then by continuity F will have a constant sign on some subinterval (c, d) of $[a, b]$ containing x.

▷ **Exercise 3–15.**   Recall again that σ in $C^2(I, \mathbf{R}^n)$ is an extremal for the functional F if it satisfies the Euler-Lagrange Equations. You should now be able to prove the following result in your head.

3.7.5. Theorem.   *A necessary and sufficient condition for a path σ in $C^2(I, \mathbf{R}^n)$ to be an extremal for the functional $F = F_\mathcal{L}$ is that $DF(\sigma, h) = 0$ for all variations h in $C^2(I, \mathbf{R}^n)$ that vanish at the endpoints a and b.*

And now we give the result we have been promising for so long.

3.7.6. Theorem.   *Let $\sigma \in C^2([a,b], \mathbf{R}^n)$, and suppose that the functional F assumes a maximum (or minimum) at σ among all paths with the same endpoints as σ. Then σ is an extremal of F.*

Proof.   Given any variation h in $C^2([a,b], \mathbf{R}^n)$ vanishing at a and b, $\gamma(s) = \sigma + sh$ is a straight line in $C^2([a,b], \mathbf{R}^n)$ and each $\gamma(s)$ clearly has the same endpoints as σ. Since $\gamma(0) = \sigma$, it follows that the function of s, $s \mapsto F(\sigma + sh)$, has a maximum at $s = 0$, and hence

$$DF(\sigma, h) = (\frac{d}{ds})_{s=0} F(\sigma + sh) = 0,$$

so by the preceding theorem σ is an extremal of F.   ∎

3.8. More General Variations

For s near zero let $\sigma_s(t) = (x_1^s(t), \ldots, x_n^s(t))$ be a path in $C^2(I, \mathbf{R}^n)$. If the $x_i^s(t)$ are C^2 functions of (t, s), then we call σ_s a smooth variation of $\sigma = \sigma_0$, and we define $\delta\sigma$ in $C^2(I, \mathbf{R}^n)$ by $\delta\sigma(t) = (\frac{d}{ds})_{s=0} \sigma_s(t)$. (Note that for the straight line variation $\sigma_s = \sigma + sh$, $\delta\sigma$ is just h.) By the equality of cross-derivatives, $\delta\sigma'(t) = (\frac{d}{ds})_{s=0} \sigma_s'(t)$.

3.8.1. Proposition. $(\frac{d}{ds})_{s=0} F(\sigma_s) = DF(\sigma, \delta\sigma)$.

Proof. Define $q_i(t,s) = x_i^s(t)$ and $\dot{q}_i(t,s) = \frac{dx_i^s(t)}{dt}$, so that $\delta\sigma_i(t) = (\frac{d}{ds})_{s=0} q_i(t,s)$ and $\delta\sigma_i'(t) = (\frac{d}{ds})_{s=0} \dot{q}_i(t,s)$. Then as before,

$$F(\sigma_s) = \int_a^b \mathcal{L}(q_1(t,s), \ldots, q_n(t,s), \dot{q}_1(t,s), \ldots, \dot{q}_n(t,s)) \, dt,$$

and the proposition follows by a computation similar to that for the previous variational formula. ∎

3.9. The Theorem of E. Noether

Let ϕ be a diffeomorphism of $\mathbf{R}^n$. Then ϕ induces a diffeomorphism $D\phi$ of $\mathbf{TR}^n$, $(q,\dot{q}) \mapsto (\phi(q), D\phi_q(\dot{q}))$, where $D\phi_q$ is the linear map of $\mathbf{R}^n$ whose matrix is the Jacobian matrix of ϕ at q. The characteristic property of $D\phi$ is that if γ is any smooth curve in $\mathbf{R}^n$, then $(\phi \circ \gamma)' = D\phi \circ \gamma'$. (This is a simple consequence of the chain rule.)

We say that ϕ is a *symmetry* of a Lagrangian $\mathcal{L} : \mathbf{TR}^n \to \mathbf{R}$ (or that ϕ preserves $\mathcal{L}$) if $\mathcal{L} \circ D\phi = \mathcal{L}$. If $V : \mathbf{R}^n \to \mathbf{R}^n$ is a smooth vector field on $\mathbf{R}^n$ that generates a flow ϕ_t, we call V an infinitesimal symmetry of $\mathcal{L}$ if each element ϕ_t of the flow is a symmetry of $\mathcal{L}$.

We associate to the vector field V a smooth real-valued function $\hat{V}$ on $\mathbf{TR}^n$, called the *conjugate momentum* of V, by $\hat{V}(q,\dot{q}) := \sum_i \frac{\partial \mathcal{L}}{\partial \dot{q}_i}(q,\dot{q}) V_i(q)$. In particular if σ is a smooth path in $\mathbf{R}^n$, then $\hat{V}(\sigma'(t)) = \sum_i \frac{\partial \mathcal{L}}{\partial \dot{q}_i}(\sigma'(t)) V_i(\sigma(t))$.

3.9.1. E. Noether's Principle. *If V is an infinitesimal symmetry of a Lagrangian $\mathcal{L}$, then the conjugate momentum $\hat{V}$ is a constant of the motion for the associated Euler-Lagrange Equations; i.e., $\hat{V}$ is constant along the canonical lifting σ' of every solution σ of the Euler-Lagrange Equations.*

Proof. Define a variation σ_s of σ by $\sigma_s = \phi_s(\sigma)$, where ϕ_t is the flow generated by V, so that $\sigma_s' = D\phi_s(\sigma')$. Note that since $\phi_s(x)$ is the solution curve of the vector field V with initial condition x, it follows that $\delta\sigma(t) = V(\sigma(t))$. Since V is an infinitesimal symmetry of $\mathcal{L}$, $D\phi_s$ preserves $\mathcal{L}$, so $\mathcal{L} \circ \sigma_s' = \mathcal{L}(\sigma')$, and hence $F(\sigma_s)$ is a constant, and therefore $DF(\sigma, \delta\sigma) = (\frac{d}{ds})_{s=0} F(\sigma_s) = 0$. Then by Corollary 3.7.2 of

82   3. Second-Order ODE and the Calculus of Variations

the Variational Formula and the above formula for $\hat{V}(\sigma'(t))$ it follows that $\hat{V}(\sigma'(t))$ is constant. ∎

▷ **Exercise 3–16.** Given a Lagrangian $\mathcal{L} : \mathbf{TR}^n \to \mathbf{R}$, define an associated Hamiltonian function $H_\mathcal{L} : \mathbf{TR}^n \to \mathbf{R}$ by $H_\mathcal{L}(q,\dot{q}) := \sum_i \frac{\partial \mathcal{L}}{\partial \dot{q}_i}(q,\dot{q})\dot{q}_i - \mathcal{L}(q,\dot{q})$, and show that it is constant along the canonical lifting of every solution of the Euler-Lagrange Equations. Hint: Imitate the proof of Noether's Principle, but use the variation $\sigma_s(t) = \sigma(t+s)$.

3.10. Lagrangians Defining the Same Functionals

A Lagrangian is just a smooth real-valued function on $\mathbf{TR}^n$, the tangent bundle of $\mathbf{R}^n$, and an especially simple class of such functions is the family of differential forms (more precisely, differential 1-forms). These are smooth functions $\omega : \mathbf{TR}^n \to \mathbf{R}^n$ that are linear on each tangent space, $\mathbf{T}_p\mathbf{R}^n$, or, in other words, having the property that $\omega(q,\dot{q})$ is linear in $\dot{q}$ for each choice of q. Clearly the most general such ω is of the form $\omega(q,\dot{q}) = \sum_{i=1}^n \omega_i(q)\dot{q}_i$, where the ω_i are n smooth real-valued functions on $\mathbf{R}^n$ called the components of the differential form ω. The associated functional F_ω of a differential form ω is the well-known line integral of elementary analysis: given a smooth path $x : [a,b] \to \mathbf{R}^n$, the line integral of ω along x, denoted by $\int_x \omega$, is defined by $\int_a^b \sum_{i=1}^n \omega_i(x(t))\frac{dx_i}{dt} dt$, which of course is also just the definition of $F_\omega(x)$.

A smooth real-valued function ϕ on $\mathbf{R}^n$ defines a differential form $\omega = d\phi$, its differential, with components $\omega_i = \frac{\partial \phi}{\partial x_i}$. (More intrinsically, if (p,v) is a tangent vector to $\mathbf{R}^n$ at p, then $d\phi(p,v)$ is the directional derivative of ϕ at p in the direction v.) A differential form ω is called *exact* if there exists a smooth function ϕ with $\omega = d\phi$, so by equality of cross-derivatives the components of an exact form clearly satisfy the identities $\frac{\partial \omega_i}{\partial x_j} = \frac{\partial \omega_j}{\partial x_i}$. Any differential form satisfying these identities is called a *closed* form. If $\omega = d\phi$ is an exact form and x as above is a smooth path in $\mathbf{R}^n$, then, by the chain rule, $\omega(x(t), \frac{dx}{dt}) = \frac{d}{dt}\phi(x(t))$, so that the line integral $\int_x \omega$ evaluates to

3.10. Lagrangians Defining the Same Functionals

$\phi(x(b)) - \phi(x(a))$. In particular its value depends only on the endpoints of the path x, and we can recover ϕ from ω by the formula $\phi(p) := \phi(0) + \int_x \omega$, where x is any path from 0 to p (for example the straight line path $x(t) := tp$).

3.10.1. Proposition. *For a differential form ω on $\mathbf{R}^n$ the following are equivalent:*

1) ω *is exact.*

2) ω *is closed.*

3) *The Euler-Lagrange Equations for the Lagrangian $\mathcal{L} = \omega$ are satisfied identically (i.e., by all smooth paths), or equivalently every smooth path is an extremal for the functional F_ω.*

4) *The line integral of ω along a path depends only on the endpoints of the path.*

Proof. As already remarked, the implication 1) $\Longrightarrow$ 2) is trivial. The Euler-Lagrange Equations for $\mathcal{L}(q,\dot{q}) := \sum_{k=1}^n \omega_k(q)\dot{q}_k$ are $\frac{d}{dt}\omega_i(q) = \sum_{k=1}^n \frac{\partial \omega_k}{\partial q_i}\frac{dq_k}{dt}$, and by the chain rule this is an identity if ω is closed, so 2) $\Longrightarrow$ 3). Next suppose that $\sigma_0 : [a,b] \to \mathbf{R}^n$ and $\sigma_1 : [a,b] \to \mathbf{R}^n$ are smooth paths with the same endpoints p_1, p_2 and let $\sigma_s(t) := \sigma_0(t) + s(\sigma_1(t) - \sigma_0(t))$ be the straight line in $\Sigma(p_1, p_2)$ joining them. Then $(\frac{d}{ds})_{s=s_0} F_\omega(\sigma_s) = 0$ if and only if σ_{s_0} is an extremal of F_ω, so that if 3) holds, then $\frac{d}{ds} F_\omega(\sigma_s)$ is identically zero and it follows that $F_\omega(\sigma_0) = F_\omega(\sigma_1)$, proving 3) $\Longrightarrow$ 4). Finally, define $\phi(p) := \int_{x_p} \omega$ where x_p is the straight line $x(t) := tp$, $0 \le t \le 1$, i.e., $\phi(p) := \int_0^1 \omega(tp, p)\,dt$. If 4) holds, then we can evaluate $\phi(p+sv)$ by first integrating ω along x_p and then adding the integral of ω along the straight line γ_{sv} from p to $p+sv$ (i.e., $\gamma_{sv}(t) = p + tv$, $0 \le t \le s$). Thus $\frac{1}{s}(\phi(p+sv) - \phi(p)) = \frac{1}{s}\int_{\gamma_{sv}} \omega = \frac{1}{s}\int_0^s \omega(p+tv, v)\,dt$. Letting s approach zero gives $d\phi(p, v) = \omega(p, v)$, so 4) $\Longrightarrow$ 1). ∎

Suppose $\mathcal{L}(q, \dot{q})$ is **any** Lagrangian such that every smooth path in $\mathbf{R}^n$ is an extremal of $F_\mathcal{L}$, i.e., such that the Euler-Lagrange Equations associated to $\mathcal{L}$ are automatically satisfied. Does it follow that $\mathcal{L}$ is necessarily a differential form and hence by the above proposition the differential of a function? Not quite, since we can certainly add a constant to $\mathcal{L}$. But that is all.

3.10.2. Proposition. *Let $\mathcal{L}$ be a Lagrangian function on $\mathbf{TR}^n$ for which the Euler-Lagrange Equations hold identically (or equivalently for which the functional $F_\mathcal{L}$ is constant on each of the subspaces $\Sigma(p_1, p_2)$). Then $\mathcal{L}$ has the form $d\phi + C$, where ϕ is a smooth function on $\mathbf{R}^n$ and C is a constant.*

Proof. One direction of the equivalence of the two forms of the hypothesis is trivial and the other is just the implication 3) $\Longrightarrow$ 4) of the preceding proposition. Given such a Lagrangian $\mathcal{L}$ and given three points $q, \dot{q}, \ddot{q}$ in $\mathbf{R}^n$, let $x(t)$ be a smooth path in $\mathbf{R}^n$, defined for t near 0, such that $x(0) = q$, $x'(0) = \dot{q}$, and $x''(0) = \ddot{q}$ (e.g., $x(t) := q + t\dot{q} + (t^2/2)\ddot{q}$). Since $x(t)$ satisfies the Euler-Lagrange Equation, at time $t = 0$ we get the identity

$$\frac{\partial \mathcal{L}}{\partial q_i}(q, \dot{q}) - \sum_{j=1}^n \frac{\partial^2 \mathcal{L}}{\partial q_j \partial \dot{q}_i}(q, \dot{q}) \dot{q}_j = \sum_{j=1}^n \frac{\partial^2 \mathcal{L}}{\partial \dot{q}_j \partial \dot{q}_i}(q, \dot{q}) \ddot{q}_j. \quad (*)$$

Note that since the left-hand side of $(*)$ does not involve $\ddot{q}$, it follows that the right-hand side must have the same value for all choices of the $\ddot{q}$ as it has for $\ddot{q} = 0$, i.e., $\sum_{j=1}^n \frac{\partial^2 \mathcal{L}}{\partial \dot{q}_j \partial \dot{q}_i}(q, \dot{q}) \ddot{q}_j = 0$. Choosing $\ddot{q}_j = 1$ for $j = k$ and $\ddot{q}_j = 0$ for $j \neq k$, it follows that $\frac{\partial^2 \mathcal{L}}{\partial \dot{q}_k \partial \dot{q}_i}(q, \dot{q}) = 0$ for all k and j, and hence $\mathcal{L}(q, \dot{q}) = \sum_{k=1}^n \omega_k(q) \dot{q}_k + C(q)$. Substituting this into the left-hand side of $(*)$ (which recall is identically zero) gives $\sum_{k=1}^n \frac{\partial \omega_k}{\partial q_i}(q) \dot{q}_k + \frac{\partial C}{\partial q_i} = \sum_{j=1}^n \frac{\partial \omega_i}{\partial q_j}(q) \dot{q}_j$. If we first take all the $\dot{q}_k$ equal to zero, it follows that $\frac{\partial C}{\partial q_i} = 0$, so C is a constant, and if we now take $\dot{q}_\ell = 1$ and $\dot{q}_j = 0$ for $j \neq \ell$, we get $\frac{\partial \omega_\ell}{\partial q_i}(q) = \frac{\partial \omega_i}{\partial q_\ell}(q)$, so that ω is a closed form and hence exact. ∎

3.10.3. Corollary. *Let $\mathcal{L}_1$ and $\mathcal{L}_2$ be two Lagrangian functions on $\mathbf{TR}^n$. A neccessary and sufficient condition for the associated functionals $F_{\mathcal{L}_1}$ and $F_{\mathcal{L}_2}$ to differ by at most a constant on each of the spaces $\Sigma(p_1, p_2)$ is that $\mathcal{L}_1$ and $\mathcal{L}_2$ differ by the sum of an exact form and a constant. In particular, if this condition is satisfied, then*

3.11. Riemannian Metrics and Geodesics

$\mathcal{L}_1$ and $\mathcal{L}_2$ give the same Euler-Lagrange Equations and so they have same extremals.

Proof. Given the proposition, this is immediate from the linearity of $\mathcal{L} \mapsto F_\mathcal{L}$. ∎

Caution! While Lagrangians whose difference is of the form $d\phi + C$ have the same extremals, the converse is **not** valid; i.e., from the fact that $\mathcal{L}_1$ and $\mathcal{L}_2$ have the same extremals, one cannot conclude that $F_{\mathcal{L}_1}$ and $F_{\mathcal{L}_2}$ differ by constants on the spaces $\Sigma(p_1, p_2)$. For example, multiplying a Lagrangian by a nonzero constant clearly does not change its extremals. (This is no more mysterious than the fact that although two functions from **R** to **R** that differ by a constant have the same critical points, the converse is clearly false.)

When we introduced Lagrangian functions $\mathcal{L}$, we noted that their essential function was to define the associated functionals $F_\mathcal{L}$ on the spaces $\Sigma(p_1, p_2)$. It is these functionals, rather than the Lagrangians themselves, that should be regarded as the primary objects of study in the Calculus of Variations, and this suggests that we should consider two Lagrangians functions to be in some sense "the same" if their associated functionals agree, or differ by at most a constant, on each $\Sigma(p_1, p_2)$.

3.10.4. Definition. Two Lagrangian functions on $\mathbf{TR}^n$ are called *quasi-equivalent* if their difference is of the form $d\phi + C$. If $\mathcal{L}$ is a Lagrangian function on $\mathbf{TR}^n$ and Φ is a diffeomorphism of $\mathbf{R}^n$, then Φ is called a *quasi-symmetry* of $\mathcal{L}$ if $\mathcal{L} \circ D\Phi$ is quasi-equivalent to $\mathcal{L}$, i.e., if $\mathcal{L} \circ D\Phi = \mathcal{L} + d\phi + C$ for some smooth real-valued function ϕ on $\mathbf{R}^n$ and some constant C. If Φ_s is the smooth flow on $\mathbf{R}^n$ generated by a vector field V, then V is called an *infinitesimal quasi-symmetry* of $\mathcal{L}$ if there exists a smooth one-parameter family of functions $\phi_s : \mathbf{R}^n \to \mathbf{R}$ and constants C_s such that $\mathcal{L} \circ D\Phi_s = \mathcal{L} + d\phi_s + C_s$.

3.11. Riemannian Metrics and Geodesics

Recall that if we fix q in $\mathbf{R}^n$, then the set of all $(q, \dot{q})$ in $\mathbf{TR}^n$ is denoted by $\mathbf{T}_q\mathbf{R}^n$ and is called the tangent space to $\mathbf{R}^n$ at q. We are now going to contemplate having a different inner-product (or dot-product) on each tangent space. Note that the standard inner product of two vectors $\dot q^1 = (\dot q^1_1, \ldots, \dot q^1_n)$ and $\dot q^2 = (\dot q^2_1, \ldots, \dot q^2_n)$ is $\langle \dot q^1, \dot q^2 \rangle = \sum_i \dot q^1_i \dot q^2_i = \sum_{i,j} \delta_{ij} \dot q^1_i \dot q^2_j$, where of course δ_{ij} is the identity matrix. To define an inner-product in $\mathbf{T}_q \mathbf{R}^n$, we will use $\langle \dot q^1, \dot q^2 \rangle_q = \sum_{i,j} g_{ij}(q) \dot q^1_i \dot q^2_j$, where $g_{ij}(q)$ is a symmetric positive definite matrix whose entries are smooth functions of q. Positive definite means that $\|\dot q\|^2_q = \langle \dot q, \dot q \rangle_q$ is positive unless $\dot q$ is the zero vector. We note that this clearly implies that the matrix $g_{ij}(q)$ is invertible, and we will denote its inverse by $g^{ij}(q)$, so that $\sum_k g_{ik}(q) g^{kj}(q) = \delta_{ij}$.

We will use this so-called Riemannian metric tensor $g_{ij}(q)$ to define two Lagrangians on $\mathbf{TR}^n$. The first is $\mathcal{L}^1(q, \dot q) = \|\dot q\|_q = \sqrt{\sum_{i,j} g_{ij}(q) \dot q_i \dot q_j}$, and the corresponding functional $F_{\mathcal{L}^1}(\sigma)$ is called the length of σ. The second is $\mathcal{L}^2(q, \dot q) = \frac{1}{2} \|\dot q\|^2_q = \frac{1}{2} \sum_{i,j} g_{ij}(q) \dot q_i \dot q_j$, and the corresponding functional $F_{\mathcal{L}^2}(\sigma)$ is called the action of σ.

▷ **Exercise 3–17.** Show that $\mathcal{L}^2$ is nondegenerate. What is the matrix M_{ij} in this case?

▷ **Exercise 3–18.** Derive the Euler-Lagrange Equations for $\mathcal{L}^1$ and $\mathcal{L}^2$. Extremals of $\mathcal{L}^1$ are called *geodesics* of the Riemannian metric. Since the length is easily seen to be independent of parametrization, the reparametrization of a geodesic is again a geodesic. Show that extremals of $\mathcal{L}^2$ are just geodesics parametrized proportionally to arc-length.

3.12. A Preview of Classical Mechanics

In this final section of the present chapter we will have a preliminary look at three important special classes of differential equations, called, respectively, Newton's Equations, Lagrange's Equations, and Hamilton's Equations. As we shall see, they are three equivalent ways of looking at essentially the same mathematics, but each has special advantages. The analysis of these equations is the core of what is called Classical Mechanics, a subject of considerable beauty that lies at an important interface of mathematics and physics. The following discussion is intended as an introduction to and preview of things to

3.12. A Preview of Classical Mechanics

come, describing some of the more salient mathematical features of these equations. In Chapter 4 we will study them in considerably more detail.

"Newton's Equations" have already been introduced above. They are in fact a whole class of second-order differential equations on $\mathbf{R}^n$. Each member of this class is defined by giving n positive constants $m_1, m_2, \ldots, m_n$ (called the masses) and a C^2 real-valued function $U : \mathbf{R}^n \to \mathbf{R}$, called the potential energy function. From U we define a C^1 vector field ∇U, the gradient of U, whose ith component is $\frac{\partial U}{\partial x_i}$, and the "force" vector field F is defined to be $-\nabla U$. Then Newton's Equations read force = mass × acceleration, or $m_i \frac{d^2 x_i}{dt^2} = -\frac{\partial U}{\partial x_i}$. As usual for a second-order system, we can replace Newton's Equations by an equivalent system of first-order equations in $\mathbf{R}^n \times \mathbf{R}^n$. Namely, using coordinates $(x, v) = (x_1, \ldots, x_n, v_1, \ldots, v_n)$, we get the first-order system: $\frac{dx_i}{dt} = v_i$ and $m_i \frac{dv_i}{dt} = -\frac{\partial U}{\partial x_i}$.

But it turns out to be more convenient to make a simple linear change of coordinates from $(x, v) = (x_1, \ldots, x_n, v_1, \ldots, v_n)$ to $(q, p) = (q_1, \ldots, q_n, p_1, \ldots, p_n)$, where $q_i = x_i$ and $p_i = m_i v_i$. The old v_i are called velocity (or "tangent bundle") coordinates, whereas the new p_i are called momenta (or "cotangent bundle") coordinates, and the transformation from the coordinates (x, v) to the coordinates (q, p) is called the Legendre Transformation. (There is no real difference between the x_i and the q_i, and it is mainly to conform with tradition that we have changed names.) In terms of the q's and p's, Newton's Equations take the form $m_i \frac{dq_i}{dt} = p_i$ and $\frac{dp_i}{dt} = -\frac{\partial U}{\partial q_i}$.

It will be convenient to define three real-valued functions on $\mathbf{R}^n \times \mathbf{R}^n$: the kinetic energy, K; the Lagrangian, $\mathcal{L}$; and the Hamiltonian (or total energy), H. By definition, $K(x, v) := \frac{1}{2} \sum_i m_i v_i^2$, or $K(q, p) := \sum_i \frac{p_i^2}{2m_i}$, and then $\mathcal{L}$ and H are, respectively, the difference $\mathcal{L} := K - U$ and the sum $H := K + U$ of the kinetic and potential energies. (Here we are regarding U as a function on $\mathbf{R}^n \times \mathbf{R}^n$ that is independent of its second argument: $U(x, v) = U(x)$.)

In the context of Newton's Equations, the fact that a certain function f is a constant of the motion is usually called a *conservation law* (and f is called a *conserved quantity*).

3. Second-Order ODE and the Calculus of Variations

▷ **Exercise 3–19.** Prove by direct computation that the total energy, H, is a conserved quantity—i.e., that if $(q(t), p(t))$ is any solution of Newton's equations, then the time derivative of $H(q(t), p(t))$ is identically zero.

Note that the partial derivative of the Lagrangian function with respect to the ith velocity v_i is just the ith momentum p_i: $\frac{\partial \mathcal{L}}{\partial v_i} = \frac{\partial K}{\partial v_i} = \frac{\partial}{\partial v_i}\left(\frac{1}{2}\sum_j m_i v_j^2\right) = m_i v_i = p_i$. On the other hand, the partial derivative of $\mathcal{L}$ with respect to x_i is just the ith component of the force: $\frac{\partial \mathcal{L}}{\partial x_i} = -\frac{\partial U}{\partial x_i}$. It follows that we can rewrite Newton's Equations in the form $\frac{d}{dt}\left(\frac{\partial \mathcal{L}}{\partial v_i}\right) - \frac{\partial \mathcal{L}}{\partial x_i} = 0$, which we recognize as the Euler-Lagrange Equations for the Lagrangian $\mathcal{L}$.

From our study of the Calculus of Variations above, we now see that this Lagrangian reformulation of Newton's Equations, which might otherwise seem unmotivated (and gratuitously complicated) has a conceptual advantage. Namely, define a real-valued function A (called the *action functional*) on the space of C^2 paths σ joining two points p and q in $\mathbf{R}^n$ by $A(\sigma) = F_\mathcal{L}(\sigma) := \int_a^b \mathcal{L}(\dot\sigma(t))\,dt = \int_a^b \mathcal{L}(q_i(t), \dot q_i(t))\,dt$. Then Newton's Equations express a remarkable and nonobvious fact ("The Principle of Least Action"): the extremals of A are precisely the solutions of Newton's Equations. One immediate consequence of this is that by the general Noether Principle, one-parameter groups of symmetries of the Lagrangian function give rise to conservation laws.

The Lagrangian formulation also has a more mundane use (one that we considered above in a special case). Namely, it is often more convenient in studying a particular example of Newton's Equations to use a coordinate system other than Cartesian coordinates. It can be messy to convert Newton's Equations to these new coordinates, and it is often easier to rewrite the Lagrangian function in the new coordinates. And because Lagrange's Equations express a fact that is independent of coordinates, they also must be equivalent to Newton's Equations in the new coordinate system.

A classic example of this is the following so-called "Central Force Problem". We take $n = 2$, $m_1 = m_2 = m$ and consider a potential

3.12. A Preview of Classical Mechanics

U that is a function of the distance from the origin $r = \sqrt{x_1^2 + x_2^2}$. This models the dynamics of a particle of mass m in the plane that is attracted towards the origin with a force $-\frac{\partial U}{\partial r}$, and we shall consider it in more detail later.

What you should notice first is that both the kinetic energy and the potential energy are invariant under rotation about the origin, and hence so too are the Lagrangian and the Hamiltonian. For this reason, it is natural in studying this problem to use polar coordinates (r, θ) where as usual $\theta = \arctan(x_2/x_1)$. We have already looked at this in the special case that $U = 0$, and let us now consider the more general situation of an arbitrary potential $U(r)$.

We recall that $x_1 = r\cos\theta$ and $x_2 = r\sin\theta$. As previously, we write $(r, \theta, \dot{r}, \dot{\theta})$ for the corresponding canonical coordinates in the space $\mathbf{R}^2 \times \mathbf{R}^2$ of positions and velocities (i.e., if a curve is given parametrically in polar coordinates by $(r(t), \theta(t))$, then its canonical lifting is given parametrically by $(r(t), \theta(t), \dot{r}(t), \dot{\theta}(t))$ where $\dot{r}(t) = \frac{d}{dt}r(t)$ and $\dot{\theta}(t) = \frac{d}{dt}\theta(t)$). The kinetic energy K is expressed in Cartesian coordinates by $K(x_1, x_2, v_1, v_2) = \frac{m}{2}(v_1^2 + v_2^2)$ and so in polar coordinates it is given by $K(r, \theta, \dot{r}, \dot{\theta}) = \frac{m}{2}(\dot{r}^2 + r^2\dot{\theta}^2)$. Thus, in polar coordinates, $\mathcal{L}(r, \theta, \dot{r}, \dot{\theta}) = \frac{m}{2}(\dot{r}^2 + r^2\dot{\theta}^2) - U(r)$, so Lagrange's Equations become $\frac{d}{dt}\left(\frac{\partial \mathcal{L}}{\partial \dot{r}}\right) - \frac{\partial \mathcal{L}}{\partial r} = 0$ and $\frac{d}{dt}\left(\frac{\partial \mathcal{L}}{\partial \dot{\theta}}\right) - \frac{\partial \mathcal{L}}{\partial \theta} = 0$, i.e., $\frac{d}{dt}(m\dot{r}) = -U'(r)$ and $\frac{d}{dt}(mr^2\dot{\theta}) = 0$. As we know, we must adjoin to these two equations the two further equations $\frac{dr}{dt} = \dot{r}$ and $\frac{d\theta}{dt} = \dot{\theta}$, so finally, as consequences of Lagrange's Equations we have $m\frac{d^2r}{dt^2} = -U'(r)$ and $r^2\frac{d\theta}{dt} = \text{constant}$.

The quantity $mr^2\frac{d\theta}{dt}$ is called the angular momentum, and its constancy is the law of conservation of angular momentum, also referred to as "Kepler's Law of Equal Areas in Equal Times". We shall see later that it is a consequence of Noether's Principle and the fact that $\mathcal{L}$ is invariant under rotation.

▷ **Exercise 3–20.** Transform Newton's Equation from Cartesian to polar coordinates in the straightforward, pedestrian way.

A function $H(q_1, \ldots, q_n, p_1, \ldots, p_n)$ on $\mathbf{R}^n \times \mathbf{R}^n$ gives rise to a system of differential equations $\frac{dq_i}{dt} = \frac{\partial H}{\partial p_i}$, $\frac{dp_i}{dt} = -\frac{\partial H}{\partial q_i}$. Systems that

90   3. Second-Order ODE and the Calculus of Variations

arise in this way are called *Hamiltonian*, and H is called the Hamiltonian function for the system. These systems share many remarkable and characteristic properties, and in particular they have important invariance properties, a couple of which we will look at now. In fact, the study of Hamiltonian systems is a subject in and of itself, one that we will consider in more detail later.

▷ **Exercise 3–21.** Show that any system of Hamiltonian differential equations is volume preserving. (This is called Liouville's Theorem.) Show conversely that if $\frac{dq}{dt} = V_1(q,p)$ and $\frac{dp}{dt} = V_2(q,p)$ is any area-preserving autonomous system of differential equations in $\mathbf{R} \times \mathbf{R}$, then there is a Hamiltonian function $H(q,p)$ such that $V_1 = \frac{\partial H}{\partial p}$ and $V_2 = -\frac{\partial H}{\partial q}$. (This does **not** generalize to higher dimensions.)

▷ **Exercise 3–22.** Check that Newton's Equations are Hamiltonian differential equations, with the total energy as the Hamiltonian function. Show that the Hamiltonian function is always a constant of the motion for a Hamiltonian system. (Together these facts give still another proof of the conservation of energy for Newtonian systems.)

Chapter 4

Newtonian Mechanics

4.1. Introduction

The world we live in is a complex place, and we must expect any theory that describes it accurately to share that complexity. But there are three assumptions, satisfied at least approximately in many important physical systems, that together lead to a considerable simplification in the mathematical description of systems for which they are valid.

The first of these assumptions is that the system is "isolated", or "closed", meaning that all forces influencing the behavior of the system are accounted for within the system. The second assumption is that the system is "nonrelativistic", meaning that all velocities are small compared to the speed of light. The third assumption is that the system is "nonquantum", meaning that the basic size parameters of the system are large compared with those of atomic systems (or, more precisely, that the actions involved are large multiples of the fundamental Planck unit of action).

These assumptions put us into the realm of "classical" physics, where dynamical interactions of material bodies are adequately described by the famous three laws of motion of Newton's Principia. Of course, such systems can still exhibit great complexity, and in fact even the famous "three body problem"—to describe completely the motions of three point particles under their mutual gravitational attraction—is still far from "solved". Moreover, at least the latter two of these assumptions are quite sophisticated in nature, and even explaining them carefully requires some doing. Later we shall see that a comparatively unsophisticated fourth assumption—that a system is "close to equilibrium"—cuts through all the complexity and reduces a problem to one that is completely analyzable (using an algorithm called the "method of small vibrations"). This magical assumption, which in effect linearizes the situation, is far from universally valid—magic after all only works on special occasions. But when it does hold, its power is much too valuable to ignore, and we will look at it in some detail at the end of this chapter, after developing the basic theory of Newtonian mechanics and illustrating it with several important examples.

We commence our study of Classical Mechanics with a little history.[1]

4.2. Newton's Laws of Motion

We have already referred several times to "Newton's Laws of Motion". They are a well-recognized milestone in intellectual history and could even be said to mark the beginning of modern physical science, so it is worth looking at them in more detail. They were first published in July of 1686 in a remarkable treatise, usually referred to as Newton's *Principia*,[2] and it is not their mere statement that gives them such importance but rather the manner in which Newton was able to use them in Principia to develop a mathematically rigorous theory of particle dynamics.

Let us look first at Newton's original formulation of his Laws of Motion:

AXIOMATA SIVE LEGES MOTUS

Lex I. Corpus omne perseverare in statuo suo quiescendi vel movendi uniformiter in directum, nisi quatenus a viribus impressis cogitur statum illum mutare.

---

[1] We are grateful to Professor Michael Nauenberg of UCSC for his critical reading of this section and for correcting several inaccuracies in these historical remarks.

[2] The full Latin title is "Philosophiae Naturalis Principia Mathematica", or in English, "Mathematical Principles of Natural Philosophy". This first edition is commonly referred to as the 1687 edition, since it was not distributed until a year after it was printed.

4.2. Newton's Laws of Motion

Lex II. Mutationem motus proportionalem esse vi motrici impressae, & fieri secundum lineam rectam qua vis illa imprimitur.

Lex III. Actioni contrariam semper & qualem esse reactionem: sive corporum duorum actiones in se mutuo semper esse quales & in partes contrarias dirigi.

Even though we are sure you had no difficulty with the Latin, let's translate that into English:

AXIOMS CONCERNING LAWS OF MOTION

Law 1. Every body remains in a state of rest or of uniform motion in a straight line unless compelled to change that state by forces acting on it.

Law 2. Change of motion is proportional to impressed motive force and is in the same direction as the impressed force.

Law 3. For every action there is an equal and opposite reaction, or, the mutual actions of two bodies on each other are always equal and directed to opposite directions.

The first thing to remark is that, mathematically speaking, there are only **two** independent laws here—the First Law is clearly a special case of the second, obtained by setting the "impressed motive force" to zero.[3]

Another point worth mentioning is that the Second Law does not really say "$F = ma$". Newton was developing the calculus at the same time he was writing the Principia, and no one would have understood his meaning if he had written the Second Law as we do today. In fact, if one reads the Principia, it becomes clear that what Newton intended by the Second Law is something like, "If you strike an object

---

[3] However, as we shall see later, the First Law does have physical content that is independent of and prior to the Second Law: it asserts the existence of so-called "inertial frames of reference", and it is only in inertial frames that the Second Law is valid. Moreover, the First Law also has great historical and philosophical importance, as we shall explain in more detail at the end of this section.

with a hammer, then the change of its momentum is proportional to the strength with which you hit it and is in the same direction as the hammer moves." That is, Newton is thinking about an instantaneous impulse rather than a force applied continuously over time. So how did Newton deal with a nonimpulsive force that acted over an interval of time, changing continuously as it did so? Essentially he worked out the appropriate differential calculus details each time. That is, he broke the interval into a large number of small subintervals during which the force was essentially constant, applied the Second Law to each subinterval, and then passed to the limit.

The Third Law does **not** say that the force (or "action") that one body exerts on another is directed along the line joining them. However this is how it usually gets used in the Principia, and so it is often considered to be part of the Third Law. We will distinguish between the two versions by referring to them as the weak and strong forms of Newton's Third Law.

It is pretty clear that these Laws of Motion by themselves are insufficient to predict how physical objects will move. What is missing is a specification of what the forces actually are that objects exert upon each other. However, later in the Principia Newton formulated another important law of nature, called The Law of Universal Gravitation. It states that there is an attractive force between any two particles of matter whose magnitude is proportional to the product of their masses and inversely proportional to the square of the distance separating them. One of the most remarkable achievements of the Principia was Newton's derivation of the form of his law of gravitation from the Laws of Motion together with Kepler's laws of planetary motion, and we well give an account of how Newton accomplished this later, after we have developed the necessary machinery.

If one takes Newton's law of gravitation seriously, it would appear that a small movement of a massive object on the Earth would be instantaneously felt as a change in the gravitational force at arbitrarily great distances—say on Jupiter. This "action at a distance" was something that made Newton and many of his contemporaries quite uncomfortable. Today we know that gravitation does **not** work precisely the way that Newton's law suggests. Instead, gravitation

4.2. Newton's Laws of Motion 95

is described by a field, and changes in this field propagate with the speed of light. The force on a test particle is not a direct response to the many far-off particles that together generate the field, but rather it is caused by the interaction of the test particle with the gravitational field in its immediate location. To a good approximation, the gravitational field is described by a potential function that gives Newton's law of gravity, but the detailed reality is more complicated, and accounting for small errors observed in certain predictions of Newtonian gravitation requires the more sophisticated theory of Einstein's General Relativity.

Newton's Laws of Motion themselves are now known to be only an approximation. In situations where all the velocities involved are small compared to the speed of light, Newton's Laws of Motion are highly accurate, but at very high velocities one needs Einstein's more refined theory of Special Relativity. Newton's Laws of Motion also break down when dealing with the very small objects of atomic physics. In this realm the more complex Quantum Mechanics is needed to give an accurate description of how particles move and interact.

But even though Newton's Laws of Motion and his Law of Gravitation are not the ultimate description of physical reality, it should not be forgotten that they give an amazingly accurate description of the dynamics of massive objects over a vast range of masses, velocities, and distances. In particular, in the two hundred years following the publication of Principia, the consequences of Newton's Laws of Motion were developed into a mathematical theory of great elegance and power that among other successes made predictions concerning the motions of the planets, moons, comets, and asteroids of our own solar system that were verified with remarkable accuracy. We will cover some of this theory below.

We will end this mainly historical section with an explanation of why the First Law of Motion has such great historical and philosophical importance. We quote Michael Nauenberg (with permission) from part of a private exchange with him on this subject:

> Newton made it clear in the Principia that he credited
> Galileo with the Second Law. What should be pointed

out is that the great breaktrough in dynamics in Galileo and Newton's time came about with an understanding of the First Law. Before then, it was understood that to initiate motion required an external force, but the idea that motion could be sustained without an external force seems to have escaped attention. Even stones and arrows somehow had to be continuously pushed during their flight by the surrounding air, according to Aristotles and later commentators, until Galileo finally showed that the air only slows them down, and in the absence of air friction, they travelled along a parabolic path. In earlier manuscripts Newton spoke also of "inertial forces". Apparently even he could not free himself completelely from millenia of confusion.

4.3. Newtonian Kinematics

As has become traditional, we will begin our study of the Newtonian worldview with a discussion of the *kinematics* of Newtonian physics, i.e., the mathematical formalism and infrastructure that we will use to describe motion, and only then will we go on to consider the *dynamics*, that is, the nature of the forces that express the real physical content of Newton's theory of motion.

A Newtonian (Dynamical) System (V, F) consists of an orthogonal vector space V, called the *configuration space* of the system, together with a vector field F on V, i.e., a smooth map $F : V \to V$, called the *force law* of the system. (By an orthogonal vector space we just mean a real vector space with a positive definite inner product.) For the time being V will be finite dimensional and its dimension, N, is called the number of degrees of freedom of the system (V, F). Later we will also consider the infinite-dimensional case. If you want to think of V as being $\mathbf{R}^N$ with the usual "dot-product", that is fine, but we will write $\langle u, v \rangle$ to denote the inner product of two elements u and v in V and $\|v\|$ to denote the "length" of a vector v (defined by $\|v\|^2 = \langle v, v \rangle$).

The reason why we call V configuration space is that the points of V are supposed to be in bijective correspondence with all the possible

4.3. Newtonian Kinematics

configurations of some physical system. An important example, and one that we will return to repeatedly in the sequel, is the description of a system that consists of n point particles $P_1, \ldots, P_n$ in $\mathbf{R}^3$. If we denote the location of P_i by $x^i = (x_1^i, x_2^i, x_3^i)$, then a configuration of the system S is specified by giving the n vectors x^i in $\mathbf{R}^3$, or equivalently by giving the $3n$ real numbers x_j^i. Thus in this case we can identify V with $(\mathbf{R}^3)^n$, or equivalently with $\mathbf{R}^N$, where $N = 3n$.

The force law F specifies the dynamics of the system via the second-order ODE

$$\frac{d^2 x}{dt^2} = F(x), \qquad (\text{NE})$$

called "Newton's Equations of Motion"—or simply "Newton's Equations".[4] The reason that there is no "mass" multiplying the left-hand side is that, to simplify notation, we have already divided it into the right-hand side—that is, our F does not really represent "force" but rather "specific force", or force per unit mass.

4.3.1. Remark. One can more generally consider forces that depend on time and/or velocity as well as on position. For example, in a nonisolated system, a particle may be acted upon by forces that arise from its interactions with particles that are not part of the system. If these external particles are themselves in motion, such external forces will normally be time-dependent. And there are important physical systems in which the force depends on the velocity with which a particle moves as well as on its position. For example, this is so for objects moving in a viscous medium and also for electrically charged particles moving in a magnetic field. While we will have occasion to bring in time-dependent forces, we will not consider any velocity-dependent force laws.

By definition, a "solution" of (NE) is a twice differentiable curve $\sigma : I \to V$ defined on some "time" interval I, such that for all t in I, $\sigma''(t) = F(\sigma(t))$ (where σ'' denotes the second derivative of σ).

---

[4]By this point you probably will not be too surprised to learn that Newton never used or even wrote down "Newton's Equations". The first person to do so was probably Leonhard Euler in his 1736 book "Mechanica". (This was the first mechanics textbook to be based on differential equations, and so in a sense it is the progenitor of the book you are reading!)

Note that if V is $\mathbf{R}^N$—or, equivalently, if we choose an orthonormal basis for V—and if the curve $\sigma(t)$ is given in components by $\sigma(t) = (x_1(t), \ldots, x_N(t))$, then the condition for σ to be a solution is that the $x_i(t)$ satisfy the second-order system of ODE:

$$\frac{d^2 x_i(t)}{dt^2} = F_i(x_1(t), \ldots, x_N(t)).$$

If $\sigma : I \to V$ is a solution of (NE) defined on an interval I containing 0, then we call the pair $(\sigma(0), \sigma'(0)) \in V \times V$ the *initial condition* for this solution. We call $x_0 = \sigma(0)$ the *initial position* and $v_0 = \sigma'(0)$ the *initial velocity*.

The following is an immediate consequence of the general existence, uniqueness, and smoothness theorems for solutions of ODE discussed in Chapter 1.

4.3.2. Theorem. *Given any initial condition $p = (x_0, v_0) \in V \times V$, there is a uniquely determined "maximal" solution $\sigma_p : (\alpha_p, \omega_p) \to V$ of (NE) with initial condition p.*

We now give more detail:

1) The sense in which σ_p is maximal is that every solution of (NE) with initial conditions p is a restriction of σ_p to some subinterval of (α_p, ω_p).

2) The functions $\alpha : V \times V \to (-\infty, 0)$ and $\omega : V \times V \to (0, \infty)$ are semi-continuous in the sense that given $p_0 = (x_0, v_0) \in V \times V$ and $\epsilon > 0$ there is a neighborhood N of p_0 in $V \times V$, such that for $p \in N$, $\alpha(p) < \alpha(p_0) + \epsilon$ and $\omega(p) > \omega(p_0) - \epsilon$.

3) If $\omega(q) < \infty$, then as t approaches $\omega(q)$, either $\|\sigma(t)\|$ or $\|\sigma'(t)\|$ approaches infinity. Similarly if $\alpha(q) > -\infty$, then as t approaches $\alpha(q)$, either $\|\sigma(t)\|$ or $\|\sigma'(t)\|$ approaches infinity.

4) If $O = \{(q, t) \in V \times V \times \mathbf{R} \mid \alpha(q) < t < \omega(q)\}$ is the open subset of $V \times V \times \mathbf{R}$ where $\sigma_q(t)$ is defined, then the mapping $(q, t) \mapsto \sigma_q(t)$ of O into V is smooth.

4.3.3. Remark. Sometimes the force field F is undefined or "blows up" on a closed subset C of V, in which case it is really $V \setminus C$ that is the configuration space. In this case, the initial position of a solution must be taken in $V \setminus C$, but otherwise the above theorem still holds—except that in 3) it may happen that as t approaches the endpoints of the interval (α_q, ω_q), $\sigma(t)$ may have a limit point in C.

- **Example 4–1. Free Systems.** The simplest Newtonian systems are the "free" systems, i.e., the systems (V, F) with V arbitrary and F the identically zero vector field on V. The solutions for this are clearly the straight lines in V—more precisely, the solution with initial position x_0 and initial velocity v_0 is $t \mapsto x_0 + tv_0$. This of course encapsulates Newton's First Law of Motion. Note that the time parameter t is proportional to the arclength measured from the initial point x_0, the proportionality constant being the length of the velocity vector v_0.

▷ **Exercise 4–1.** If you know basic Riemannian Geometry, generalize Newton's Equations so that they make sense for an arbitrary Riemannian manifold M as the configuration space (and not just a "flat" one). Clearly, the force F should be a vector field on M—i.e., a cross-section of the tangent bundle TM, so the problem is how to suitably generalize the concept of acceleration. Here is a hint: in the flat case, we just saw that the solutions for a free system are the straight lines, with a parametrization that is proportional to arclength. In the general case, a solution for a free system on M should be a geodesic of M parametrized proportionally to arclength, and as you perhaps know, this is equivalent to setting the covariant derivative of the velocity to zero—which of course suggests that the covariant derivative of velocity is the correct generalization of acceleration.

4.4. Classical Mechanics as a Physical Theory

Mathematics is a deductive science, physics experimental, but although this distinction is real, it should not be exaggerated. Mathematics after all started life as Geometry which, as that name suggests, was based on axioms mirroring real world experience, and even in branches as pure as number theory, mathematicians have always carried out numerical "experiments" to suggest appropriate conjectures. Conversely, until the early seventeenth century, physics, then called Natural Philosophy, was considered a deductive science[5], and while later experience has taught us that physics must start from experiments and constantly return to experiments to validate any new theoretical predictions, in between these visits to the laboratory physics remains very much a deductive science, its progress dependent on the discovery and manipulation of abstract theoretical constructs both to organize the results of prior experiments and to suggest what to look for at the next stage of experimentation.

While there appears little reason *a priori* to expect that we should be able to create simple mathematical models that reflect the real world with precision, it is a seemingly miraculous fact that we are nevertheless able to do so. Eugene Wigner referred to this as "the unreasonable effectiveness of mathematics in the natural sciences" in a famous paper with that title[6] and the name has stuck. Classical Mechanics was perhaps the first good example of an "effective theory" in Wigner's sense, and it remains one of the most striking. It stands on its own as a beautiful piece of pure mathematics, and yet it is also a remarkably successful physical theory; for two hundred years following its inception it passed *every* experimental test to which it was subjected, and still today it remains the theory of choice within its domain of applicability. While we may not be able to explain why this should be so, in this section we will at least try to explain just what the precise connection is between the mathematical and the physical aspects of Classical Mechanics.

For a mathematical theory to serve as a "model" for some part of physical reality, at least two things must be specified:

i) how to interpret the undefined mathematical abstractions of the theory in terms of concrete, physically meaningful concepts and

ii) how to determine the numerical quantities that enter into the mathematical theory by performing specific physical experiments.

We will refer to these as the Interpretation Problem and the Measurement Problem, respectively. Philosophers of science have written

---

[5]For a narrative of the heroic life of the man who perhaps did most to introduce the experimental approach into physics, see Dava Sobel's inspiring "Galileo's Daughter", Walker, 1999.

[6]Cf. Communications in Pure and Applied Mathematics, vol. 13, 1960

4.4. Classical Mechanics as a Physical Theory

whole books on the general theory of this process, but for the case of interest to us, Classical Mechanics, the solution to both of these problems can be described in fairly intuitive terms, and we will give only an informal and abbreviated account, leaving many details to the imagination of the reader.

The basic undefined abstraction of Classical Mechanics is that of a *particle*. Although the word as used today usually connotes some sort of indivisible object, for purposes of Classical Mechanics all that is required of a particle is that it be a piece of matter of known mass that (in the context in which it is considered) can be considered as being located at a point of space (rather than being a distributed object). For example, if the problem at hand is to derive the orbits of the planets about the Sun, then the Earth is usually considered to be a particle, but if the problem is to describe the rotation of the Earth on its axis or to compute the precise orbit of, say, a weather satellite around the Earth, then while the satellite can be treated as a particle, the Earth cannot.

A somewhat trickier concept is that of an *isolated* particle, which is meant to formalize the notion of a particle that is not subject to any external force. But this is clearly an idealization—since every particle of matter is the source of a gravitational field that exerts a force on every other particle in the universe, it would seem that no real particle could be considered rigorously isolated. What saves the concept from being vacuous is that if a particle is sufficiently removed from other matter, the gravitational field in its neighborhood causes an acceleration that is (very nearly) the same for itself and all nearby particles, independent of their masses or other properties[7]. This implies (using a nice metaphor that goes back to Einstein) that if we get into an elevator and cut the cable, then a particle inside the elevator will experience no gravitational force[8]. What about other forces? The fact is that the only other long-range force besides gravitation that intervenes in classical mechanical considerations is electromagnetism,

---

[7] This is Einstein's "Principle of Equivalence".

[8] Until the elevator hits the bottom of the shaft! If he were writing today, Einstein would no doubt have used the example of an orbiting space station, where this phenomenon of weightlessness has become a familiar one seen in countless TV broadcasts from space.

so that inside a freely falling laboratory, a particle that is electrically neutral, magnetically shielded, and not in contact with other matter can effectively be considered as isolated.

Let us now take up the measurement problem for Classical Mechanics. The three basic kinds of measurements that we have to make are of mass, time, and the positions of particles. The first two are relatively easy and we consider them first.

Although physics texts caution students—correctly—that mass and weight are logically distinct concepts that should not be confused, Einstein's Equivalence Principle implies that for measurement purposes they are always proportional, so that we can use a simple pan balance for measuring the mass of small bits of matter.[9] For large conglomerations of matter (such as the Earth) this clearly does not work, and we must use indirect methods, and later we shall explain how Cavendish experimentally "weighed the Earth" by measuring Newton's gravitational constant G.

As explained earlier, time is measured by choosing a periodic process and using its primitive period as the unit of time. We have also seen that there are many possible choices—for example, we can choose any normal mode of a system near equilibrium, or we can choose the motion of either one of an isolated pair of gravitating particles about their center of mass. The miracle is that these various methods are consistent with each other! In other words, **the ratio of the periods of two different natural cyclic processes remains constant in time**, giving a well-defined conversion factor between the time units using one process or the other. This is an experimental fact, and if it were not so, then physics would be a very different science from the one we know, and Newton's Laws of Motion would not model the real world. In what follows we will assume that we have chosen a *clock*, meaning a particular choice of cyclic process defining a unit of time.

---

[9]While this approach is legitimate and simplifies the discussion, it is as Michael Nauenberg remarked, "completely ahistorical". At the end of this section we shall comment briefly on how Newton handled (or mishandled) the notion of mass.

4.4. Classical Mechanics as a Physical Theory

We will call a system for assigning numerical coordinates to the positions of particles in space a *reference frame*, and what we would like is a choice of reference frame for which particles will obey Newton's Laws of Motion, and in particular the First Law.

4.4.1. Definition. A reference frame is called *inertial* if the positions of all isolated particles as measured in the frame are either stationary in time or else move in straight lines with constant velocity.

We will assume that the space in which we live is Euclidean[10]. It follows that to specify a reference frame what we have to do is

1) choose three orthogonal directions,
2) choose an origin, and
3) describe how to measure distances between two points.

The algorithm for making these choices so as to end up with an inertial frame is the most interesting and nontrivial part of solving the Measurement Problem for Classical Mechanics. We obviously must exercise some care. For example, if we were to set up a laboratory on a rotating merry-go-round, then a reference frame rigidly attached to the laboratory would certainly not be inertial, since there will be "fictitious" centrifugal and Coriolis forces resulting from the rotation.

But wait a minute! We **are** on a rotating merry-go-round, aren't we—one that makes a full rotation once in twenty-four hours? And the fictitious forces due to this rotation are very visible in the Foucault Pendulum Experiment.

The differential equations governing the motion of the Foucault pendulum are

$$\frac{d^2x}{dt^2} = -\omega^2 x + 2\Omega \frac{dy}{dt} \sin \phi,$$
$$\frac{d^2y}{dt^2} = -\omega^2 y + 2\Omega \frac{dx}{dt} \sin \phi,$$

---

[10] While we now know that this is not rigorously true, it does follow from General Relativity Theory that if we keep away from regions like black holes where there are very strong gravitational fields, then the geometry of space will be well approximated by Euclidean geometry over appreciable regions.

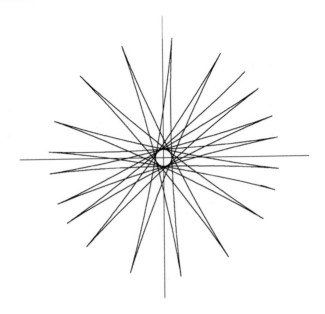

Figure 4.1. The Foucault pendulum.

where ω is the angular frequency of the pendulum, Ω is the rotational frequency of the Earth, and ϕ is the latitude of the pendulum. The terms $2\Omega \frac{dy}{dt} \sin\phi$ and $2\Omega \frac{dx}{dt} \sin\phi$ are the x and y components of the so-called Coriolis force—the "fictitious force" referred to above.

▷ **Exercise 4–2.** If you are not familiar with the notion of Coriolis force, then find out about it by looking it up on Google or in Wikipedia or in your favorite physics book. Then see if you can derive the above equations for the Foucault pendulum.

So the first problem we face is how to choose a reference frame that is "nonrotating". But nonrotating with respect to what? To find the answer, we only need to look up at the sky on a clear, dark night. We should choose our three orthogonal directions in space so that they point in constant directions with respect to the "fixed stars" (or better, the fixed galaxies). As for the choice of origin, all that is important is that it be in free fall, as discussed earlier, so that leaves open only the manner in which we measure distances, and the easiest

4.4. Classical Mechanics as a Physical Theory

way to do that is to use our clock and see how long it takes light to make a round trip.

We can now give a succinct prescription for setting up an ideal laboratory for the study of Classical Mechanics in an inertial reference frame. Launch a space station using a booster rocket, and after burnout use gas jets to stabilize the orientation of the station with respect to the stars. Take the center of mass of the station as the origin and choose any three orthogonal directions that are fixed relative to the station as coordinate axes. Measure time with a cesium vapor clock and measure distances by radar.

Now that we have given precise physical interpretations to all the undefined notions that occur in the statement of Newton's Laws of Motion, their validity becomes a matter open to experimental verification, and needless to say, countless experiments have verified their validity with great precision.

It is clear from the above discussion that there is a lot of indeterminacy in the choice of an inertial reference frame. First, we can translate the origin by any fixed vector x_0. Then, we can apply any orthogonal transformation T to the three directions defining the coordinate axes. And finally, since any particle moving with constant velocity V relative to a freely falling particle is also in free fall, we can add a variable translation tV. In other words, if x is an inertial coordinate system, then so is $x' = Tx + tV + x_0$. The group of transformations of $\mathbf{R}^3 \times \mathbf{R}$ (or "space-time") that have the form $(x, t) \mapsto (x', t)$ where $x' = Tx + tV + x_0$ is called the *Galilean group*, so we may rephrase this by saying that Galilean transformations carry inertial frames to inertial frames.[11] It is clear that if we apply a Galilean transformation to the parametric equation $x(t) = p + tv$ that expresses motion with constant velocity in a straight line, then we get another such equation, $x'(t) = (Tp + x_0) + t(Tv + V)$, and it is an easy exercise to show that Galilean transformations are the only invertible transformations $(x, t) \mapsto (\Phi(x, t), t)$ with this property.

---

[11]The fact that the coordinate t is required to remain the same in a Gallilean transformation is usually expressed by saying that time is considered an "absolute" quantity in Classical Mechanics.

Thus the Galilean group is the natural maximal symmetry group of Classical Mechanics.

4.5. Potential Functions and Conservation of Energy

Suppose $U : V \to \mathbf{R}$ is a smooth real-valued function on V. We recall that dU_p, the differential of U at a point p of V, is the linear functional on V defined by $dU_p(v) = (d/dt)_{t=0} U(p+tv)$—or in words, $dU_p(v)$ is the directional derivative of U at p in the direction v. Since V is an orthogonal vector space, any linear functional f on V is given by taking the inner product with a uniquely determined fixed vector in V "dual" to f. In particular, the vector dual to dU_p is called the *gradient* of U at p and is denoted by ∇U_p. It is then easily seen that for any smooth curve $\gamma : [a,b] \to V$, $(d/dt) U(\gamma(t)) = \langle \gamma'(t), \nabla U_{\gamma(t)} \rangle$.

▷ **Exercise 4–3.** Prove this, and show that, with respect to any orthonormal system of coordinates $(x_1, \ldots, x_n)$ for V, the components of ∇U are $\partial U / \partial x_i$.

We shall say that a force field F is "derived from a potential" if there exists a smooth function U such that $F = -\nabla U$, in which case we call U a potential for F. The reason for the minus sign is that this way $U(p)$ represents the work required to move the system from a standard configuration p_0 to the configuration p, and it also represents the energy available in the system to do work. We will see below that there is a close connection between the existence of a potential for the force F and Newton's Third Law of Motion.

If F is derived from a potential, then (NE) has the form

$$\frac{d^2 x}{dt^2} = -\nabla U, \qquad (\text{NE}')$$

so, with respect to an orthonormal coordinate system, a solution satisfies

$$\frac{d^2 x_i(t)}{dt^2} = -\frac{\partial U}{\partial x_i}(x_1(t), \ldots, x_N(t)).$$

It is clear that if U is a potential for F, then so is U plus a constant, and it is equally easy to see that **any** other potential for F is obtained by adding a constant to U.

4.5. Potential Functions and Conservation of Energy 107

4.5.1. Remark. Since our force field F is really "specific force", i.e., force per unit mass, our potential function U represents potential energy per unit mass.

Which force fields F are derived from a potential? The answer to this is closely bound up with the notion of work.

If $\gamma : [a,b] \to V$ is any smooth curve in V, then we define the work done by the force field F along the path γ to be the line integral $\int_\gamma F \cdot ds = \int_a^b \langle F(\gamma(t)), \gamma'(t) \rangle \, dt$. If we choose an orthonormal coordinate system in which the curve γ is given parametrically by $\gamma(t) = (x_1(t), \ldots, x_N(t))$, then the line integral defining the work can be expressed as $\int_a^b \sum_i F_i(x_1(t), \ldots, x_N(t)) x_i'(t) \, dt$. The force field F is called *conservative* if the work done along every closed path is zero. This is easily seen to be equivalent to the work done along any two paths joining the same endpoints being equal. If U is a potential for F, then an easy calculation using the Fundamental Theorem of Calculus shows that the work done by F along γ is equal to $U(\gamma(a)) - U(\gamma(b))$.

▷ **Exercise 4–4.** Carry out the details. Conversely, if F is conservative, then choose an arbitrary "basepoint" p_0 in V and define a real-valued function U on V by letting $U(p)$ be the work done by F along any path joining p_0 to p. Show that U is a potential for F.

Thus,

4.5.2. Theorem. *A force field is derived from a potential if and only if it is conservative.*

▷ **Exercise 4–5.** Suppose that in some orthonormal coordinate system $(x_1, \ldots, x_N)$ for V the field F has components $F_i(x_1, \ldots, x_N)$. Show that a necessary and sufficient condition for F to be derived from a potential is that for all $1 \leq i, j \leq N$, $\partial F_i / \partial x_j = \partial F_j / \partial x_i$.

4.5.3. Remark. There is an equivalent way to formulate the above, avoiding any reference to work and using instead the language of differential forms. (Skip the following if you are **not** familiar with forms.) Let ω_F denote the one-form on V dual to the vector field F. With respect to an orthonormal coordinate system x_i as above,

$\omega_F = \sum_i F_i \, dx_i$ (so the work done by F along γ is $\int_\gamma \omega_F$). Clearly F is derived from a potential if and only if ω_F is an exact form, i.e., there exists a smooth function U so that $\omega_F = dU$, and of course it is a standard result that a one-form ω is exact if and only if its integral around every closed loop is zero. And by the Poincaré Lemma, ω_F is exact if and only if $d\omega_F = 0$ (or as a physicist might say it, F is conservative if and only if the curl of F is zero).

If (V, F) is a Newtonian system and $\sigma : I \to V$ is a solution curve, then we define two real-valued functions K_σ and P_σ on I, called, respectively, the kinetic energy and power functions of the solution, by $K_\sigma(t) = \frac{1}{2} \|\sigma'(t)\|^2$ and $P_\sigma(t) = \langle F(\sigma(t)), \sigma'(t) \rangle$. If the force F is derived from a potential $U : V \to \mathbf{R}$, then the sum of the kinetic and potential energies along a solution σ defines a third function $H_\sigma : I \to \mathbf{R}$ called the total energy (or Hamiltonian) function of the solution, i.e., $H_\sigma(t) = \frac{1}{2} \|\sigma'(t)\|^2 + U(\sigma(t))$.

4.5.4.   Remark. If the force F is derived from a potential U, then it is immediate from the definition of the power function (and the meaning of the gradient operator) that P_σ is minus the rate of change of U along the solution σ.

▷ **Exercise 4–6.** Show that the power function is the time rate of change of the kinetic energy.

4.5.5.   Energy Conservation Theorem. *If (V, F) is a Newtonian system and if F is derived from a potential U, then the total energy is constant along any solution curve.*

Proof. The proof is immediate from the preceding remark and exercise. ∎

4.5.6.   Definition. If $(V, -\nabla U)$ is a Newtonian system, then we call a point p_0 of V an *equilibrium point* (or simply an equilibrium) if p_0 is a strict local minimum of the potential function U.

4.5.7.   Remark. Frequently "equilibrium point" is used as a synonym for critical point—i.e., any point where ∇U vanishes, and what

4.5. Potential Functions and Conservation of Energy 109

we have called an equilibrium point is then referred to as a stable equilibrium. Physicists often refer to an equilibrium point as a "vacuum state".

At *any* critical point p_0 of U, the solution of Newton's Equations with initial position p_0 and initial velocity zero is the constant curve $\sigma(t) = p_0$. But if p_0 is an equilibrium point, then something much stronger is true. Informally, any solution of Newton's Equations that starts near enough to p_0 with sufficiently small velocity will stay near p_0. To make this precise, we define $B_r(p_0) = \{p \in V \mid \|p - p_0\| \leq r\}$, the closed ball of radius r about p_0, and we denote its boundary by $S_r(p_0) = \{p \in V \mid \|p - p_0\| \leq r\}$, the sphere of radius r about p_0. If p_0 is an equilibrium point, then for r sufficiently small, $U(q) - U(p_0) > 0$ for all q on $S_r(p_0)$, and since $S_r(p_0)$ is compact, there will be a positive δ so that $U(q) - U(p_0) > \delta$ for q in $S_r(p_0)$.

4.5.8. Stability Theorem for Equilibria. *If p_0 is an equilibrium point of $(V, -\nabla U)$, then for any positive r there is a positive ϵ such that for all $p \in B_\epsilon(p_0)$ and $u \in B_\epsilon(0)$ the solution σ of Newton's Equations with initial position p and initial velocity u exists for all time and remains inside $B_r(p_0)$.*

Proof. We can replace r by any smaller positive number, so we can assume as above that for some positive δ, $U(q) - U(p_0) > \delta$ for q in $S_r(p_0)$. So to prove that $\sigma(t)$ stays inside $B_r(p_0)$, it will suffice to show that $U(\sigma(t)) - U(p_0) < \delta$ for all t. By continuity of U, we can choose ϵ with $\epsilon^2 < \delta$ such that for p in $B_\epsilon(p_0)$, $U(p) - U(p_0) < \frac{1}{2}\delta$. By energy conservation, $U(\sigma(t)) + \frac{1}{2}\|\sigma'(t)\|^2 = U(p) + \frac{1}{2}\|u\|^2$, or $U(\sigma(t)) - U(p_0) = U(p) - U(p_0) + \frac{1}{2}\|u\|^2 - \frac{1}{2}\|\sigma'(t)\|^2$, and therefore $U(\sigma(t)) - U(p_0) < \frac{1}{2}\delta + \frac{1}{2}\epsilon^2 < \delta$. ∎

▷ **Exercise 4–7.** What is the (one line) proof that these solutions do in fact exist for all time?

At a critical point p_0 of a potential function U, the so-called Hessian matrix—the symmetric matrix $K_{ij} = \frac{\partial^2 U}{\partial x_i \partial x_j}$ of second partial derivatives of U—describes the behavior of U to second-order at p_0. The operator K with matrix K_{ij} is called the Hessian operator of

U at p_0 and we recall that if it is positive-definite (i.e., if all of its eigenvalues are positive, say $\omega_1^2, \ldots, \omega_n^2$), then p_0 is an equilibrium point, and such an equilibrium is called a *nondegenerate* minimum of U. At a nondegenerate minimum, the Hessian operator has a unique positive definite square root, Ω (namely the operator having eigenvalues ω_i at the ω_i^2-eigenvectors of K). In this situation we have a more refined version of the above stability theorem. To make the statement simpler, we translate p_0 to the origin and assume that $U(0) = 0$.

4.5.9. Stability Theorem for Nondegenerate Equilibria. *Assume the origin is a nondegenerate equilibrium point of $(V, -\nabla U)$ and that $U(0) = 0$. Let $x(t, x_0, u_0)$ denote the solution of Newton's Equations with initial position x_0 and initial velocity u_0. Then for ϵ sufficiently small, $x(t, \epsilon x_0, \epsilon u_0)$ is defined for all t and there is a positive constant C such that $\|x(t, \epsilon x_0, \epsilon u_0)\|^2 \le C|\epsilon|^2(\|x_0\|^2 + \|u_0\|^2)$ and $\|x'(t, \epsilon x_0, \epsilon u_0)\|^2 \le C|\epsilon|^2(\|x_0\|^2 + \|u_0\|^2)$.*

▷ **Exercise 4–8.** Use the following steps and hints to prove this theorem. Note first that by Taylor's Theorem with Remainder, near the origin we can write $U(x) = \frac{1}{2}\|\Omega x\|^2 + \|x\|^2 \rho(x)$ where ρ is continuous and $\rho(0) = 0$.

1) Let ω_m and ω_M denote the smallest and largest eigenvalues of Ω. Show that near the origin $\frac{\omega_m^2}{4}\|x\|^2 < U(x) < \frac{3\omega_M^2}{4}\|x\|^2$. Hint: On one hand, choosing orthonormal coordinates defined by an eigenbasis for Ω, it follows that $\omega_m^2 \|x\|^2 \le \|\Omega x\|^2 \le \omega_M^2 \|x\|^2$, and on the other hand, near the origin, $-\frac{\omega_m^2}{4} < \rho(x) < \frac{\omega_M^2}{4}$.

2) Show using 1) that if E is the total energy function on $V \times V$, defined by $E(x, u) = \frac{1}{2}\|u\|^2 + U(x)$, then for any x_0 and u_0 the inequality $E(\epsilon x_0, \epsilon u_0) \le A\epsilon^2(\|x_0\|^2 + \|u_0\|^2)$ is satisfied for some positive A and all sufficiently small ϵ. Hint: Take $A = \min(\frac{1}{2}, \frac{3\omega_M^2}{4})$.

3) $E(x(t, \epsilon x_0, \epsilon u_0), x'(t, \epsilon x_0, \epsilon u_0)) = E(\epsilon x_0, \epsilon u_0)$ by conservation of energy. Since the kinetic energy $\frac{1}{2}\|x'(t, \epsilon x_0, \epsilon u_0)\|^2$ is positive, it follows that the potential energy is less than the total energy,

so by 2) $U(x(t, \epsilon x_0, \epsilon u_0)) \leq A\epsilon^2(\|x_0\|^2 + \|u_0\|^2)$. Now use 1) to derive $\|x(t, \epsilon x_0, \epsilon u_0)\|^2 \leq C|\epsilon|^2(\|x_0\|^2 + \|u_0\|^2)$.

4) Since the potential energy $U(x(t, \epsilon x_0, \epsilon u_0))$ is positive (for small ϵ) the kinetic energy $\frac{1}{2}\|x'(t, \epsilon x_0, \epsilon u_0)\|^2$ is less than $E(\epsilon x_0, \epsilon u_0)$, and so the inequality $\|x'(t, \epsilon x_0, \epsilon u_0)\|^2 \leq C|\epsilon|^2(\|x_0\|^2 + \|u_0\|^2)$ follows from 2).

4.6. One-Dimensional Systems

For a one-dimensional Newtonian system, we can take V to be $\mathbf{R}$, and the force law and potential function are then just "real-valued functions of one real variable"—i.e., exactly the domain of elementary Calculus. These systems are interesting to analyze for their own sake, and in addition they play an important rôle in analyzing higher-dimensional systems.

One-dimensional systems are also very easy to visualize using a computer. There are many computer programs, such as Maple, Mathematica, MatLab, and 3D-XplorMath , that make it easy for a user to enter interactively a second-order ODE and initial conditions and that will then solve the ODE numerically and display the solution curve $x(t)$ on the computer screen. (Usually what these programs display is the parametric curve $x = x(t)$, $y = x'(t)$ in the (x, y)-plane.) In the Web Companion there is a movie, created in 3D-XplorMath, for the case of the Pendulum Equation $\frac{d^2 x}{dt^2} = -\sin(x)$ showing the direction field and a few solution curves. (See Figure 4.2 for a static version.)

▷ **Exercise 4–9.** Show that, in a one-dimensional system, F is **always** derived from a potential U. In fact, show that we can take U to be minus the indefinite integral of F.

A major simplification in the one-dimensional case is that we can use conservation of energy to reduce Newton's Equation to a first-order equation. If $x(t)$ is the solution of Newton's Equations with initial position x_0 and initial velocity v_0, then the total energy for this solution is $E = U(x_0) + v_0^2/2$. The energy conservation equation is then $\frac{1}{2}(\frac{dx}{dt})^2 + U(x(t)) = E$.

If we rewrite this as $\frac{dx}{dt} = \sqrt{2(E - U(x))}$, then it follows that we can find the solution $x(t)$ explicitly in two computable steps—first a quadrature and then a function inversion. Indeed, the time t as a function of the position x is given by the definite integral

$$t(x) = \int_{x_0}^{x} \frac{ds}{\sqrt{2(E - U(s))}}$$

and we then only have to invert this function.

4.6.1. Remark. Hey, wait a minute! Something must be wrong with this! For a periodic orbit, the position $x(t)$ takes the same value for many different values of the time t, so the "function" $t(x)$ is **not** single-valued. So how is it that we have derived an explicit formula for it? Aha!—what happens if $U(t) = E$? And shouldn't that square root have a sign? The velocity isn't always positive after all.

▷ **Exercise 4–10.** We'll come back to this later for a closer look—and try to make rigorous sense out of it—but for now see if you can figure out for yourself what is going on. Try to analyze what happens in the harmonic oscillator example below (where in fact **all** of the solutions are periodic).

We next introduce two important one-dimensional systems. (Actually, the first one, the harmonic oscillator, was already considered in Section 2.3.)

• **Example 4–2. The One-Dimensional Harmonic Oscillator.** This models a particle of mass m attached to an ideal spring with spring constant (or Young's Modulus) k. That is, the spring has an unstretched length of 0, and when stretched to position x, there is a restoring force $-kx$ acting on the particle ("Hooke's Law"). Newton's Equation is $\frac{d^2x}{dt^2} = -\omega^2 x$, where we have set $\omega = \sqrt{\frac{k}{m}}$, and the potential function is $U(x) = \frac{1}{2}\omega^2 x^2$. Notice that the force law is linear, so that Newton's Equations are linear differential equations, and it is therefore easy to solve them explicitly. In fact, the general solution is $x(t) = a\cos(\omega t) + b\sin(\omega t)$, where a and b are arbitrary constants. In particular, the solution with initial position x_0 and initial velocity v_0 is $x(t) = x_0 \cos(\omega t) + (v_0/\omega)\sin(\omega t)$. We note that all solutions are periodic, with the same period $\frac{2\pi}{\omega} = 2\pi\sqrt{\frac{m}{k}}$ and

4.6. One-Dimensional Systems

frequency $\omega/2\pi = \frac{1}{2\pi}\sqrt{\frac{k}{m}}$. (The parameter ω is called the angular frequency of the oscillator and represents the reciprocal of the time it takes the "phase" ωt to increase by one radian.)

▷ **Exercise 4–11.** Show that the general solution of the harmonic oscillator can also be expressed in the form $x(t) = A\cos(\omega(t - \tau))$. The two parameters A and τ are called the amplitude and phase shift of the solution. What is their geometric significance?

▷ **Exercise 4–12.** Find the general solution for the linear system $\frac{d^2x}{dt^2} = \omega^2 x$, i.e., Newton's Equation when the force is again linear, but repulsive rather than attractive. Why do you think this is less physically interesting? (Well, perhaps it *is* interesting—the potential energy is not bounded below, so this force could do an unlimited amount of work. Hmm, perhaps one should call *that* an attractive force.)

• **Example 4–3. The Pendulum.** This models the oscillations of a particle of mass m attached to the end of a weightless, rigid rod of length L. The other end of the rod is attached to a pivot, and the rod swings under the influence of gravity. The coordinate x is the angle that the rod is displaced from the equilibrium configuration— straight down. The height of the particle above the equilibrium height is clearly $L\cos(x)$, and since the force of gravity is of constant magnitude mg and directed downward, the work done to move the particle from its equilibrium configuration is $mgL\cos(x)$ and so the potential function is $U(x) = gL\cos(x)$. (Of course, g is the "acceleration of gravity". For a pendulum on the surface of the Earth, if we measure time in seconds and distance in feet, then g is approximately 32 ft/sec^2.) Newton's Equation is now $\frac{d^2x}{dt^2} = -gL\sin(x)$. The general solution cannot be expressed in terms of elementary functions (it involves elliptic functions). But if x is close to zero, then $\cos(x)$ is approximately $\frac{1}{2}x^2$, so if we displace a pendulum of length L only slightly from its equilibrium position, then we might expect it to behave like a harmonic oscillator whose angular frequency ω is $\sqrt{gL}$. In particular, we might expect it to swing back and forth with a period of approximately $2\pi\sqrt{gL}$.

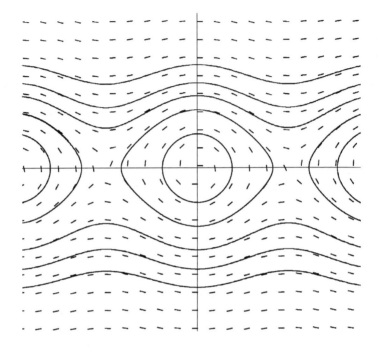

Figure 4.2. The pendulum.

Figure 4.2 shows the direction field and several orbits for the Pendulum Equation, $\frac{d^2x}{dt^2} = -gL\sin(x)$—or rather for the equivalent first-order system $\frac{dx}{dt} = \omega$, $\frac{d\omega}{dt} = -gL\sin(x)$. The horizontal x coordinate represents the pendulum angle, and the vertical coordinate ω represents its angular velocity. The closed orbits are those with insufficient kinetic energy (or initial velocity) to swing "over the top".

The Web Companion contains instructions for carrying out computer experiments involving the harmonic oscillator and pendulum, using various programming systems or 3D-XplorMath.

4.6.2. A Universal One-Dimensional Model. It is helpful to have a good intuitive physical picture in mind when dealing with a Newtonian system, and another nice feature of one-dimensional systems is that there is a "universal" method for constructing a physical model that realizes any abstract potential function U.

4.6. One-Dimensional Systems

Let us start with a smooth curve $y = Y(x)$ in the plane. We imagine an ideal infinitely thin wire positioned along this curve and strung with a single bead that slides without friction along the wire under the action of a constant downward gravitational acceleration $-g$. The (specific) force is of course derived from the potential $gY(x)$. We will assume that for x large and positive, Y tends monotonically to infinity, and similarly for x large and negative. We will also assume that at each critical point of Y the second derivative is nonzero, so that it is either a strict local maximum or a strict local minimum.

▷ **Exercise 4–13.** Show that these assumptions imply there can be only a finite number of critical points of Y. Deduce that for any real number E, the equation $Y(x) = E$ has only a finite number of solutions.

We reparametrize our curve by arclength $s(x) = \int_0^x \sqrt{1 + Y'(x)^2}$, measured from the point $p_0 = (0, Y(0))$ since it is easier to describe the motion of the bead in terms of s than of x. If we solve for x as a function of arclength, we get a relation $x = X(s)$, and we write $U(s) = gY(X(s))$ for the potential expressed as a function of s. (Note that it is still just g times the height of the bead.) If we denote the position of the bead along the curve at time t by $s(t)$, the velocity will be $v(t) = s'(t)$. (If we had used x as our parameter, we would instead have to deal instead with the more complicated expression $x'(t)\sqrt{1 + Y'(x(t))^2}$.) The component of the gravitational force on the bead in the direction of the wire is just $-U'(s)$, so Newton's Equation of motion for the bead is $s''(t) = -U'(s(t))$.

Now let's see if we can describe the solution $s(t)$ with some arbitrary initial position s_0 and initial velocity v_0. Before doing this rigorously, let's consider it intuitively in terms of the sliding bead picture. We place the bead on the wire at s_0 and then give it a shove so that its velocity is v_0. Assuming v_0 is positive, it will keep travelling to the right until it reaches a position s_M where it has gone so high that all its kinetic energy is converted to potential energy and its total energy E is equal to $U(s_M)$. At this point it will start sliding to the left and will continue doing so until it reaches the next point, x_m, to the left, where $U(x_m) = E$. It will then keep oscillating periodically between x_m and x_M.

OK, now let's try to do this carefully. One possibility is that s_0 is a critical point of U and $v_0 = 0$. Clearly the solution for these initial conditions is $s(t) \equiv s_0$. (And, conversely, any such "stationary" solution must be located at a critical point.)

To describe the other less trivial solutions, let's introduce some auxiliary quantities. First we define the total energy E of the solution by $E = \frac{1}{2}v_0^2 + U(s_0)$. There are now two cases to consider. First, if $v_0^2 > 0$, then $U(s_0) < E$, and since U tends to $+\infty$ in both directions, there will (by the preceding exercise) be a first point s_m to the left of s_0 where U has the value E and similarly a first point s_M to the right of s_0 where U has the value E. On the other hand, if $v_0 = 0$, (so that $U(s_0) = E$), then s_0 cannot be a critical point of U (since we are assuming that $s(t)$ is **not** a constant solution). If $U'(s_0) < 0$, then we define $s_m = s_0$ and define s_M to be the next largest solution of $U(x) = E$, while if $U'(s_0) > 0$, then we define $s_M = s_0$ and let s_m be the next smallest solution of $U(t) = E$. Thus, in any case we have $U(s_m) = U(s_M) = E$, $s_m < s_M$, $s_m \leq s_0 \leq s_M$, and $U(s) = E$ has no solution strictly between s_m and s_M—i.e., we have bracketed s_0 between two successive solutions of $U(s) = E$.

▷ **Exercise 4–14.** Show that the solution $s(t)$ is periodic with period

$$T = \sqrt{2} \int_{s_m}^{s_M} \frac{ds}{\sqrt{E - U(s)}}$$

and that it oscillates between s_m and s_M.

▷ **Exercise 4–15.** Let's see if we can describe more precisely the periodic solution that oscillates between two successive roots s_m and s_M of $U(x) = E$. Without loss of generality we can assume that the solution starts at s_m with velocity zero at time 0. Let T be as in the preceding exercise, and define $t : [0, T/2] \to \mathbf{R}$ by $t(s) = \sqrt{2} \int_{s_m}^{s} d\sigma/\sqrt{E - U(\sigma)}$. Show that this is a one-to-one map of $[0, T/2]$ onto $[s_m, s_M]$, and let $s : [s_m, s_M] \to [0, T/2]$ denote its inverse. Extend s to the interval $[0, T]$ by defining $s(T/2 + t) = s(T/2 - t)$, and then extend s to a function on the whole real line by making it

4.6. One-Dimensional Systems

periodic of period T. Show that this is in fact the desired solution. Carry this out for the harmonic oscillator and (if you are ambitious) carry it out for the pendulum, expressing the period T as a "complete elliptic integral" and expressing the general solution in terms of the Jacobi elliptic function sn.

▷ **Exercise 4–16.** Show that in the two previous exercises we have missed an important special situation! Namely, if either (or both) of the points x_m and x_M are critical points of U, show that the integral that is supposed to represent the semi-period is infinite. Deduce that in this case, instead of reaching this critical point in finite time, the solution approaches the critical point asymptotically as t tends to infinity. (Such a solution is called a "brake orbit".)

▷ **Exercise 4–17.** Show that the sliding bead model really is "universal", in the sense that given any smooth function $U(s)$ (satisfying the above restrictions on Y), there is a $Y(x)$ for which $U(s) = gY(X(s))$.

▷ **Exercise 4–18. Classical Inverse Scattering.** In this exercise we will for simplicity assume that U assumes its minimum value, zero, at $s = 0$ and that it is strictly increasing for $s > 0$. The goal will be to recover the function $U(s)$ (for s positive) by performing "scattering experiments". That is, we imagine that we place the bead on the wire at $x = s = 0$ and that for each positive energy, E, we give it a shove to the right with kinetic energy E. The bead will slide up the wire until it reaches the point at distance $S(E)$ along the curve where its potential energy is E and its kinetic energy (and hence its velocity) is zero. This will take a time $T(E) = \sqrt{2} \int_0^{S(E)} \frac{ds}{\sqrt{E-U(s)}}$, and the bead will then turn around and slide back down to the point $s = 0$ in an equal time, so the time for it to make the round trip is $2T(E)$ and is what we experimentally measure. We would like to recover the potential function U from these measured values $T(E)$, but we will instead look at the computationally easier problem of recovering the inverse function, i.e., the value $S(E)$ where U assumes a particular value E. Show that in fact

$$S(E) = \frac{\sqrt{2}}{2\pi} \int_0^E \frac{T(e)\,de}{\sqrt{E-e}}.$$

Hint: Evaluate $\int_0^\alpha \frac{T(e)\,de}{\sqrt{\alpha-e}}$, using the above integral formula for $T(E)$ to evaluate $T(e)$, but using $u = U(s)$ rather than s as the variable of integration. Then interchange the order of the du and de integrations, and note that $\int_e^\alpha \frac{du}{\sqrt{(\alpha-u)(u-e)}} = \pi$. For details, see page 27 of [LL].

4.7. The Third Law and Conservation Principles

We return now to the consideration of n point particles $P_1, \ldots, P_n$ in $\mathbf{R}^3$. In the Newtonian worldview, every particle P_i exerts a force, say f_{ij}, on each other particle P_j, and f_{ij} is a smooth function of the positions x^i and x^j of the two particles. Thus (in an isolated system) the total force f_i acting on the particle P_i is just the vector sum $f_i = \sum_{j=1}^n f_{ij}$ (where for convenience we have defined $f_{ii} = 0$). Thus f_i is a smooth function of $x^1, \ldots, x^n$ and Newton's Second Law of Motion ("force equals mass times acceleration") says,

$$\frac{d^2 x^i}{dt^2} = (1/m_i) f_i,$$

where m_i denotes the mass of P_i, while the weak form of the Third Law says f_{ij} is skew-symmetric in i and j:

$$f_{ij} = -f_{ji}.$$

We define the linear momentum p_i of the particle P_i to be the vector function of t, $m_i \frac{dx^i}{dt}$ (i.e., its mass times its velocity), so Newton's Equations become $\frac{dp_i}{dt} = f_i$. Hence if we define the total linear momentum p of the system of n particles to be the vector sum of the p_i, then $\frac{dp}{dt} = \sum_{i=1}^n f_i = \sum_{i=1}^n \sum_{j=1}^n f_{ij}$, and by the antisymmetry of the f_{ij} (Newton's Third Law) it follows that $\frac{dp}{dt} = 0$, so

4.7.1. The Principle of Conservation of Linear Momentum. *If an isolated system of particles obeys Newton's Second Law and the weak form of Newton's Third Law, then its total linear momentum is a constant of the motion.*

4.7. The Third Law and Conservation Principles

▷ **Exercise 4–19.** Recall that the center of mass of the particles P_i is the point X of $\mathbf{R}^3$ given by $X = \frac{1}{M}\sum_i m_i x^i$ (where $M = \sum_i m_i$ is the total mass of the system). Show that an equivalent statement of conservation of total linear momentum is that the center of mass moves in a straight line with constant velocity (or remains stationary).

For nonisolated systems of particles we will assume that the total force f_i on the particle P_i is the sum of an internal part f_i^{int} and an external force f_i^{ext}. The internal force is, as above, the sum of contributions f_{ij} from the other particles of the system, while the external force is some given function of x_i and t. The vector sum $f^{\text{ext}} = \sum_i f_i^{\text{ext}}$ is called the total external force on the system.

▷ **Exercise 4–20.** Show that, for a nonisolated system, the rate of change of the total linear momentum is equal to the total external force. Equivalently, X, the center of gravity, satisfies the Newton's Equation $M\frac{d^2 X}{dt^2} = f^{\text{ext}}$. (So for certain purposes, we can represent the system of particles by a fictitious particle of mass M located at the center of mass and moving as if the total external force were acting on it.)

The vector cross-product of the position vector x^i of the particle P_i and its momentum $p_i = m_i \frac{dx^i}{dt}$ is called the *angular momentum*, j_i, of P_i,

$$j_i = m_i \left(x^i \times \frac{dx^i}{dt} \right),$$

and the vector sum $j = \sum_i j_i$ is called the total angular momentum. Note that $\frac{dj_i}{dt} = m_i(x^i \times \frac{d^2 x^i}{dt^2})$ (since the cross-product of anything with itself is zero). Thus by Newton's Equation, $\frac{dj_i}{dt} = x^i \times f_i$. The quantity $x^i \times f_i$ is called the *torque* on P_i and we will denote it by τ_i, so the above equation says that the rate of change of the angular momentum of a particle equals the torque on the particle. The sum $\tau = \sum_i \tau_i$ is called the total torque acting on the system and is the sum of the total internal torque $\sum_i \sum_j x^i \times f_{ij}$ and the total external torque $\sum_i x_i \times f_i^{\text{ext}}$. But now suppose that the strong form of Newton's Third Law holds, i.e., the direction of f_{ij} is along the line joining P_i to P_j, i.e., parallel to $(x^i - x^j)$. Then since $x^i \times f_{ij} + x^j \times f_{ji} = (x^i - x^j) \times f_{ij} = 0$, it follows that the total internal torque is zero, and so the total torque coincides with the total external torque and in particular is zero for an isolated system. This proves

4.7.2. The Principle of Conservation of Angular Momentum. *If an isolated system of particles obeys Newton's Second Law and the strong form of Newton's Third Law, then its total angular momentum is a constant of the motion. More generally, for a nonisolated system the rate of change of angular momentum is equal to the external torque.*

Now let's try to find conditions under which a system of forces f_{ij}, as above, are conservative and hence derivable from a potential function. It will be convenient to introduce quantities $r_{ij} = x^i - x^j$, their lengths $\rho_{ij} = \left\| x^i - x^j \right\|$, and also their normalizations, the unit vectors $\hat{r}_{ij} = \frac{r_{ij}}{\rho_{ij}}$.

▷ **Exercise 4–21.** Suppose that the force $f_{ij}(x^i, x^j)$ depends only on the distance $r_{ij} = \left\| x^i - x^j \right\|$ separating P_i from P_j. Suppose moreover that f_{ij} and f_{ji} are "equal in magnitude and opposite in direction and directed along the line joining P_i and P_j" (the strong form of Newton's Third Law of Motion). Show that in this case F is conservative.

• **Example 4–4. The Gravitational n-Body Problem.** We return to the system of n particles in $\mathbf{R}^3$, but now we assume that the force f_{ij} has a specific form, namely that its magnitude is $Gm_i m_j / r_{ij}^2$ (where G is a positive constant, called the gravitational coupling constant) and that it acts in the direction from P_i to P_j. So, explicitly,

$$f_{ij} = \frac{Gm_i m_j}{\|x_j - x_i\|^2} \frac{(x_j - x_i)}{\|x_j - x_i\|}.$$

We note that this is an example of the sort mentioned earlier, where the force field blows up on a certain closed subset C which must therefore be removed from V. In this case the subset C is the so-called "collision set", namely the union of the hyperplanes $x_i = x_j$.

4.7. The Third Law and Conservation Principles

▷ **Exercise 4–22.** Show that in this example we can take as the potential

$$U = \sum_{i<j} \frac{-Gm_i m_j}{\|x_j - x_i\|}.$$

4.7.3. Remark. In the Web Companion, we show an animation of Keplerian motion that contains a graphical proof that Kepler's Laws of Motion (in particular the fact that each planetary orbit is an ellipse with the radius vector sweeping out equal areas in equal times—i.e., conservation of angular momentum) imply that the potential energy of the central force must be inversely proportional to the distance r from the center and hence that the force is proportional to $\frac{1}{r^2}$.

You might wonder what would happen if the force instead of being proportional to $\frac{1}{r^2}$ were proportional to $\frac{1}{r^e}$ where e is close to 2. One might guess that the planets would still move in closed orbits that were very close to ellipses. But, no. While the shape of the orbits would be quite elliptical locally, they would not close up but would rather precess around the sun. Figure 4.3 shows the precession for $e = 2.05$.

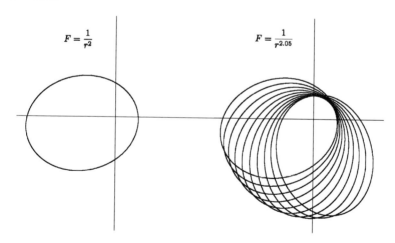

Figure 4.3. Precession.

In actuality, the orbit of each planet **does** precess. However, this is not because e is different from 2, but rather it is a consequence of the perturbing effects of the attraction of the other planets (plus a little more due to general relativity).

4.7.4. Definition. An *equivalence* between two Newtonian systems $(V_1, -\nabla U_1)$ and $(V_2, -\nabla U_2)$ is an orthogonal linear transformation $\phi : V_1 \to V_2$ such that $U_1 = U_2 \circ \phi$. If an equivalence exists, then we call the systems equivalent.

▷ **Exercise 4–23.** Show that an equivalence between $(V_1, -\nabla U_1)$ and $(V_2, -\nabla U_2)$ maps solutions of Newton's Equations for $(V_1, -\nabla U_1)$ to solutions of Newton's Equations for $(V_2, -\nabla U_2)$.

▷ **Exercise 4–24.** Show that $(V_1, -\nabla U_1)$ and $(V_2, -\nabla U_2)$ are equivalent if and only if there exist orthogonal coordinates for V_1 and orthogonal coordinates for V_2 with respect to which Newton's Equations are "the same".

▷ **Exercise 4–25.** Let G be a group of orthogonal transformations of V. We call G a group of symmetries of the Newtonian system $(V, -\nabla U)$ if U is an "invariant" of G, i.e., if $U \circ g = U$ for all g in G. Of course G acts on parametrized curves in V by defining $(g\sigma)(t) = g(\sigma(t))$. Show that the space of solutions of Newton's Equations is carried into itself under this action when G is a group of symmetries.

4.8. Synthesis and Analysis of Newtonian Systems

Given two Newtonian systems (V_1, F_1) and (V_2, F_2), we can form the direct sum $V = V_1 \oplus V_2$ of the two configuration spaces. A curve in V, $\sigma : I \to V$, is of course just a pair of curves $\sigma_1 : I \to V_1$ and $\sigma_2 : I \to V_2$. Can we find a force law F for V so that σ is a solution curve for (V, F) if and only if σ_i is a solution curve of (V_i, F_i) for $i = 1, 2$? Of course we can—clearly $F(x_1, x_2) = (F_1(x_1), F_2(x_2))$ is the unique such F and we denote it by $F_1 \oplus F_2$.

If we think of V_1 and V_2 as parametrizing the possible configurations of two "particles" P_1 and P_2, respectively, then V parametrizes a system that is the composite of P_1 and P_2, and the force law $F_1 \oplus F_2$ just says that the two particles are "noninteracting" or "uncoupled",

4.8. Synthesis and Analysis of Newtonian Systems

since the motion of each particle in the composite system is unaffected by the behavior of the other. If we completely understand the dynamics of (V_1, F_1) and (V_2, F_2), then we also completely understand the dynamics of (V, F).

▷ **Exercise 4–26.** If U_i is a potential for F_i for $i = 1, 2$, define a function $U_1 \oplus U_2$ on $V_1 \oplus V_2$ by $(U_1 \oplus U_2)(x_1, x_2) = U_1(x_1) + U_2(x_2)$ and show that it is a potential for $F_1 \oplus F_2$.

Now let's look at things from the other end: instead of synthesizing a larger system out of two other systems, let us try to analyze a Newtonian system (V, F) into smaller "subsystems". But what should we mean by a subsystem (V_1, F_1) of (V, F)? Clearly V_1 should be a linear subspace of V with the induced inner-product.

▷ **Exercise 4–27.** Let (V, F) be a Newtonian system and let V_1 be a linear subspace of V. Show that the following are equivalent:

1) $F(v) \in V_1$ for all $v \in V_1$; i.e., the restriction of F to V_1 is the vector field $F_1 : V_1 \to V_1$.

2) If σ is any solution curve of (V, F) such that both the initial position and initial velocity lie in V_1, then the whole solution curve σ lies in V_1.

4.8.1. Definition. If either and hence both of conditions 1) and 2) hold, then we call V_1 (or (V_1, F_1)) a *subsystem* of (V, F).

▷ **Exercise 4–28.** If U is a potential for F, show that its restriction U_1 to V_1 is a potential for F_1.

Suppose now that (V, F) is the orthogonal direct sum of two subsystems (V_1, F_1) and (V_2, F_2). In general the two subsystems will be coupled, i.e., F will not equal $F_1 \oplus F_2$, and we define the *coupling force*, or *interaction force*, $F_{12} = F - (F_1 \oplus F_2)$ and think of the force law F for V as being the sum of two terms: the uncoupled force law $F_1 \oplus F_2$ and the coupling term F_{12}.

If it should happen that the interaction is zero, i.e., if the two subsystems are in fact uncoupled, then we have reduced the study of (V, F) to the study of the two smaller systems (V_1, F_1) and (V_2, F_2).

If the interaction is not exactly zero, but the subsystems are only "weakly coupled", i.e., F_{12} is in some sense small, then there is an approach that physicists often use to derive knowledge of the coupled system from its components. The technique is sometimes referred to as "turning on the interaction", or "perturbation theory" (although the latter term has a more general meaning). It works as follows. Frequently there is a natural parameter ϵ in a physical theory (called the coupling constant) and the interaction force F_{12} has the form $\epsilon\Phi_{12}$. If we denote by $x_\epsilon(t)$ the solution of (V, F) corresponding to some fixed initial conditions and a particular value of the coupling constant, then $x_0(t)$ will be the known solution of the uncoupled system, and we can try to find a power series in ϵ, $x_0(t) + \epsilon\xi_1(t) + \epsilon^2\xi_2(t) + \cdots$, that converges to (or is asymptotic to) the actual solution $x_\epsilon(t)$. In good situations, where the actual value of the coupling constant is small, it may be enough for practical purposes to calculate only a few terms of this perturbation theory expansion, and even knowing just ξ_1 can be very helpful.

Everything we have said about direct sums of two systems goes over in a completely obvious way to direct sums of any finite number of systems, and if you have ever wondered how astronomers go about making precise calculations of the orbits of the planets, it is by using just such perturbation methods. Initially each planet together with the sun is considered as a two-body system and the planet's elliptical orbit is calculated using the explicit solution of the Kepler problem. This gives a first approximation for the predicted positions of the planets—one that is already quite accurate for short periods of time. In the next approximation, a time-dependent external gravitational force is calculated for each planet, based on its position and the positions of all the other planets from the first approximation, and the orbit of each planet is then re-calculated (numerically) adding this force to the gravitational attraction of the sun. This process can then be iterated if still greater accuracy is required.

4.9. Linear Systems and Harmonic Oscillators

As we saw, the one-dimensional harmonic oscillator is a system that we can solve explicitly, essentially because the force law (and hence

4.9. Linear Systems and Harmonic Oscillators

Newton's Equation) is linear. That suggests that we investigate linear force laws in higher dimensions.

Suppose we have a linear Newtonian system (V, F). Since we can always choose an orthonormal basis for V, there is no loss of generality in assuming that V is $\mathbf{R}^N$, so F is given by an $N \times N$ matrix of real numbers, F_{ij}, and Newton's Equations become

$$\frac{d^2 x_i}{dt^2} = \sum_{j=1}^{N} F_{ij} x_j.$$

The situation simplifies further when the linear force F is derived from a potential, U.

4.9.1. Theorem. *A necessary and sufficient condition for a linear Newtonian system (V, F) to be derived from a potential is that F is a self-adjoint operator on V, and in this case we can take for the potential $U(x) = -\frac{1}{2} \langle Fx, x \rangle$.*

Proof. Let's look at the condition for U to be a potential for F, namely $-\frac{\partial U}{\partial x_i} = \sum_j F_{ij} x_j$. It follows that $\frac{\partial^2 U}{\partial x_i \partial x_j} = -F_{ij}$ (so that U is a quadratic function of the x's) and by "equality of cross-derivatives" it follows that the matrix F_{ij} is symmetric, or equivalently that the linear operator F is self-adjoint. Conversely, if $F : V \to V$ is any self-adjoint operator on an orthogonal vector space, the it is easy to check that F is derived from the potential function $U : V \to \mathbf{R}$ defined by $U(x) = -\frac{1}{2} \langle Fx, x \rangle = -\frac{1}{2} \sum_{ij} F_{ij} x_i x_j$. ∎

For the remainder of this section we assume that (V, F) is a linear system derived from a potential. It will be more convenient to make the primary self-adjoint operator we work with $K = -F$, the negative of the force, rather than the force F itself. The potential now becomes $U(x) = \frac{1}{2} \langle Kx, x \rangle$, and Newton's Equations are $\frac{d^2 x}{dt^2} = -Kx$.

4.9.2. Definition. *A linear Newtonian system (V, F) that is derived from a potential is called a harmonic oscillator system if the operator $K = -F$ is strictly positive.* We choose an orthonormal basis $v_1, \ldots, v_N$ for V consisting of eigenvectors of K such that the corresponding sequence of eigenvalues $\omega_1^2, \ldots, \omega_N^2$ is nondecreasing.

The set of real numbers ω_i is called the frequency spectrum of the system and we call the v_i the *normal modes* of the system.

4.9.3. Theorem. *Every N-dimensional harmonic oscillator system is the uncoupled direct sum of N one-dimensional harmonic oscillators, namely its normal modes, and two harmonic oscillator systems are equivalent if and only if they have the same frequency spectrum.*

▷ **Exercise 4–29.** Prove this.

We can write down explicitly the general solution of the above harmonic oscillator system, namely $x(t) = \sum_i x_i(t) v_i$ where $x_i(t) = \sum_{i=1}^{N} a_i \cos(\omega_i t) + b_i \sin(\omega_i t)$ and the a_i and b_i are arbitrary constants. To get the solution of the initial value problem with $x(0) = x^0$ and $x'(0) = v^0$, take $a_i = \langle x^0, v_i \rangle$ and $b_i = \frac{1}{\omega_i} \langle v^0, v_i \rangle$.

▷ **Exercise 4–30.** Let's assume that we can make "observations" on a harmonic oscillator system (V, F) by using certain measuring devices, D_x. These devices are indexed by elements x of V, and taking a measurement with D_x when the system is in the configuration v will give the measurement $\langle x, v \rangle$. If we "listen in" on the system with D_x when its solution is as above and record the results, the signal we will see is $S_x(t) = \sum_{i=1}^{N} a_i \langle x, v_i \rangle \cos(\omega_i t) + b_i \langle x, v_i \rangle \sin(\omega_i t)$. Now of course the numbers we would like to learn about are the elements ω_i of the frequency spectrum (after all, they completely determine the system). Can you think of a way to pull these out from the $S_x(t)$? (This is a typical "signal analysis" question.)

4.10. Small Oscillations about Equilibrium

In the introduction to this chapter we promised to show how the dynamics of a Newtonian system simplifies greatly when the system is "close to equilibrium", and we are at last in a position to make good on this.

We begin with an approximate, intuitive description of what it means for a Newtonian system to be close to equilibrium and an explanation of why we might expect physical systems to normally be in such a state. As we have seen, a Newtonian system is described

4.10. Small Oscillations about Equilibrium

by a real-valued function of its configuration—the potential energy function—and the sum of kinetic and potential energies (the total energy) remains constant during any dynamical motion of the system **if the system is truly isolated**. But real macroscopic systems are never completely isolated, and when they are not at rest, there will be external frictional forces that drain energy from the system at a rate that is essentially proportional to the velocities of its components. Now the kinetic energy is positive, so if the potential energy is bounded below (as is the case for realistic systems), then the velocities of the system must eventually tend to zero and the system will asymptotically approach a rest point p. At this rest point the force acting on the system is zero, and since the force is the gradient of the potential, p must be a critical point of the potential energy. In fact, generically p will be a local minimum of the potential function—i.e., an "equilibrium point" of the system. (If this paragraph seems too abstract, then try re-reading it with the following concrete systems in mind—a marble rolling in a bowl, a swinging pendulum, a weight vibrating at the end of a spring, and a vibrating guitar string.)

Recall that perhaps the simplest of any nontrivial Newtonian system is a one-dimensional harmonic oscillator, and only slightly more complicated is an N-dimensional harmonic oscillator—for as we saw in the preceding section, this can be decomposed into a direct sum of one-dimensional oscillators, its normal modes. Now the remarkable fact is that when an arbitrary N-dimensional system is perturbed only slightly from an equilibrium state, not only does it remain close to that equilibrium, but moreover its motion relative to the equilibrium looks very similar to that of a certain N-dimensional harmonic oscillator relative to the origin, the approximation getting better as the perturbation from equilibrium gets smaller.

What this means is that we can find local orthonormal coordinates $(x_1, \ldots, x_N)$ centered at the equilibrium such that the motion is approximately given in these coordinates by $x_i(t) = A_i \cos(\omega_i(t - \tau_i))$, where the amplitudes A_i and phase shifts τ_i depend on the particular perturbation from equilibrium, but the N angular frequencies ω_i are characteristic of the equilibrium. (They are square roots of the eigenvalues of the Hessian matrix of the potential at the equilibrium.) For reasons coming from quantum field theory, physicists refer to an equilibrium configuration of a system as a "vacuum state", and they call the N harmonic oscillators associated to the equilibrium the "normal modes" associated to the vacuum state.

This approach to describing dynamics near equilibrium is called the method of small vibrations, and it continues to hold even for continuum systems that are described by infinite-dimensional configuration spaces. A particularly intuitive model is a stretched guitar string that is plucked gently from its equilibrium state and then allowed to vibrate. In this case the frequencies of the normal modes are exactly the principal frequency and various overtones of the vibrating string, transmitted to our ears as sound by pressure waves in the surrounding air.

To make our discussion of the method of small vibrations precise, we must describe carefully the harmonic oscillator that approximates the nonlinear system near a given equilibrium point and also give a rigorous definition of the sense in which solutions of this linearized system approximate the solutions of the nonlinear system.

Recall that if V and W are orthogonal vector spaces and $f : V \to W$ is a differentiable map, then the differential of f at a point p of V is a linear map Df_p of V to W that maps v to the directional derivative of f at p in the direction v; that is, $Df_p(v) = (\frac{d}{dt})_{t=0} f(p+tv)$. If we pick bases for V and W, then the mapping f is described by giving its components $f_i(x_1, \ldots, x_N)$, and the matrix for Df_p is the Jacobian matrix $\frac{\partial f_i}{\partial x_j}$ evaluated at p.

Now suppose that $U : V \to \mathbf{R}$ is twice differentiable, so that the vector field $\nabla U : V \to V$ is differentiable. Then for each p in V we define a linear map $\text{hess}(U)_p$, of V to itself, called the Hessian of U at p, by $\text{hess}(U)_p = D(\nabla f)_p$. By what we have just noted, $\text{hess}(U)_p(u) = (\frac{d}{ds})_{s=0} \nabla U_{p+sv}$. But the definition of ∇U gives $\langle \nabla U_{p+su}(u), v \rangle = (\frac{d}{dt})_{t=0} U(p+su+tv)$, and putting these together we get $\langle \text{hess}(U)_p(u), v \rangle = (\frac{\partial}{\partial t})_{t=0} (\frac{\partial}{\partial s})_{s=0} U(p+su+tv)$. Now if U has continuous second partial derivatives, we know that cross-derivatives are equal, so that $\langle \text{hess}(U)_p(u), v \rangle$ is clearly symmetric in u and v.

4.10. Small Oscillations about Equilibrium

This proves

4.10.1. Proposition. *If V is an orthogonal vector space and if $U : V \to \mathbf{R}$ is C^2, then $\mathrm{hess}(U)_p$ is self-adjoint for all p in V.*

▷ **Exercise 4–31.** Give a (slightly) different proof of this proposition by showing that the matrix of $\mathrm{hess}(U)_p$ in an orthonormal coordinate system is just the matrix $\frac{\partial^2 U}{\partial x_i \partial x_j}$ of second partial derivatives of f at p in these coordinates, and hence it is symmetric.

Now suppose that $(V, -\nabla U)$ is a conservative Newtonian system, and let p_0 be a nondegenerate equilibrium point. Without loss of generality, in what follows, we can assume that p_0 is the origin, and we can also assume that $U(0) = 0$. Denote the Hessian of U at 0 by K, so by the nondegeneracy assumption, the eigenvalues of K are positive real numbers ω_i^2, and we denote by v_i an orthonormal base of eigenvectors with $K v_i = \omega_i^2 v_i$. We will denote by Ω the positive square root of K, defined by $\Omega v_i = \omega_i v_i$. Then Taylor's Theorem with Remainder applied to ∇U gives $\nabla U(x) = \Omega^2 x + \|x\|^2 R(x)$, where $R : V \to V$ is continuous and in particular is bounded near the origin. Newton's Equation is then $\frac{d^2 x}{dt^2} = -\nabla U(x) = -\Omega^2 x - \|x\|^2 R(x)$. Near the origin, the second term will be small compared to the first, and if we ignore it, then we have just the harmonic oscillator equation $\frac{d^2 x}{dt^2} = -\Omega^2 x$. We will let $\xi(t, x^0, u^0)$ denote the solution of this harmonic oscillator equation satisfying the initial conditions $\xi(0, x^0, u^0) = x^0$ and $\xi'(0, x^0, u^0) = u^0$, while $x(t, x^0, u^0)$ will denote the solution of the full Newton's Equation $\frac{d^2 x}{dt^2} = -\nabla U(x)$ with these same initial conditions.

Recall that we know the harmonic oscillator solution $\xi(t, x^0, u^0)$ explicitly; in fact we saw (see 2.3.1) that $\xi(t, x^0, u^0) = \cos(t\Omega)x^0 + \sin(t\Omega)(\Omega^{-1}u^0)$. An obvious property of this solution (one that follows immediately from the linearity of the harmonic oscillator equation) is that $\frac{1}{\epsilon}\xi(t, \epsilon x^0, \epsilon u^0) = \xi(t, x^0, u^0)$. Of course the analogous identity does not hold for the solution of the nonlinear equation, and we will call $\frac{1}{\epsilon}x(t, \epsilon x^0, \epsilon u^0)$ the ϵ-rescaled solution of the nonlinear equation. The sense in which solutions ξ of the harmonic oscillator equation approximate solutions x of the nonlinear equation as we approach an equilibrium is expressed by the following theorem.

4.10.2. Rescaling Approximation Theorem. *The ϵ-rescaled solution of the nonlinear equation*

$$\frac{d^2x}{dt^2} = -\nabla U(x) = -\Omega^2 x - \|x\|^2 R(x)$$

converges, as ϵ tends zero, to the solution of the harmonic oscillator equation $\frac{d^2x}{dt^2} = -\Omega^2 x$, with the same initial conditions (x_0, u_0). Moreover the convergence is uniform for t, x_0, and u_0 bounded. In fact the norm of their difference is bounded by a constant times $\epsilon |t|(\|x_0\|^2 + \|u_0\|^2)$.

Proof. Define

$$g(t, x^0, u^0) = -\|x(t, x^0, u^0)\|^2 R(x(t, x^0, u^0))$$

so that $x(t, x^0, u^0)$ is a solution of the forced harmonic oscillator equation $\frac{d^2x}{dt^2} = -\Omega^2 x + g(t)$. From the Variation of Parameters Formula for solutions of such equations (see 2.4.1) it follows that the "deviation" $\Delta(t, x^0, u^0) := x(t, x^0, u^0) - \xi(t, x^0, u^0)$ is given by the integral $\int_0^t G(t-s)g(s, x^0, u^0)\, ds$, with $G(t) = \Omega^{-1}\sin(t\Omega)$. Since ξ is scaling invariant, $\frac{1}{\epsilon}x(t, \epsilon x^0, \epsilon u^0) - \xi(t, x^0, u^0) = \frac{1}{\epsilon}\Delta(t, \epsilon x^0, \epsilon u^0)$, so what remains to be shown is that $\frac{1}{\epsilon}\Delta(t, \epsilon x^0, \epsilon u^0)$ converges to zero as $\epsilon \to 0$ and that the convergence is uniform for t, x_0, and u_0 bounded. We shall see that in fact $\left\|\frac{1}{\epsilon}\Delta(t, \epsilon x^0, \epsilon u^0)\right\| < C\epsilon |t|(\|x_0\|^2 + \|u_0\|^2)$ for some positive constant C. Since $G(t)$ is uniformly bounded, it follows from the above integral representation for Δ that it will suffice to show that

$$\|g(t, \epsilon x^0, \epsilon u^0)\| = \|x(t, \epsilon x^0, \epsilon u^0)\|^2 \|R(x(t, \epsilon x^0, \epsilon u^0))\|$$

is bounded by a constant times $\epsilon^2(\|x_0\|^2 + \|u_0\|^2)$, and since R is bounded near the origin, that in turn will follow if $\|x(t, \epsilon x^0, \epsilon u^0)\|^2$ is bounded by a constant times $\epsilon^2(\|x_0\|^2 + \|u_0\|^2)$. But that is just the essential content of the Stability Theorem for Equilibria (see 4.5.8).

4.10. Small Oscillations about Equilibrium

4.10.3. Remark. Most of the above theorem can be derived more directly from the differentiability of solutions of ODE with respect to parameters (see 1.4.5). However it would only follow from that approach that sup of the difference over a time interval of length T grows like e^{KT}, rather than CT. The much better bound here is a direct consequence of conservation of energy.

Chapter 5

Numerical Methods

5.1. Introduction

In the previous chapters we have developed a theoretical understanding of initial value problems for ODEs. Only rarely can these problems be solved in closed form, and even when closed-form solutions do exist, their behavior may still be difficult to understand. To gain greater insight, solutions are most commonly approximated numerically using discretization methods. This chapter is intended as a survey of these methods focusing on how they are designed, implemented, and analyzed.

Numerical methods are designed to approximate solutions of locally well-posed initial value problems

$$\mathbf{y}' = f(t, \mathbf{y}), \quad \mathbf{y}(t_o) = \mathbf{y}_o, \quad \mathbf{y} \in \mathbf{R}^d. \tag{5.1}$$

Well-posedness means that there exists a unique solution $\mathbf{y}(t; t_o, \mathbf{y}_o)$ that satisfies (5.1) on a maximal interval of existence $[t_o, t_o + T_*)$, $0 < T_* \leq +\infty$, and that depends continuously on $(t_o, \mathbf{y}_o) \in \mathbf{R}^{d+1}$. We will assume that $\mathbf{f}(t, \mathbf{y})$ is continuous in its first argument, t, and locally uniformly Lipschitz continuous in its second argument, $\mathbf{y}$, i.e.,

$$||\mathbf{f}(t, \mathbf{y}_1) - \mathbf{f}(t, \mathbf{y}_2)|| \leq L ||\mathbf{y}_1 - \mathbf{y}_2||, \tag{5.2}$$

for some $L > 0$ and any $\mathbf{y}_1, \mathbf{y}_2$ in a neighborhood of $\mathbf{y}_o$. As we have shown in previous chapters, these assumptions guarantee local well-posedness.

Discretization methods employ approximations of (5.1) to construct a discrete set of **y**-values, $\mathbf{y}_n$, $n = 0, 1, \ldots$, in such a manner that $\mathbf{y}_n$ should approximate $\mathbf{y}(t_n)$ at a corresponding set of t-values, t_n, called *time-steps*, as the separation of the time-steps, $h_n = t_{n+1} - t_n$, tends uniformly to zero. For most purposes, we will restrict ourselves to h_n that do not vary with n and call their common value $h = t_{n+1} - t_n$ the *step-size* or *discretization parameter*. Variable step-size methods are also useful, but we will only discuss them briefly in the context of automatic error control. We are often interested in the behavior of a discretization method as the discretization parameter decreases to zero, in which case the meaning of $\mathbf{y}_n$ becomes ambiguous. When it is required for clarity in such situations, we will write $\mathbf{y}_{n,h}$ to indicate both the step number and the step-size and otherwise suppress the explicit dependence on h.

Discretization methods are broadly categorized as explicit or implicit. Briefly, an *explicit method* obtains the successive values of y_{n+1} parametrically in terms of given or previously computed quantities and is represented symbolically in the form

$$\mathbf{y}_{n+1} = \mathbf{H}(\mathbf{f}, t_n, \ldots, t_{n+1-m}, \mathbf{y}_n, \ldots, \mathbf{y}_{n+1-m}).$$

In contrast, an *implicit method* defines $\mathbf{y}_{n+1}$ as the solution of an equation:

$$\mathbf{G}(\mathbf{f}, t_{n+1}, \ldots, t_{n+1-m}, \mathbf{y}_{n+1}, \ldots, \mathbf{y}_{n+1-m}) = 0$$

that cannot in general be put in the explicit form above.

Discretization methods are also characterized by the number m of previously computed quantities, or *steps*, that the method uses to compute each subsequent approximate value of the solution and by the number of evaluations of the vector field $\mathbf{f}$, or *stages*, that are used per time-step. In the next section, we introduce some basic examples that illustrate the considerations involved in choosing between explicit or implicit methods, single- or multistep, and one- or multistage methods, in order to obtain the greatest computational efficiency in different situations.

All of the r-*stage one-step methods* we will consider can be written in the form that characterizes methods known as *Runge-Kutta*

5.1. Introduction

Methods. Given a numerical initial value y_0, these methods take the specific form

$$\mathbf{y}_{n+1} = \mathbf{y}_n + h \sum_{i=1}^{r} \gamma_i \mathbf{y}'_{n,i}, \quad n = 0, 1, \ldots \tag{5.3}$$

where

$$\mathbf{y}'_{n,i} = \mathbf{f}(t_n + \alpha_i h, \mathbf{y}_n + h \sum_{j=1}^{r} \beta_{ij} \mathbf{y}'_{n,j}) \quad \text{and} \quad \alpha_i = \sum_{j=1}^{r} \beta_{ij}. \tag{5.3'}$$

If $\beta_{ij} = 0$ for $j \geq i$, the method is explicit; otherwise it is implicit. The strategy behind these methods is to obtain better approximations of $\mathbf{y}(t_{n+1})$ by sampling the vector field $\mathbf{f}(t, \mathbf{y})$ at r points near the solution curve emanating from $(t_n, \mathbf{y}_n)$. Each additional sample provides cumulatively better estimates of the solution curve, and thus subsequent samples can also be chosen more usefully. The analytical initial value is sufficient to initialize a one-step method, and no storage of previously computed values is required.

All of the m-step one-stage methods we will consider can be written in the form that characterizes methods known as *linear m-step methods*. Given numerical initial values $\mathbf{y}_0, \ldots, \mathbf{y}_{m-1}$, these methods take the specific form

$$\mathbf{y}_{n+1} = \sum_{j=0}^{m-1} a_j \mathbf{y}_{n-j} + h \sum_{j=-1}^{m-1} b_j \mathbf{y}'_{n-j}, \tag{5.4}$$

$$n = 0, 1, \ldots, \text{ where } \mathbf{y}'_j = \mathbf{f}(t_j, \mathbf{y}_j).$$

If $b_{-1} = 0$, the method is explicit; otherwise $\mathbf{y}_{n+1}$ appears on the right-hand side in the form $\mathbf{f}(t_{n+1}, \mathbf{y}_{n+1})$ and the method is implicit. The strategy behind these methods is to obtain better approximations of $\mathbf{y}(t_{n+1})$ by using information from m prior approximations and vector field evaluations, $t_j, \mathbf{y}_j, \mathbf{f}(t_j, \mathbf{y}_j)$, $j = n, \ldots, n - (m-1)$ that have been stored or generated for initialization. In contrast to multistage methods, only one evaluation of the vector field $\mathbf{f}$ defining the ODE is required per time-step. Discussions of more general methods that combine both Runge-Kutta and multistep characteristics can be found in [GCW] and other references listed in the Web Companion.

Even a discussion of numerical methods must address "theory" as well as "practice". First and foremost, one needs to answer the theoretical question of whether the values obtained by applying a method converge to the analytical solution $\mathbf{y}(t_o+T)$, as the discretization parameter tends to zero and the number of steps increases in such a way that the time interval they represent remains fixed. We call a method *convergent* if and only if, for any IVP (5.1) satisfying (5.2) and any $T > 0$ such that $t_o + T \in [t_o, t_o + T_*)$, the values $\mathbf{y}_{n,h}$ obtained from the method satisfy

$$||\mathbf{y}(t_o + T) - \mathbf{y}_{n,h}|| \to 0 \tag{5.5}$$

as $n \to \infty$ and $h = T/n$. Note that (5.5) implies that $y_{n,h}$ exists for sufficiently large n, an issue for implicit methods. If a method is explicit, $\mathbf{y}_{n,h}$ is defined for any $h > 0$ and $n > 0$. The existence of $\mathbf{y}_{n,h}$ is only an issue for implicit methods since they are defined for any $h > 0$ and $n > 0$ if the method is explicit.

We will analyze both the theoretical convergence and practical efficiency of a numerical method in terms of two essential concepts, accuracy and absolute stability. The order of accuracy of a (convergent) method refers to how rapidly errors decrease in the limit as the step-size tends to zero. We say that such a method converges with *order of accuracy P*, or, simply, is a *Pth-order accurate* method, if and only if there exists a $C > 0$ depending only on $\mathbf{y}$, its derivatives, and T, such that

$$||\mathbf{y}(t_o + T) - \mathbf{y}_{n,h}|| \leq Ch^P = C\left(\frac{T}{N}\right)^P \tag{5.6}$$

as $n \to \infty$ and no such estimate holds for any greater value of P. The dependence of C on T can be removed by considering closed subintervals of the maximal interval of existence. The potential significance of accuracy is immediate: if the increase in computational effort per step required to achieve higher-order accuracy is outweighed by reducing the number of steps required to obtain an approximation within a desired tolerance, the overall computation can be performed more efficiently.

5.1. Introduction

Different notions of stability for numerical methods refer to its tendency 1) to dissipate, 2) to not amplify, or 3) to not uncontrollably amplify perturbations introduced into an approximation. It is well known that there is conflicting nomenclature for certain numerical methods. Less well known is the fact that the term used to describe one of the most essential characteristics of a numerical method, its *absolute stability*, is defined by property 1) in some treatments but by property 2) in others! (See the Web Companion for some examples.) Both properties rule out unbounded growth of perturbations when applied to a problem or class of problems using a particular time-step, so that any systematic amplification is prohibited. Because we wish to encompass the two main types of well-posed initial value problems, those modeling oscillation, transport, and waves, along with those modeling dissipation and diffusion, we shall follow the convention that says that a method is absolutely stable with respect to a particular ODE and step-size h if the numerical solution is bounded as $t_n \to \infty$. Specifically, there exists a $C > 0$ depending only on the initial value such that

$$||\mathbf{y}_{n,h}|| \leq C \qquad (5.7)$$

for all $n \geq 0$. There are also reputable treatments whose definition of absolute stability also requires that $||\mathbf{y}_{n,h}|| \to 0$ as $t_n \to \infty$.

For the theoretical considerations of convergence, only the much weaker notion of stability corresponding to property 3) is a necessary condition. This minimal form of stability is called either *0-stability*, or just plain *stability*. A method is 0-stable with respect to a given problem if, for sufficiently small h, the growth of perturbations introduced in each step, representing errors made in prior approximations, can be controlled by some (possibly growing and problem-dependent) function of time. Formally, we say that a method is 0-stable with respect to a particular ODE (5.1), (5.2) if there exists a step-size $h_o > 0$ such that for any N making $0 < h = T/N < h_o$, and all $0 \leq n \leq N$, the difference between the numerical solution $\mathbf{y}_{n,h}$ and any numerical solution $\mathbf{y}_{n,h;\delta}$, defined by the same method with the same step-size, but with perturbations of magnitude no greater than $\delta > 0$ introduced initially and added to the resulting, cumulatively perturbed values $\mathbf{y}_{n+1,h;\delta}$ at each subsequent step, satisfies

$$||\mathbf{y}_{n,h} - \mathbf{y}_{n,h;\delta}|| \leq C(T)\delta. \tag{5.8}$$

for some positive function of the interval length, $C(T)$. Calling such a method 'not uncontrollably unstable' might be more appropriate. We might wonder if all approximation methods that formally approximate (5.1) satisfy this condition. Indeed, the Runge-Kutta Methods (5.3) are 0-stable by construction. However, we will see examples of linear multistep methods that arise from approximations that are formally quite accurate but that violate 0-stability. A method that is not 0-stable and thus nonconvergent should only be considered in order to understand the causes of its failure, never for actual computation. The facts that 0-stability only involves sufficiently small step-sizes and that it is associated with absolute stability for the problem $y' = 0$, or equivalently for the step-size $h = 0$, explain the terminology.

We can now describe how we aim to understand the theoretical behavior, convergence and higher-order convergence, of these classes of numerical methods. Various equivalent conditions characterize the order of accuracy a method will attain *if* it is also 0-stable. The simplest example of such a condition is based on applying a method to the scalar ODE $y' = 1$, $y(t_o) = y_o$ (using exact initialization $y_j = y_o + (t_j - t_o)$, $j = 0, \ldots, m-1$, in the multistep case). If the resulting solution is exact at all subsequent time-steps, $y_n = y_o + (t_n - t_o)$ for all $n > 0$, we say that the method is *consistent*. A fundamental result is that 0-stability and consistency are not only necessary, but together they are sufficient for a method to be convergent with order $P = 1$. Higher-order potential (or formal) accuracy of a linear multistep method, subject to 0-stability, is equivalent to exactness on ODEs whose solutions are polynomials of degree $\leq P$. For each successive order, this corresponds to one condition that can either be expressed as a linear equation in the coefficients in (5.4), in terms of the asymptotic behavior of a certain polynomial formed from these coefficients, or in terms of dependence on h of the magnitude of errors the method introduces at each step. And while we have noted that 0-stability is inherent in the form of Runge-Kutta Methods, the conditions on their coefficients required for each additional degree of

5.1. Introduction

accuracy are nonlinear and their number grows exponentially, while the number of coefficients only grows quadratically with the number of stages.

Convergence is not a guarantee that a method will perform even adequately in practice. The error bound that guarantees convergence and describes its rate includes contributions reflecting both the accuracy and absolute stability of a method. Growth of perturbations does not prevent high-order convergence, but the growth of perturbations that 0-stability permits can easily dominate these estimates until the step-size h is far smaller than accuracy considerations alone would otherwise require, and so render high-order accuracy irrelevant. For this reason, absolute stability for moderate step-sizes with respect to the problem at hand is the more practically significant property. So no matter how high the order of convergence that theory predicts, we will see that absolute stability analysis is often the deciding factor in performance. There are two situations in particular where these phenomena are especially important. One occurs in the case of multistep methods, whose solutions depend on initial conditions not present in the analytical problem. Because of this, multistep methods have more modes of potential amplification of perturbations than one-step methods. The other occurs in systems whose modes have widely separated temporal scales and applies to both one-step and multistep methods. This separation is quantified through a ratio of magnitudes of eigenvalues of the linearized system. As we shall see, such modes inevitably appear and are excited when we attempt to use discretized systems to accurately approximate the behavior of interest in certain important systems of ODE. Such systems are usually referred to as being *stiff* in the numerics literature. Absolute stability of a method with respect to those peripheral modes and step-size used ensures that small errors arising in these modes do **not** amplify and overwhelm phenomena in modes we wish to and can otherwise resolve.

In the next section, we will introduce seven basic numerical methods in order to motivate and illustrate the theory, and the role of absolute stability in making the theory useful in practice. These methods have been chosen as prototypes for multiple reasons. First, they exemplify a number of the important different features and properties numerical methods can possess. Second, their derivations are motivated by approximation techniques that can be generalized to obtain entire families of methods with useful characteristics. And third, they are simple enough that their behavior on two important classes of model problems can be rigorously analyzed and completely understood.

The first class of model problems is the Pth-order accuracy model problem, $M_A(P)$:
$$y' = f(t), \quad y(0) = y_o, \qquad (M_A(P))$$
where $f(t)$ is any polynomial of degree $\leq P-1$, so that the analytical solution $y(t)$ is a polynomial of degree P satisfying the initial condition. This class of model problems can be used to understand the order accuracy of any linear multistep method, and explicit Runge-Kutta Methods for $P \leq 2$. Exact solutions of this model problem for comparison with numerical solutions are easily obtained by antidifferentiation. For each example method, we will obtain an explicit formula for the approximations y_n that it generates when applied to $(M_A(P))$ for some appropriate degree P.

The second class of model problems is the absolute stability model problem, $(M_S(\lambda))$:
$$y' = \lambda y, \quad y(0) = 1, \qquad (M_S(\lambda))$$
where we call λ the *model parameter*. We will eventually allow both λ and $y(t)$ to be complex, but to begin with, we will take both to be real scalars. These model problems can be used to understand the stability properties, of a method, especially absolute stability, which is why we refer to it as $(M_S(\lambda))$. For these homogeneous, first-order, linear, constant coefficient, scalar ODEs, amplification of perturbations amounts to the same thing as amplification of solutions, and therefore absolute stability with respect to $(M_S(\lambda))$, sometimes called linearized absolute stability, forbids unbounded growth of the numerical solutions themselves. For Runge-Kutta Methods this is equivalent to restricting $w = \lambda h$ so that the *amplification factor* of the method $a(w) = y_{n+1}/y_n$ satisfies $|a(w)| \leq 1$. (As noted above, some authors require $|a(w)| < 1$.) We call the set $\{w \in \mathbb{C} \mid |a(w)| < 1\}$ the

5.1. Introduction

region of absolute stability of the method. The set $\{w \in \mathbf{C} \mid |a(w)| < 1\}$ has been called the linearized stability domain, or lsd [IA1, p. 68]. Though it may seem surprising, we will also consider the analytical solution, $y_{n+1} = e^{\lambda h} y_n$ as a numerical method. We will when we solve systems of ODEs designed to approximate constant coefficient PDEs, where it is known as the Fourier, or *spectral*, method. The spectral method is often used in conjunction with nonexact methods via a technique known as splitting. The region of absolute stability of the analytical solution method is $\{w \in \mathbf{C} \mid \text{Re}(w) \leq 0\}$. We are primarily interested in the absolute stability of other methods for these values of w although nonlinear stabilization can lead us to consider problems with $\text{Re}(w) > 0$ as well.

The model problems $(M_S(\lambda))$ are universal, in the sense that their solutions form a basis for the space of homogeneous solutions of any diagonalizable systems of first-order constant coefficient ODEs. Such systems arise upon linearizing more general systems of ODEs about an equilibrium. They also appear in the spatial discretizations of PDEs of evolution discussed in Section 5.6. The absolute stability model problems for negative real λ arise directly from eigenfunction expansions of solutions of the diffusion equations mentioned above, and for purely imaginary λ in expansions of solutions of wave equations. The model parameter corresponds to the characteristic exponent (eigenvalue) of a mode. Our analysis will demonstrate rather extreme consequences when numerical methods lack absolute stability with respect to the approximating ODEs and the step-size employed.

We shall invite the reader to implement each example method on $(M_S(\lambda))$ and experiment with its behavior for certain combinations of λ and h. We can use these programs to perform accuracy studies of each method by fixing $\lambda = -1$ while we successively halve h, keeping $Nh = T$, and absolute stability studies by fixing h and successively doubling λ. To understand our results, we will describe the amplification factor and region of absolute stability for each method.

The results regarding theoretical convergence can be formulated quite nicely in terms of the model problems as follows. A method is 0-stable if and only if it is absolutely stable applied to $M_S(0)$, a condition that is automatically satisfied by Runge-Kutta Methods.

We say that a numerical method has *formal accuracy* (or *polynomial accuracy*) of order P if it can be applied to every problem in the class $(M_A(P))$ using exact initial values and the resulting numerical solution is exact ($y_n = y(t_n)$ for all time-steps t_n, $n \geq 0$). Because of this, a method is consistent if and only if it has polynomial accuracy of order $P \geq 1$. Therefore a method is convergent if and only if it is exact in the sense above on $M_A(1)$ and absolutely stable on $M_S(0)$. In terms of its coefficients, the Runge-Kutta Method (5.3), (5.3′) is consistent if and only if $\sum_{i=1}^{r} \gamma_i = 1$, and the the linear multistep method (5.4) is consistent if and only if $\sum_{j=0}^{m} a_j = 1$ and $\sum_{j=-1}^{m} b_j - \sum_{j=0}^{m} j a_j = 1$.

Each additional degree of polynomial accuracy depends on one additional algebraic condition on the coefficients. Each degree of formal, polynomial accuracy implies another order of actual accuracy for 0-stable linear multistep methods. For Runge-Kutta Methods, beyond second order, this polynomial accuracy of degree P is insufficient to guarantee general accuracy. For $P = 3$, an additional condition on the accuracy of a Runge-Kutta Method when applied to $M_S(P)$ is sufficient to guarantee general Pth-order accuracy, but even this is insufficient for any greater P. For this reason, a more correct name for the model problems $(M_A(P))$ might be the linear multistep accuracy model problems, and for $P = 1$, the consistency model problem. Further details and discussion of issues pertaining individually to Runge-Kutta Methods and linear multistep methods may be found in Appendices H and I, respectively.

Although the model problems $M_S(0)$ and $M_A(0)$ both refer to the same ODE, $y' = 0$, 0-stability only excludes unbounded growth; it does not require exactness. Numerical solutions of $M_S(\lambda)$ obtained from the linear multistep method (5.4) remain bounded as $n \to \infty$ if and only if the roots of the characteristic polynomial of the method

$$p_w(r) = \rho(r) - w\sigma(r), \text{ where}$$

$$\rho(r) = r^m - \sum_{j=0}^{m-1} a_j r^{m-(j+1)}, \text{ and } \sigma(r) = \sum_{j=-1}^{m-1} b_j r^{m-(j+1)} \quad (5.9)$$

satisfy the following root condition: All roots are either inside the unit

5.1. Introduction

circle in the complex plane or on the unit circle and simple. In fact, one definition of absolute stability for a linear multistep method with respect to $M_S(\lambda)$ is that the roots of $p_w(r)$ satisfy the root condition when $w = \lambda h$. Under that definition of absolute stability, 0-stability is immediately equivalent to absolute stability with respect to $M_S(0)$, and to the fact that $\rho(r)$ satisfies the root condition.

So for methods of the form (5.3), (5.3'), and (5.4) the classes of model problems $M_A(P)$ and $M_S(\lambda)$ are sufficient to determine the theoretical issue of convergence, the practical issue of linearized absolute stability, and even higher-order accuracy, except for Runge-Kutta Methods beyond $P = 3$. But even for linear multistep methods, we will discover that attaining high-order accuracy tends to compete with retaining absolute stability when h is not vanishingly small. Only by increasing the per-step computational effort, including the number of evaluations of the vector field required for each time-step can both be increased independently. In particular, implicit methods that permit greater accuracy without compromising stability require additional effort to solve the nonlinear equations that define each step.

In conclusion, many factors must be considered in choosing or designing an efficient method for a specific problem and its parameters. Our understanding of the practical performance of numerical methods can be guided effectively in terms of the order of accuracy and absolute stability of a method. These two concepts interact to determine the step-size required to obtain an approximation within a specified tolerance, when using a particular method on a particular ODE. Along with the problem-specific computational effort required per step, reflecting the number of stages and steps in the method and the work required to evaluate the vector field, they determine the relative efficiency of different methods in different situations. It is important to realize that no method is universally superior to all others, and the selection of an effective method depends upon careful consideration of features of the problem or class of problems one wishes to solve and the accuracy of approximation required.

5.2. Fundamental Examples and Their Behavior

Now we introduce several working examples of numerical methods for IVPs that are motivated by relatively elementary principles. Then we will apply them to the model problems we introduced above. We will focus on how their behavior depends on the nature of the problem and the step-size. We do not claim that a method that performs well on such simple problems will necessarily perform well on more challenging problems. However, methods that perform poorly on simple problems with certain features will likely *not* perform well on more complex problems with similar features.

• **Example 5–1. Euler's Method.** The most familiar and elementary method for approximating solutions of an initial value problem is Euler's Method. Euler's Method approximates the derivative in (5.1) by a finite difference quotient $\mathbf{y}'(t) \approx (\mathbf{y}(t+h) - \mathbf{y}(t))/h$. We shall usually discretize the independent variable in equal increments:

$$t_{n+1} = t_n + h, \quad n = 0, 1, \ldots, t_0 = t_o. \tag{5.10}$$

Henceforth we focus on the scalar case, $N = 1$. Rearranging the difference quotient gives us the corresponding approximate values of the dependent variable:

$$y_{n+1} = y_n + hf(t_n, y_n), \quad n = 0, 1, \ldots, y_0 = y_o. \tag{5.11}$$

Euler's Method is an r-stage Runge-Kutta Method (5.3) with $r = 1$, $\gamma_1 = 1$, and $\beta_{11} = 0$. It is also a linear m-step method (5.4) with $m = 1$, $a_0 = 1$, $b_{-1} = 0$, and $b_0 = 1$. Since $b_{-1} = 0$, it is explicit. However, it is too simple to capture essential features that occur for m or $r > 1$ and that we will find present in our next examples.

Geometrically, Euler's Method follows the tangent line approximation through the point (t_n, y_n) for a short time interval, h, and then computes and follows the tangent line through (t_{n+1}, y_{n+1}), and so on, as shown in Figure 5.1.

▷ **Exercise 5–1.** Write a program that implements Euler's Method, the values of $f(t, y)$ coming from a function defined in the program. Test the results on the model problem $(M_S(\lambda))$,

$$y' = \lambda y, \quad y(0) = 1, \quad t \in [0, T], \quad y, \lambda \in \mathbf{R},$$

5.2. Fundamental Examples and Their Behavior

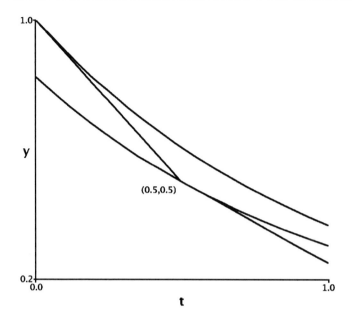

Figure 5.1. Interpretation of Euler's Method.

whose analytical solution is $y(t) = 1e^{\lambda t}$. Use $T = 1$ and combinations of $\lambda = \pm 2^l, l = 0, 1, \ldots, L$, and $h = 2^{-m}, m = 0, 1, 2, \ldots, M$, $L = 5$, $M = 5$.

The anticipated results for $l = 0$ ($\lambda = -1$) and $m = 0, \ldots, 4$ are displayed along with the exact solution in Figure 5.2.

Strictly speaking, Euler's Method generates a sequence of points in the (t, y)-plane, but we conventionally associate this sequence with the piecewise-linear curve obtained by joining consecutive points with line segments. Figure 5.2 shows three of these approximating curves and illustrates an important distinction involved in analyzing the convergence of these approximations. We call the difference between the exact solution and an approximate solution at a certain value of $t_o + T$ a *global error*, since it is the cumulative result of local errors introduced in each of N steps of size $h = T/N$ and the propagation of errors accumulated in earlier steps to later steps. These errors may either be amplified or attenuated from earlier steps to later steps.

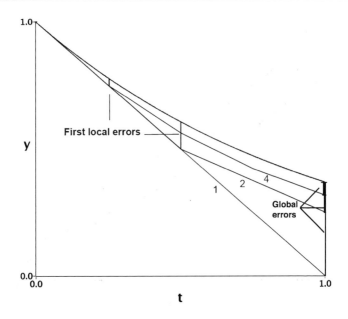

Figure 5.2. Behavior of Euler's Method: Accuracy.

Global errors corresponding to $N = 1, 2,$ and 4 are represented by vertical lines at $T = 1$. As this implies, a *local error* does not include the effect of prior errors but is the difference between one step of an approximate solution and the exact solution sharing the same initial value and time interval over that one step. Local errors for one step of Euler's Method starting at $(0, 1)$, with $N = 1, 2,$ and 4, are represented by vertical lines at $T = 1$, $1/2$, and $1/4$, respectively. Two kinds of local errors are discussed and depicted in the final section of the chapter on convergence analysis. Local truncation errors arise when a step of a method is initialized using the exact solution values. Another local error involves local solutions $\hat{y}_n(t)$ passing through values of the numerical solution (t_n, y_n). In Figure 5.2, h decreases by factors of $1/2$ while the number of steps doubles. The figure indicates that local errors for Euler's Method are on the order of h^2. At $t_N = T = 1$ it appears that the difference between the approximate solution and the analytical solution also decreases by a factor of $1/2$.

5.2. Fundamental Examples and Their Behavior

This suggests that for Euler's Method, global errors are on the order of h.

According to the description in the introduction, the near proportionality of errors at $y(1)$ to h^1 suggests that Euler's Method has order of accuracy 1, or in other words that it is *first-order accurate*.

To perform an analytical *accuracy* study of Euler's Method, we apply it to the class of accuracy model problems (M_A^2) written in the form

$$y' = \frac{d}{dt}(c_1(t - t_o) + c_2(t - t_o)^2), \quad y(t_o) = y_o, \tag{5.12}$$

whose analytical solution is

$$y(t) = y_o + c_1(t - t_o) + c_2(t - t_o)^2. \tag{5.12'}$$

When Euler's Method is applied to (5.12), it reduces to $y_{n+1} = y_n + h(c_1 + 2c_2nh)$. Using $\sum_{n=1}^{N-1} 2n = N(N-1)$, we find that $y_N = y_0 + c_1Nh + c_2h^2(N^2 - N)$. In terms of $t_n - t_o$,

$$y_N = y_0 + c_1(t_n - t_o) + c_2(t_n - t_o)^2 - c_2h(t_n - t_o)$$

and the global error at time $T = Nh$ satisfies

$$y(t_o + T) - y_N = (y_o - y_0) + c_2Th.$$

Setting $c_2 = 0$ shows that Euler's Method is exact when $y(t)$ is a polynomial of degree 1. For a polynomial of degree 2, its error at a fixed T is proportional to the first power of the time-step h. When we estimate the global error in the general case to prove convergence, the bound will involve a factor $\max_{t \in [t_o, t_o+T]} \frac{y''(t)}{2}$ that reduces to the factor of c_2 above.

To perform an analytical *absolute stability* study of Euler's Method, we apply it to the class of stability model problems $(M_S(\lambda))$. When Euler's Method is applied to these model problems, it reduces to $y_{n+1} = (1+w)y_n$, where we have combined the model parameter λ and discretization parameter h into a single parameter $w = \lambda h$. The exact solution with the same initial condition y_n and time interval h is $y_n e^{\lambda h} = y_n(1 + \lambda h + (\lambda h)^2/2 + (\lambda h)^3/3! + \cdots)$. In this context, one step of Euler's Method captures the terms of order h^1 in the exact solution correctly, and the remainder is bounded by a multiple of h^2. Successive iterates of Euler's Method can be written as $y_n = (1+w)^n y_o$. Heuristically, N errors of order h^2 accumulate to give an error of order h. Stability is necessary in order to make this argument rigorous.

The absolute stability properties of Euler's Method are illustrated by results of the exercise for $m = 3$ ($h = 1/8$) and $\lambda = -2^l$, $l = 0, \ldots, 5$, displayed in Figure 5.3. The value of λ appears adjacent to the corresponding exact and approximate solutions. As λ grows progressively more negative to $\lambda = -16$ where $w = -2$ and $y_n = (1+w)^n y_o = (-1)^n$, the approximate solution does not decay but simply oscillates. Beyond this value, e.g., $\lambda = -32$ so $w = -4$ and $(1+w)^n = -3^n$, the oscillations grow exponentially as shown in Figure 5.4. Note that some of the approximations in the accuracy and stability figures share the same values of w, e.g., $\lambda = -8$, $h = 1/8$, and $\lambda = -1$, $h = 1$; the independent variable must still be rescaled in order to make them correspond exactly. According to the description in the introduction, it follows from the form of the above solution that Euler's Method is absolutely stable with respect to the model problem when $|1 + \lambda h| \leq 1$. We will also be examining the model problem in the complex plane—that is, we will interpret λ as complex and replace the real scalar y with its complex equivalent, z. Thus we call $\{w \in \mathbf{C} \mid |1+w| \leq 1\}$, the closed disc of radius 1 about the point $w = -1$, the *region of absolute stability* for Euler's Method. The region of absolute stability of Euler's Method is depicted in Figure 5.14, together with the corresponding regions for the remaining example methods after they too have been analyzed.

In some important applications we will consider later, the instability exhibited by Euler's Method in Figure 5.4 has unavoidable negative consequences that can only be resolved by resorting to implicit methods such as those that we will derive below. In other circumstances, the rate of convergence exhibited by Euler's Method in Figure 5.2 is also unsatisfactory, and practical efficiency considerations require methods with higher-order accuracy, obtained from either multistep or multistage strategies. Along with performance improvements, each of these modifications brings with it important implementation considerations that do not appear in Euler's Method

5.2. Fundamental Examples and Their Behavior

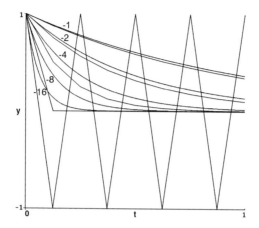

Figure 5.3. Behavior of Euler's Method: Stability.

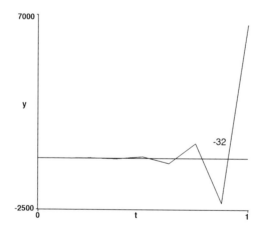

Figure 5.4. Behavior of Euler's Method: Instability.

due to its simplicity. We shall see that these considerations provide both challenges and opportunities.

To obtain the most basic examples of these kinds of methods, we provide another interpretation of Euler's Method and then modify it. In addition to the tangent line and difference quotient interpretations above, Euler's Method can be viewed as arising from the left endpoint

5. Numerical Methods

approximation of the integral of y' over an interval of width h:

$$y(t+h) - y(t) = \int_t^{t+h} y'(s)\, ds \approx hy'(t). \tag{5.13}$$

To improve upon Euler's Method, we can use more symmetric approximations:

$$\int_t^{t+h} y'(s)\, ds \approx hy'(t + \frac{h}{2}) \tag{5.14}$$

and

$$\int_t^{t+h} y'(s)\, ds \approx h\frac{y'(t) + y'(t+h)}{2}. \tag{5.15}$$

Both are exact if y is a polynomial of degree ≤ 2 (see Appendix J).

We expect that the methods known as the midpoint method and the leapfrog method, both obtained from (5.14) in the form

$$y(t+h) - y(t) \approx hy'(t + \frac{h}{2}), \tag{5.16}$$

and the trapezoidal method, obtained from (5.15) in the form

$$y(t+h) - y(t) \approx h\frac{y'(t) + y'(t+h)}{2}, \tag{5.17}$$

would lead to more accurate approximations of (5.1) than Euler's Method. In the next section we will show rigorously that they do. The geometric interpretation of these approximations and of the Euler's Method approximation, (5.13), are depicted in Figure 5.5.

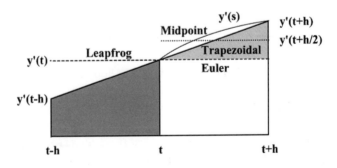

Figure 5.5. Interpretation of some basic methods.

5.2. Fundamental Examples and Their Behavior

▷ **Exercise 5–2.** The mean value theorem guarantees that the difference quotient $(y(t+h) - y(t))/h$ is equal to $y'(\xi)$, not at $\xi = t$ or $\xi = t + h$, but rather at some point ξ strictly between t and $t + h$. Show that for any polynomial $p_2(t)$ of degree ≤ 2, (5.14)–(5.17) are exact, that is,

$$\frac{p_2(t+h) - p_2(t)}{h} = p_2'(t + \frac{h}{2})$$

and also

$$\frac{p_2(t+h) - p_2(t)}{h} = \frac{p_2'(t) + p_2'(t+h)}{2}.$$

- **Example 5–2. The midpoint method.** To obtain the midpoint method from (5.16), we discretize t_n as in (5.10) and approximate

$$y'(t + \frac{h}{2}) = f(t + \frac{h}{2}, y(t + \frac{h}{2}))$$

using one step of Euler's Method with time-step $\frac{h}{2}$ as follows:

$$y_{n+\frac{1}{2}} = y_n + hf(t_n, y_n),$$

$$y_{n+1} = y_n + hf(t_n + \frac{h}{2}, y_{n+\frac{1}{2}}), \quad n = 0, 1, \ldots. \tag{5.18}$$

The midpoint method is an explicit r-stage Runge-Kutta Method, with $r = 2, \gamma_1 = 0, \gamma_2 = 1, \beta_{11} = \beta_{12} = \beta_{22} = 0$, and $\beta_{21} = \frac{1}{2}$.

▷ **Exercise 5–3.** If we accept the fact that the global error in Euler's Method is proportional to h within $O(h^2)$, the midpoint method can be derived using a technique known as *extrapolation*. Show that applying this assumption to one Euler step of size h and two steps of size $\frac{h}{2}$ tells us

$$y_{n+1} = y_n + hf(t_n, y_n) + Ch + O(h^2)$$

and

$$y_{n+1} = y_n + \frac{h}{2}f(t_n, y_n) + \frac{h}{2}f(t_n + \frac{h}{2}, y_n + \frac{h}{2}f(t_n, y_n)) + C\frac{h}{2} + O(h^2).$$

Then form a combination of these formulas, twice the latter minus the former, to obtain the midpoint method:

$$y_{n+1} \approx y_n + hf(t_n + \frac{h}{2}, y_n + \frac{h}{2}f(t_n, y_n)) + O(h^2).$$

▷ **Exercise 5–4.** Modify the program implementing Euler's Method to implement the midpoint method on the model problem $(M_S(\lambda))$, using the same parameters, and compare the results.

For our analytical accuracy study of the midpoint method, we consider the class of initial value problems (M_A^3) written in the form

$$y' = \frac{d}{dt}(c_1(t-t_o) + c_2(t-t_o)^2 + c_3(t-t_o)^3), \quad y(t_o) = y_o, \quad (5.19)$$

whose exact solution is

$$y(t) = y_o + c_1(t-t_o) + c_2(t-t_o)^2 + c_3(t-t_o)^3. \quad (5.19')$$

This is simply the next higher-order analogue of (5.12).

If we apply the midpoint method to (5.19), it reduces to

$$y_{n+1} = y_n + h\left(c_1 + c_2(2n+1)h + 3c_3\left(\frac{(2n+1)h}{2}\right)^2\right).$$

Using $\sum_{n=0}^{N-1} 2n+1 = N^2$ and $\sum_{n=0}^{N-1} 3(2n+1)^2 = 4N^3 - N$, we find

$$y_N = y_0 + c_1 Nh + c_2(Nh)^2 + c_3(Nh)^3 - c_3 h^3 \frac{N}{4}.$$

From this, we see that the global error at time $T = Nh$ satisfies $y(t_o + T) - y_N = (y_o - y_0) + \frac{1}{4}c_3 Th^2$. The midpoint method is exact when $y(t)$ is a polynomial of degree 2, and for a polynomial of degree 3, the error at a fixed T is proportional to h^2. While formal accuracy analysis using the model problems $(M_A(P))$ does not in general tell the whole story regarding accuracy of Runge-Kutta Methods, if $P \leq 2$, formal order of accuracy P does imply order of accuracy P. We will discuss these issues in greater detail below and in Appendix H.

In the context of the model problem, with $w = \lambda h$, the midpoint method becomes $y_{n+1} = (1 + w + w^2/2)y_n$ whose solution is $y_n = (1 + w + w^2/2)^n y_o$.

Figure 5.6 depicts the results of using the midpoint method with the same parameters as in Figure 5.3. As h decreases by factors of $\frac{1}{2}$, the number of steps doubles. At $t_N = T = 1$ it now appears that the difference between the approximate solution and the analytical solution decreases by a factor of $1/4$, suggesting that the midpoint

5.2. Fundamental Examples and Their Behavior

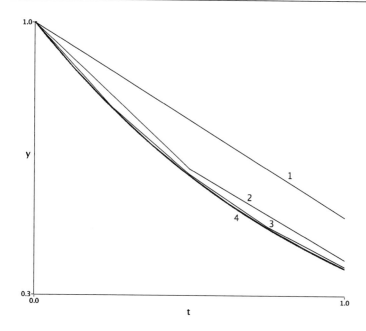

Figure 5.6. Behavior of the midpoint method: Accuracy.

method is second-order accurate; i.e., its order of accuracy is 2. The approximation with $j = 5$, $h = 1/32$ is indistinguishable from the exact solution. We might expect this behavior from the observation that one step of the midpoint method captures the terms of order $\leq w^2$ in the exact solution $y_n e^w = y_n(1 + w + w^2/2 + w^3/3! + \cdots)$ correctly, and the remainder is bounded by a multiple of w^3.

In the same fashion, Figure 5.7 corresponds to Figure 5.3, except the final value of λ has been changed from -16 to -17 to illustrate the incipient instability similar to that of Euler's Method when $w < -2$. For $\lambda = -16$, $w = -2$, the approximate solution $y_n = 1^n$ neither decays nor grows, nor does it oscillate as it did when Euler's Method was used. From the solution above, the midpoint method is absolutely stable with respect to the model problem when $|1 + \lambda h + (\lambda h)^2/2| \leq 1$. In the complex plane, the region of absolute stability for the midpoint method is then $\{w \in \mathbf{C} \mid |1 + w + w^2/2| \leq 1\}$.

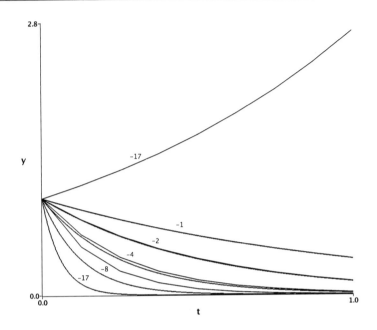

Figure 5.7. Behavior of the midpoint method: Stability.

- **Example 5–3. The Leapfrog Method.** To obtain the leapfrog method, we discretize t_n as in (5.10), but we double the time interval, h, and write the midpoint approximation (5.16) in the form

$$y'(t+h) \approx (y(t+2h) - y(t))/h$$

and then discretize it as follows:

$$y_{n+1} = y_{n-1} + 2hf(t_n, y_n). \tag{5.20}$$

The leapfrog method is a linear $m = 2$-step method, with $a_0 = 0$, $a_1 = 1$, $b_{-1} = -1$, $b_0 = 2$, and $b_1 = 0$. It uses slopes evaluated at odd values of n to advance the values at points at even values of n, and vice versa, reminiscent of the children's game of the same name. For the same reason, there are multiple solutions of the leapfrog method with the same initial value $y_0 = y_o$. This situation suggests a potential instability present in multistep methods, which must be addressed

5.2. Fundamental Examples and Their Behavior

when we analyze them—two values, y_0 and y_1, are required to initialize solutions of (5.20) uniquely, but the analytical problem (5.1) only provides one. Also for this reason, one-step methods are used to initialize multistep methods.

▷ **Exercise 5–5.** Modify the program implementing Euler's Method to implement the leapfrog method on the model problem $(M_S(\lambda))$, using the same parameters. Initialize y_1 1) using the 'constant method', $y_1 = y_0$, 2) using one step of Euler's Method, $y_1 = y_0 + hf(t_0, y_0)$, and 3) using one step of the midpoint method. Compare the results.

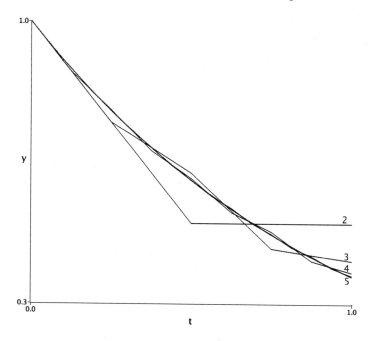

Figure 5.8. Behavior of the leapfrog method: Accuracy.

If we apply the leapfrog method to (5.19), it reduces to

$$y_{n+1} = y_{n-1} + 2h(c_1 + 2c_2nh + 3c_3(nh)^2).$$

If $N = 2K$ is even, we use $\sum_{k=1}^{K} 4(2k-1) = (2K)^2 = N^2$ and $\sum_{k=1}^{K} 6(2k-1)^2 = (2K-1)(2K)(2K+1) = N^3 - N$ to show that

$$y_N = y_0 + c_1 Nh + c_2(Nh)^2 + c_3(Nh)^3 - c_3 Nh^3.$$

If $N = 2K + 1$ is odd, we use $\sum_{k=1}^{K} 4(2k) = (2K+1)^2 - 1 = N^2 - 1$ and $\sum_{k=1}^{K} 6(2k)^2 = 2K(2K+1)(2K+2) = N^3 - N$ to show

$$y_N = y_1 - (c_1 h + c_2 h^2 + c_3 h^3) + c_1 Nh + c_2(Nh)^2 + c_3(Nh)^3 - c_3(N-1)h^3.$$

From this, we see that the global error at time $T = Nh$ satisfies

$$y(t_o + T) - y_N = (y(0) - y_0) + \frac{1}{4} c_3 T h^2$$

when N is even and

$$y(t_o + T) - y_N = (y(t_1) - y_1) + c_3(Th^2 - h^3)$$

when N is odd.

Assuming at first that $y_0 = y_o$ and $y_1 = y(t_1)$, the leapfrog method is exact when $y(t)$ is a polynomial of degree 2, and for a polynomial of degree 3, the error at a fixed T is proportional to h^2.

When the initial values are not exact, the formulas illustrate the dependence of global errors on the values used to initialize linear multistep methods. If the leapfrog method is initialized with the constant method, i.e., if we use $y_1 = y_o$, then $y(t_1) - y_1 = y(t_1) - y(t_0) = y'(\xi)h$ for some $\xi \in (t_0, t_1)$, the overall error degrades to $O(h)$ and the effort involved in employing a higher-order method is wasted. If we use Euler's Method, $y_1 = y_o + hy'(t_o)$, then $y(t_1) - y_1 = y(t_1) - (y(t_0) + hy'(t_0)) = y''(\xi)\frac{h^2}{2}$ for some $\xi \in (t_0, t_1)$ has the same order of magnitude as the largest other term contributing to the global error. The overall error achieves its maximum potential order of $O(h^2)$. The principle is the same for methods with more steps and initialization values. We only need to initialize using a method whose *global* error has order one less than the method it initializes. Only the local errors of the initialization method affect the global error of the overall method, since they are incurred over a fixed number of steps independent of h. In contrast, to reach a fixed $t_o + T$, the number of steps of the method being initialized is $N = Th^{-1}$. This factor is responsible for the different orders of magnitude of global errors, and

5.2. Fundamental Examples and Their Behavior 157

both local and initialization errors. There is no benefit gained from any additional effort devoted to computing initialization values more accurately. If we used one step of the midpoint method or Heun's Method instead of Euler's Method to compute y_1, improvement in the accuracy of the solution, if any, would be negligible.

We have implicitly assumed that we can use the analytical initial value y_o to initialize a numerical method. But even for a one-step method, sometimes initial values themselves are only computed approximately. If we imagine stopping a computation and then continuing, the results must be identical to those obtained had we not stopped in the first place. If we compare several steps of a computation with the same computation broken into two, the results are clearly the same. In the latter case, the second part begins with an inexact value. Fortunately, the new initial error is the order of the global error of the numerical method, and we have seen that this is all that is required in order for global errors to continue having the same order.

When the leapfrog method is applied to the absolute stability model problem $(M_S(\lambda))$, it takes the form $y_{n+1} = y_{n-1} + 2wy_n$. This is a linear second-order constant coefficient difference equation whose general solution is a linear combination, $y_n = c_+ y_n^+ + c_- y_n^-$, of two basic solutions, $y_n^+ = r_+^n$ and $y_n^- = r_-^n$, where $r_\pm$ are roots of $p_w(r) = r^2 - 2wr - 1$, the *characteristic polynomial* associated with the leapfrog method. In general, we find that $y_j = r^j$ is a nonzero solution of (5.4) if and only if r is a root of the characteristic polynomial of (5.4),

$$p_w(r) = \rho(r) - w\sigma(r)$$

where $\rho(r) = r^m - \sum_{j=0}^{m-1} a_j r^{m-(j+1)}$ and $\sigma(r) = \sum_{j=-1}^{m-1} b_j r^{m-(j+1)}$.

(5.21)

For any real w the characteristic polynomial of the leapfrog method has two distinct roots given by the quadratic formula, $r_\pm = w \pm \sqrt{w^2 + 1}$. When $w > 0$, $r_+ > 1$ and $-1 < r_- < 0$, and when $w < 0$, $0 < r_+ < 1$ and $r_- < -1$. (If $w = 0$, then $r_\pm = \pm 1$.)

Therefore, when $\lambda < 0$ and the analytic solution has exponentially decreasing magnitude, the leapfrog method applied to the model problem exhibits unstable exponential growth regardless of how small h may be, as long as $c_- \neq 0$. Since $c_- = 0$ implies $y_{n+1} = r_+ y_n$, if we initialize y_1 using one step of Euler's Method, or either of the other methods suggested above, we are guaranteed $c_- \neq 0$. Using a binomial expansion, $(1+u)^{1/2} = 1 + u/2 - u^2/8 + \cdots$, $|u| < 1$, $r_+ = 1 + w + w^2/2 - w^4/8 + \cdots$, $|w| < 1$; i.e., for small $|w|$, one step of the mode r_+ of the leapfrog method agrees with the terms of order $\leq w^2$ in the exact solution $y_n e^w$, and the remainder is bounded by a multiple of w^3. When w approaches zero along the negative real axis, $r_+ \approx 1 + w$ has magnitude less than 1. Since $r_+ r_- = -1$, or using the expansion above, $r_- \approx -1 + w - w^2/2$, in this situation r_- has magnitude greater than 1 and the powers of r_- explain the exponentially growing oscillations observed in solutions of the leapfrog method.

Figure 5.8 shows a series of results using the leapfrog method with the same parameters as in Figures 5.2 and 5.5—as h decreases by factors of $1/2$, the number of steps N gets doubled. (We start with $l = 2$, $h = 1/2$ since it is a 2-step method.) At $t_N = T = 1$, the difference between the approximate solution and the analytical solution decreases by a factor of $1/4$, similar to the behavior of the midpoint method. Along with the accuracy model analysis, this adds evidence that the leapfrog method is also second-order accurate. The approximation with $j = 5$, $h = 1/32$ is indistinguishable from the exact solution. Figure 5.9 is a stability study corresponding to to Figures 5.3 and 5.7, but to capture the behavior with different parameters, the time-steps h and time intervals T are varied along with λ. Starting with $\lambda = -1$ and $h = 1/8$ as before, we have extended the time interval to $T = 8$ to observe the visible onset of instability. Each time λ is doubled, we have divided h by four so $w = \lambda h$ is halved, but this only accelerates the onset of instability, and our time intervals must shrink so that further amplification does not prevent us from showing the results on a common graph. From the form of the solution above, the leapfrog method is only absolutely stable with respect to the real scalar model problem when $w = 0$. We will analyze the situation for complex w when we apply the leapfrog method to a

5.2. Fundamental Examples and Their Behavior

2×2 system below. We will see that the region of absolute stability for the leapfrog method is the open interval on the imaginary axis, $\{w \in \mathbf{C} \mid w = bi, -1 < b < +1\}$, i.e., the set of complex w such that $w = -\bar{w}$ and $|w| < 1$. The endpoints are not included because when $w = \pm i$, $p_w(r) = r^2 \pm 2ir - 1 = (r \pm i)^2$ has a multiple root on the unit circle, so the general solution of the difference equation has an algebraically growing mode.

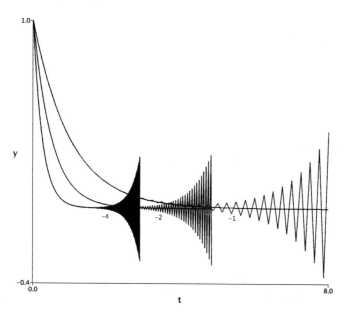

Figure 5.9. Behavior of the leapfrog method: Stability.

- **Example 5–4. The trapezoidal method.** To obtain the trapezoidal method, we define t_n as above and discretize (5.15) as follows:

$$y_{n+1} = y_n + h \frac{f(t_n, y_n) + f(t_{n+1}, y_{n+1})}{2}, \quad n = 0, 1, \ldots. \quad (5.22)$$

The trapezoidal method is classified as an *implicit method* because each step requires solving an equation to obtain y_{n+1}. In contrast, Euler's Method is called an *explicit method* because y_{n+1} is given parametrically. We will see that implicit methods can have stability advantages that make them more efficient in spite of this additional computational work needed to implement them.

The trapezoidal method is an implicit linear m-step method with $m = 1$, $a_0 = 1$, $b_0 = 1/2$, and $b_{-1} = 1/2$. It is also an implicit r-stage Runge-Kutta Method with $r = 2$, $\gamma_1 = \gamma_2 = 1/2$, $\beta_{11} = \beta_{12} = 0$, and $\beta_{21} = \beta_{22} = 1/2$. Even though the trapezoidal method is a 2-*stage* method, only one *new* evaluation of f is required per time-step after the first step. In general, if both y'_n and y'_{n+1} are among the $r > 1$ evaluations of an r-stage method, the number of new evaluations per time-step after the first step is $r - 1$.

▷ **Exercise 5-6.** Modify the program implementing Euler's Method to implement the trapezoidal method on the model problem $(M_S(\lambda))$, using the same parameters. To find y_{n+1}, you may treat (5.22) as a fixed-point problem $y_{n+1} = g(y_{n+1})$ and implement *fixed-point iteration*, $y_{n+1}^{(k+1)} = g(y_{n+1}^{(k)})$. Or you may rewrite (5.22) in the form $F(y_{n+1}) = 0$ and apply a root-finding method, e.g., *Newton's Method*, $y_{n+1}^{(k+1)} = y_{n+1}^{(k)} - F'(y_{n+1}^{(k)})^{-1} F(y_{n+1}^{(k)})$. In the case of the model problem, using the analytical solution is the simplest approach, although this would not be useful for general ODEs.

If we apply the trapezoidal method to (5.19), it reduces to

$$y_{n+1} = y_n + h(c_1 + c_2(2n+1)h + 3c_3 \frac{(nh)^2 + ((n+1)h)^2}{2}).$$

The right-hand side is close to the right-hand side we obtained when we applied the midpoint method to (5.19), only less an additional $3c_3 h^3 \frac{1}{4}$. Modifying the solution appropriately, we find the trapezoidal method yields

$$y_N = y_0 + c_1 Nh + c_2(Nh)^2 + c_3(Nh)^3 - c_3 h^3 \frac{N}{2}$$

and the global error at time $T = Nh$ satisfies $y(t_o + T) - y_N = (y_o - y_0) + \frac{1}{2} c_3 T h^2$. The trapezoidal method is exact when $y(t)$ is a polynomial of degree 2, and for a polynomial of degree 3, the error at a fixed T is proportional to the second power of the time-step, h^2.

When the trapezoidal method is applied to the absolute stability model problem $(M_S(\lambda))$, it takes the form

$$y_{n+1} = (1 + w/2)(1 - w/2)^{-1} y_n,$$

5.2. Fundamental Examples and Their Behavior

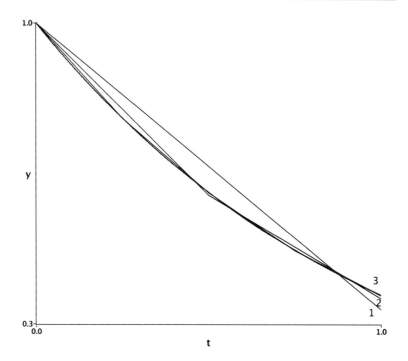

Figure 5.10. Behavior of the trapezoidal method: Accuracy.

so $y_n = ((1+w/2)/(1-w/2))^n y_o$. Using a geometric series expansion, $(1-w/2)^{-1} = 1+(w/2)+(w/2)^2+(w/2)^3\cdots$, $|w/2| < 1$, so for small w, one step of the trapezoidal method applied to the model problem, $(1+w/2)(1-w/2)^{-1} = 1 + (w/2) + (w/2)^2 + \cdots + (w/2) + (w/2)^2 + (w/2)^3 \cdots = 1 + w + w^2/2 + w^3/4 + \cdots$, captures the terms of order $\leq w^2$ in the exact solution $y_n e^w$, and the remainder is bounded by a multiple of w^3.

In Figure 5.10, the trapezoidal method is employed with the same parameters as in Figure 5.3. As usual, each time h decreases by factors of $1/2$, the number N of steps doubles. At $t_N = T = 1$, as with the midpoint and leapfrog methods, the difference between the approximate solution and the analytical solution appears to decrease by a factor of $1/4$, suggesting that the order of accuracy for the trapezoidal method is also 2. The approximations with $j \geq 4$, $h \leq 1/16$ are indistinguishable from the exact solution. In the same fashion, Figure 5.11 corresponds to Figure 5.4. At increasingly negative values of λ, we begin to observe the numerical solution becoming oscillatory and decaying less rapidly, even though the analytical solution continues to decay monotonically and more rapidly. This should be expected, since the factor $(1 + w/2)(1 - w/2)^{-1} \to -1$ as $w \to -\infty$. From the form of the solution above, we should expect absolute stability when $|(1 + w/2)(1 - w/2)^{-1}| \leq 1$, and instability otherwise. Rewriting this condition as $|w - (-2)| < |w - 2|$, we see that the trapezoidal method should be absolutely stable for any value of w that is closer to -2 than to $+2$. The trapezoidal method is absolutely stable with respect to the model problem for any $w \leq 0$. Below, when we consider the complex scalar model problem, equivalent to the real 2×2 model problem in the plane, we will see that the region of absolute stability for the trapezoidal method is the closed left half-plane $\{w \in \mathbf{C} \mid \mathrm{Re}(w) \leq 0\}$.

A method that is absolutely stable for any complex w whose real part is negative is known as an *A-stable* method.

• **Example 5–5. The modified trapezoidal method (aka Heun's Method and the improved Euler Method).** We can approximate the solution of the nonlinear equation that defines the trapezoidal method (5.22) quite easily if we approximate the value of y_{n+1} on its right-hand side by using one step of Euler's Method and solve for y_{n+1} as follows:

$$\bar{y}_{n+1} = y_n + hf(t_n, y_n),$$
$$y_{n+1} = y_n + h\frac{f(t_n, y_n) + f(t_{n+1}, \bar{y}_{n+1})}{2}, \quad n = 0, 1, \ldots \quad (5.23)$$

This is another example of the explicit Runge-Kutta Methods and is known by many names, including the modified trapezoidal method, Heun's Method, and the improved Euler Method. Heun's Method is an explicit r-stage Runge-Kutta Method, with $r = 2, \gamma_1 = \gamma_2 = 1/2$, $\beta_{11} = \beta_{12} = 0, \beta_{21} = 1$, and $\beta_{22} = 0$. When Heun's Method is applied to the stability model problem $(M_S(\lambda))$, it coincides with the midpoint method, and so the results of the exercise will be identical and their regions of absolute stability are the same. When Heun's Method is applied to the accuracy model problem (5.19), it coincides

5.2. Fundamental Examples and Their Behavior

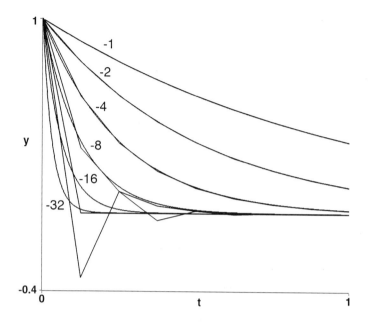

Figure 5.11. Behavior of the trapezoidal method: Stability.

with the trapezoidal method and so its formal order of accuracy is also 2.

We can also view (5.23) as a *predictor-corrector* method associated with the trapezoidal method (5.22). It may be worthwhile to solve the nonlinear equations associated with an implicit method using higher-order Newton-Raphson and quasi-Newton algorithms. These will usually require problem-specific implementations for evaluating or approximating and inverting derivatives involving considerable overhead. For universality and simplicity it is often preferable to take advantage of the natural fixed-point form

$$y_{n+1} = g(y_{n+1}; h, y_n, \ldots, y_{n+1-m})$$

of implicit numerical methods. To solve this using fixed-point iteration, we apply an explicit method called the 'predictor' to initialize $y_{n+1}^{(0)}$. For example,

$$y_{n+1}^0 = y_n + hf(t_n, y_n) \tag{5.24}$$

is called an Euler predictor step. Then we apply one or more 'corrector' steps, i.e., steps of the associated iteration algorithm

$$y_{n+1}^{(k+1)} = g(y_{n+1}^{(k)}; h, y_n, \ldots, y_{n+1-m}).$$

For (5.23), each step of

$$y_{n+1}^{(k+1)} = g(y_{n+1}^{(k)}) = y_n + h\frac{f(t_n, y_n) + f(t_{n+1}, y_{n+1}^{(k)})}{2} \qquad (5.25)$$

is called a trapezoidal 'corrector' step. When (5.24) is followed by one trapezoidal corrector step, the result is (5.23). So one more name for Heun's Method is the *Euler predictor-trapezoidal corrector* method.

The Lipschitz constant of the fixed-point iteration mapping approaches the magnitude of the derivative of the iteration function at the fixed point, $|g'(y_{n+1})|$. The iterations obtained from implicit numerical methods depend on the parameter h and have y_n as a fixed point for $h = 0$. Since $|g'(y_{n+1})|$ contains a factor of h, g will be a contraction on some neighborhood of y_n provided h is sufficiently small, and for any $y_{n+1}^{(0)}$ in this neighborhood, $y_{n+1}^{(k)} \to y_{n+1}$ as $k \to \infty$. If we denote the 'local solution' of the ODE $y' = f(t, y)$ passing through (t_n, y_n) by $\hat{y}_n(t)$, a Qth-order accurate predictor produces an initial approximation whose local error $|y_{n+1}^{(0)} - \hat{y}_n(t_{n+1})|$ has order of magnitude h^{Q+1}. A Pth-order accurate predictor produces a converged approximation whose local error $|y_{n+1}^{(\infty)} - \hat{y}_n(t_{n+1})|$ has order of magnitude h^{P+1}. By the triangle inequality, we can estimate $|y_{n+1}^{(\infty)} - y_{n+1}^{(0)}| \leq Ch^{\min\{P+1, Q+1\}}$. Additional factors of h from the iteration toward $y_{n+1}^{(\infty)}$ only decrease the magnitude of the local error for $k \leq P - Q$. For example, the Euler predictor-trapezoidal corrector method attains the full accuracy that would be obtained by iterating the trapezoidal corrector to convergence. But in problems where absolute stability is crucial, the difference in performance is substantial.

The modified trapezoidal method is only absolutely stable if h is sufficiently small, while the trapezoidal method is absolutely stable for any h. This additional labor that characterizes the implicit method makes no difference at all to the order of accuracy, but all

5.2. Fundamental Examples and Their Behavior

the difference in the world to absolute stability. The extra effort of each corrector iteration often pays itself back with interest by further relaxing the absolute stability restriction on h.

The local errors of a Pth-order accurate linear multistep method have a specific asymptotic form $C\hat{y}_n^{(P+1)}(t_n)h^{P+1}$ that can be determined from the coefficients of the method. This makes it possible to use a predictor and corrector of the same order to estimate the error of both methods efficiently, provided their constants C are distinct. For example, the leapfrog predictor coupled with one trapezoidal corrector iteration permits efficient error estimation and stabilization, even though it does not add even one degree of accuracy to the predictor.

In contrast, the difference in local errors of different Runge-Kutta Methods of the same order $P \geq 2$ will in general be different for different ODEs. This makes it difficult to use such a pair for error estimation, and methods of different order are used instead. The Euler predictor-trapezoidal corrector pair serves as a prototype of Runge-Kutta error estimation. The local error of an order $P - 1$ predictor is $|y_{n+1}^{(0)} - \hat{y}_n(t_{n+1})| \approx Ch^P$. The local error of a corrector of order P is $|y_{n+1}^{(1)} - \hat{y}_n(t_{n+1})| \approx Ch^{P+1}$. Therefore, the correction $|y_{n+1}^{(1)} - y_{n+1}^{(0)}|$ is an order h^{P+1} accurate estimate of the local error of the *lower* order method. For an Euler predictor-trapezoidal corrector pair, this technique estimates the error of the Euler step, even though we advance using the corrected Heun step. The resulting estimate is conservative, as one would want it. This approach is also efficient, since the lower-order method is *embedded*; i.e., it only involves evaluations performed by the higher-order method.

• **Example 5–6. The Backward Euler Method.** Another example of an implicit method is the Backward Euler Method,

$$y_{n+1} = y_n + hf(t_{n+1}, y_{n+1}), \quad n = 0, 1, \ldots, \qquad (5.26)$$

that arises by replacing the left endpoint approximation that characterizes Euler's Method with the right endpoint approximation. The Backward Euler Method is an implicit linear m-step method with $m = 1$, $a_0 = 1$, $b_{-1} = 1$, and $b_0 = 0$. It is also an explicit r-stage Runge-Kutta Method with $r = 1$, $\gamma_1 = 1$, and $\beta_{11} = 1$.

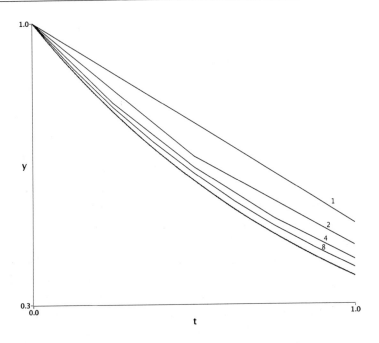

Figure 5.12. Behavior of the Backward Euler Method: Accuracy.

▷ **Exercise 5–7.** Modify the program implementing Euler's Method to implement the Backward Euler Method on the model problem $(M_S(\lambda))$, using the same parameters and using the options described above for implementing the trapezoidal method.

To understand the accuracy of the Backward Euler Method analytically, we use the same class of accuracy model problems (M_A^2) in the same form (5.12), (5.12′) that we used to analyze Euler's Method. When the Backward Euler Method is applied to (5.12), it takes the form $y_{n+1} = y_n + h(c_1 + 2c_2(n+1)h)$. Using $\sum_{n=0}^{N-1} 2(n+1) = N^2 + N$, we find that $y_N = y_0 + c_1 Nh + c_2 h^2(N^2 + N)$, or in terms of $t_n - t_o$, $y_N = y_0 + c_1(t_n - t_o) + c_2(t_n - t_o)^2 + c_2 h(t_n - t_o)$. From this, we see that the global error at time $T = Nh$ satisfies $y(t_o + T) - y_N = (y_o - y_0) - c_2 Th$. The method is exact on polynomials of degree 1, and for a polynomial of degree 2, its error at a fixed T is proportional to h. In the general case the bound involves a factor $\max_{t \in [t_o, t_o + T]} \frac{y''(t)}{2}$

5.2. Fundamental Examples and Their Behavior

that reduces to the factor of c_2 above. Note that the errors of Euler's Method and the Backward Euler Method have the same magnitude but opposite signs on these problems. We can show that the leading order errors in these methods are opposite in general and obtain a more accurate method by averaging them. In its general form, this process is known as extrapolation. Extrapolation by averaging Euler's Method and the Backward Euler Method is an alternate approach to deriving the trapezoidal method.

When the Backward Euler Method is applied to the absolute stability model problems $(M_S(\lambda))$, it takes the form

$$y_{n+1} = (1-w)^{-1} y_n$$

whose solution is $y_n = (1/(1-w))^n y_o$. Using a geometric series expansion, $(1-w)^{-1} = 1 + w + w^2 + \cdots$, $|w| < 1$, so for small w, the Backward Euler Method captures the terms of order $\leq w^1$ in the exact solution $y_n e^w$, and the remainder is bounded by a multiple of w^2.

In Figure 5.12, the Backward Euler Method is employed with the same parameters as in Figure 5.3, as h decreases by factors of $1/2$, the number of steps doubles. As with Euler's Method, at $t_N = T = 1$, the difference between the approximate solution and the analytical solution appears to decrease by a factor of $1/2$, suggesting that the Backward Euler Method has the same first-order accuracy as Euler's Method.

In the same fashion Figure 5.13 corresponds to Figure 5.4. Unlike the trapezoidal method, at increasingly negative values of λ, the Backward Euler Method does not generate oscillations. The factor $(1-w)^{-1} \to 0^+$ as $w \to -\infty$. From the form of the solution above, we should expect absolute stability when $|(1/(1-w))| \leq 1$ and instability otherwise. Rewriting this, the region of absolute stability of the Backward Euler Method is $\{w \in \mathbf{C} \mid |w - 1| \geq 1\}$; i.e., the Backward Euler Method should be absolutely stable for any complex value of w outside the circle of radius 1 centered at $w = 1$. Like the trapezoidal method, the Backward Euler Method is A-stable. Note that in the right half-plane outside this circle, the Backward Euler

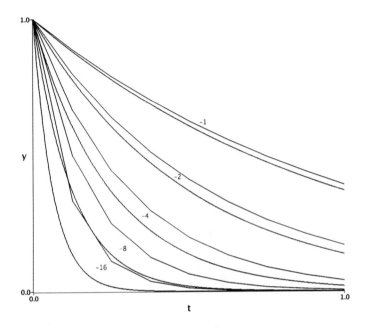

Figure 5.13. Behavior of the Backward Euler Method: Stability.

Method has decaying solutions where the analytic solution is growing exponentially, the opposite problem to what we observed previously!

Implicit linear m-step methods with $m > 1$ can be obtained using elementary means by simply doubling the discretization parameter of the trapezoidal and Backward Euler Methods:

$$y_{n+1} = y_{n-1} + 2h\frac{f(t_{n-1}, y_{n-1}) + f(t_{n+1}, y_{n+1})}{2}, \quad n = 0, 1, \ldots . \quad (5.27)$$

$$y_{n+1} = y_{n-1} + 2hf(t_{n+1}, y_{n+1}), \quad n = 0, 1, \ldots . \quad (5.28)$$

We refer to (5.27) as the $2h$ trapezoidal method and to (5.28) as the $2h$ Backward Euler Method. We do not have to repeat the accuracy studies for these methods since the results are easily obtained from and are essentially the same as that of their 1-step relatives. Stability is a different story. The approximations y_n with even and odd n are completely independent, and when these methods are applied to the

5.3. Summary of Behavior on Model Problems

model problem, both now possess an additional mode of the form $-1 + O(\lambda h)$. Continuing this approach to mh methods only make things worse. Any method of the form $y_{n+1} = y_{n+1-m} + h(\cdots)$ will have m independent modes of the form $e^{2\pi i j/m} + O(h)$ forming a basis of solutions when it is applied to the model problem, resulting in additional instability. Less contrived and better behaved examples of implicit (and explicit) 2-step methods are derived and analyzed in Appendix I.

• **Example 5–7. The y-midpoint method (also known as the implicit midpoint method).** Our final example is another implicit method that has features of midpoint and trapezoidal methods. We will call it the y-midpoint method:

$$y_{n+1} = y_n + hf(t_n + \frac{h}{2}, \frac{y_n + y_{n+1}}{2}), \quad n = 0, 1, \ldots. \tag{5.29}$$

We may also write this method in the form

$$y_{n+1} = y_n + hy'_{n,1}, \text{ where } y'_{n,1} = f(t_n + \frac{h}{2}, y_n + \frac{h}{2}y'_{n,1}) \tag{5.29'}$$

that exhibits it as an r-stage Runge-Kutta Method, with $r = 1$, $\gamma_1 = 1$, and $\beta_{11} = 1/2$. The form (5.29) shows that when the y-midpoint method is applied to the stability model problem ($M_S(\lambda)$), it coincides with the trapezoidal method and so the results of the exercise will be identical and their regions of absolute stability the same. When the y-midpoint method is applied to the accuracy model problem (5.19), it coincides with the ordinary midpoint method and so its formal order of accuracy is also 2.

5.3. Summary of Method Behavior on Model Problems

To review what we have seen, all seven example methods satisfy the condition of 0-stability, because $w = 0$ is in their regions of absolute stability. Two of the seven, Euler's Method and the Backward Euler Method have formal accuracy of order 1, and the remaining five have formal accuracy of order 2. Based on just these facts, rigorous theory will show that they are all convergent and either first- or second-order accurate, respectively. We summarize the formulas for one step of each of our example methods, along with their order of accuracy and the expressions for amplification factors that we have obtained from analytical study of the absolute stability model problem in Table 5.1.

The regions of absolute stability obtained by bounding the magnitude of the amplification factors by 1 are depicted in the shaded regions of Figure 5.14. (The form of the regions for the leapfrog method and higher-order Runge-Kutta Methods depend on analyses given below.)

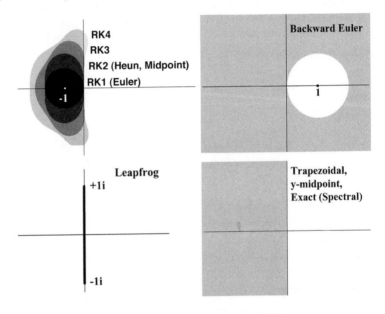

Figure 5.14. Regions of absolute stability for example methods.

The fact that $w = 0$ is in the region of absolute stability for six of the methods can be viewed as a simple consequence of the fact that they are one-step methods (and also Runge-Kutta Methods). For $y' = 0$ or just $h = 0$, they reduce to $y_{n+1} = y_n$. For a multistep method such as the leapfrog method, more analysis was required. In the introduction, we referred to a condition associated with the multistep method (5.4) when it is applied to the absolute stability model problem ($M_S(\lambda)$). In this situation, (5.4) reduces to a linear,

5.3. Summary of Behavior on Model Problems

Table 5.1

Method (order p)	$y_{n+1} =$ (m-steps, r-stages)	$a(w)$, $w = h\lambda$
Euler (1)	$y_n + hf(t_n, y_n)$ (1, 1)	$1 + w$
Backward Euler*(1)	$y_n + hf(t_{n+1}, y_{n+1})$ (1, 1)	$(1 - w)^{-1}$
Midpoint (2)	$y_n + hf(t_n + h/2, y_n + hf(t_n, y_n)/2)$ (1, 2)	$1 + w + w^2/2$
Leapfrog (2)	$y_{n-1} + 2hf(t_n, y_n)$ (2, 1)	$w \pm \sqrt{1 + w^2}$
Trapezoidal*(2)	$y_n + h(f(t_n, y_n) + f(t_{n+1}, y_{n+1}))/2$ (1, 2)	$(1 + w/2)(1 - w/2)^{-1}$
Heun (2)	$y_n + h(f(t_n, y_n) + f(t_{n+1}, y_n + hf(t_n, y_n)))/2$ (1, 2)	$1 + w + w^2/2$
y-Midpoint*(2)	$y_n + hf(t_n + h/2, (y_n + y_{n+1})/2)$ (1, 2)	$(1 + w/2)(1 - w/2)^{-1}$

*Implicit method

homogenous, constant coefficient difference equation. If we set $w = \lambda h$ and look for solutions of the form $y_n = r^n$, we find that r must be a root of the characteristic polynomial (5.9),

$$p_w(r) = \rho(r) - w\sigma(r) = r^m - \sum_{j=0}^{m-1} a_j r^{m-(j+1)} - w\Big(\sum_{j=-1}^{m-1} b_j r^{m-(j+1)}\Big).$$

Roots of $p_w(r)$ outside the unit circle guarantee exponential growth of solutions, and multiple roots on the unit circle guarantee algebraic growth. By excluding these possibilities, we guarantee that solutions remain uniformly bounded in terms of the initial values. Short-term growth is still possible if there are multiple roots inside the unit circle, or if there is cancellation of linear combinations of basic solutions at some time-steps. This leads us to formalize the definition of the stability condition described in the introduction. A polynomial $p(r)$ satisfies the condition (R) if and only if its complex roots r_j satisfy

$$|r_j| < 1 \quad \text{or} \quad |r_j| = 1 \text{ and } p'(r_j) \neq 0. \tag{R}$$

This condition guarantees that the roots are strictly inside the unit circle in the complex plane, where they may have nontrivial multiplicity, or on its boundary, where they must be simple. Some authors take the condition (R) for $p_w(r)$ as the definition of linearized absolute stability of the method (5.4) for $w = \lambda h$. The most important case this is for $w = 0$, since it determines whether or not a consistent method is convergent. The *Dahlquist root condition*, or simply the *root condition*, requires that $\rho(r) = p_0(r)$ satisfy condition (R). If the root condition is violated, a linear multistep method cannot be 0-stable, since then solutions can grow exponentially or algebraically without any dependence on h. By making h sufficiently small, the amplified solution can be made to correspond to a time interval of arbitrarily small length $T = Nh$.

A simple example exhibiting this phenomenon is the explicit two-step method

$$y_{n+1} = -4y_n + 5y_{n-1} + h(4y'_n + 2y'_{n-1}).$$

This method is derived and its formal accuracy shown to be 3 in Appendix I. Simple induction arguments can be used to show that

5.3. Summary of Behavior on Model Problems 173

for $f(t,y) = 0$ and $y_0 = y_1 = y_o$, we have $y_n = y_o$ for $n \geq 0$, and for $f(t,y) = 1$, $y_0 = y_o$, $y_1 = y_o + h$, we have $y_n = y_o + nh$ for $n \geq 0$, so the method is consistent. By applying the method to the most gentle equation possible, $y' = 0$, whose Lipschitz constant is $L = 0$, we can see that it violates the root condition and is truly unstable. For this problem, the method reduces to $y_{n+1} = -4y_n + 5y_{n-1}$. The solution y_n satisfying the initial conditions $y_0 = y_1 = \delta$ remains constant, $y_n = \delta$ for all $n \geq 0$. The solution z_n satisfying the initial conditions $z_0 = \delta$ and $z_1 = -5\delta$, grows exponentially, $z_n = (-5)^n \delta$ for $n \geq 0$. So even though $\max\{|y_0 - z_0| = 0, |y_1 - z_1|\} = 6\delta$, we have $\max_{Nh \leq T}\{|y_0 - z_0|, \ldots, |y_N - z_N|\} = (5^N - 1)\delta$. No matter how small $T > 0$ is chosen, we can make N arbitrarily large while still satisfying $Nh \leq T$ by making h correspondingly small. This shows that we cannot bound the maximum difference of solutions of this multistep method by any multiple of the maximum difference of their initial values, no matter how small the time interval or the step-size $h > 0$. This method violates the definition of stability since the numerical approximation fails to converge to the analytical solution $y(t) = 0$ as $h \to 0$ no matter how close the initializing values y_0 and y_1 are to zero. Generalizing this example shows that if (5.4) violates the root condition, it cannot be convergent. Conversely, if the root condition is satisfied, (5.4) is 0-stable. If, furthermore, (5.4) is formally accurate of order $P \geq 1$, it can be proven that the method is convergent, with order of accuracy P. The details for Euler's Method are given in the final section of the chapter. More generality is beyond the scope of the current introduction, and we refer the reader to Isaacson and Keller [IK] where convergence is proven under similar assumptions for a more general class of methods that includes the linear multistep methods (5.4).

There are both explicit and implicit examples with arbitrarily high formal accuracy that lack even this minimal form of stability, and there are those that possess it. For any $n \geq 1$, there is an implicit method with formal accuracy of order n that generalizes the Backward Euler Method known as the backward difference formula method, BDFn. For $n = 1, 2$, BDFn is A-stable, but for $n > 6$, BDFn

is not even 0-stable. For any $m \geq 1$, there is an explicit 0-stable m-step method with formal accuracy of order m known as the Adams-Bashforth Method, ABm, and an implicit 0-stable m-step method with formal accuracy of order $m+1$ known as the Adams-Moulton Method, AMm. Each of these families of methods is discussed in greater detail in Appendix I.

The accuracy model problem alone cannot be used to analyze the formal accuracy of Runge-Kutta Methods for order of accuracy greater than two. When the Runge-Kutta Method (5.3) is applied to an ODE of the form $y' = f(t)$, it takes the form

$$y_{n+1} = y_n + h \sum_{i=1}^{r} \gamma_i f(t_n + \alpha_i h)$$

that only depends on the coefficients $\alpha_i = \sum_j \beta_{ij}$ in (5.3'). For the method to have formal accuracy of order P, all that is required is that the terms to order P in the Taylor series

$$y(t_n + h) = y_n + \sum_{n=1}^{\infty} f^{n-1}(t) \frac{h^n}{n!}$$

match the Taylor series of

$$y_{n+1} = y_n + h \sum_{i=1}^{r} \gamma_i f(t + \alpha_i h).$$

This is equivalent to the conditions $\sum_i \gamma_i \alpha_i^{p-1} = \frac{1}{p}$ for $p = 1, \ldots, P$. For explicit Runge-Kutta Methods, these conditions are only equivalent to global accuracy of order P when $P \leq 2$.

To see this, consider the model problem $y' = \lambda y$. Setting $w = \lambda h$, an $r = 3$-stage method takes the form

$$y_{n+1} = y_n (1 + (\sum_{i=1}^{3} \gamma_i) w + (\sum_{i=1}^{3} \gamma_i \alpha_i) w^2 + (\gamma_3 \beta_{32} \alpha_2) w^3).$$

Third-order accuracy on this problem requires the first two formal second-order accuracy conditions $\sum_{i=1}^{3} \gamma_i = 1, \sum_{i=1}^{3} \gamma_i \alpha_i = \frac{1}{2}$, plus a different third-order condition, $\gamma_3 \beta_{32} \alpha_2 = \frac{1}{6}$. Thus there are 3-

5.3. Summary of Behavior on Model Problems

stage methods satisfying the third-order formal accuracy condition $\sum_{i=1}^{3} \gamma_i \alpha_i^2 = \frac{1}{3}$ but for which the third-order terms of the solution of $y' = \lambda y$ are not correct. Conversely, there are 3-stage methods satisfying $\gamma_3 \beta_{32} \alpha_2 = \frac{1}{6}$, but not $\sum_{i=1}^{3} \gamma_i \alpha_i^2 = \frac{1}{3}$. They are third-order accurate on $y' = \lambda y$ but not exact when applied to cubic polynomials.

The general accuracy conditions for explicit Runge-Kutta Methods are discussed in Appendix H. We will see that together the two conditions above are necessary and sufficient for an explicit 3-stage Runge-Kutta Method to have order of accuracy 3. In fact there is a two-parameter family of such methods, and we will refer to any of these methods by the collective abbreviation RK3. Further conditions not implied by either the accuracy or absolute stability model problems are required to guarantee that a Runge-Kutta Method has fourth-order accuracy. These additional conditions can still be satisfied by some explicit 4-stage methods. Again, there is a two-parameter family of such methods, and we will refer to them by the collective abbreviation RK4. The region of absolute stability of an RKp method for $p \leq 4$ is

$$\left\{ w \in \mathbf{C} \mid \left| \sum_{j=0}^{p} \frac{w^j}{j!} \right| \leq 1 \right\}.$$

These regions are depicted in Figure 5.14. For $r > 4$ however there are no explicit r-stage methods of order r. A pth order method may or may not approximate solutions of the absolute stability model problem to higher order, and an r-stage method may or may not even approximate these solutions to order r, although coefficients could be specified so that they do. For $5 \leq r \leq 7$, the maximum order of an explicit r-stage Runge-Kutta Method is $r - 1$, and for $r \geq 8$, the order is bounded by $r - 2$. Implicit Runge-Kutta Methods are more computationally intensive to implement, e.g., even for an autonomous scalar ODE they require the solution of an $r \times r$ nonlinear system of equations for every step. However, they can attain order $2r$ and at the same time maintain A-stability. For this reason, they cannot be dismissed in situations where unconditional stability and high-order accuracy are both required.

We have just noted that among explicit Runge-Kutta Methods requiring no more evaluations per step than their order, the fourth-order methods have the highest order of accuracy. Because of this fact, this section would be incomplete if we did not mention a method that is one of the most well known and that often serves as a default choice for handling problems not suspected of being stiff.

• **Example 5–8. Classical Fourth-Order Runge-Kutta (Classical RK4).** This method is defined by the following choice of parameters in (5.3): $\gamma_1 = \frac{1}{6}, \gamma_2 = \frac{2}{6}, \gamma_3 = \frac{2}{6}, \gamma_4 = \frac{1}{6}, \beta_{21} = \frac{1}{2}, \beta_{32} = \frac{1}{2}, \beta_{43} = 1$, and all other $\beta_{ij} = 0$. Specifically, we successively compute $y'_{n,1} = f(t_n, y_n)$, $y'_{n,2} = f(t_n + \frac{1}{2}h, y_n + h\frac{1}{2}y'_{n,1})$, $y'_{n,3} = f(t_n + \frac{1}{2}h, y_n + h\frac{1}{2}y'_{n,2})$, and $y'_{n,4} = f(t_n + h, y_n + hy'_{n,3})$, and then use them to form

$$y_{n+1} = y_n + h\left(\frac{1}{6}y'_{n,1} + \frac{2}{6}y'_{n,2} + \frac{2}{6}y'_{n,3} + \frac{1}{6}y'_{n,4}\right). \quad (5.30)$$

Note that the Classical Fourth-Order Runge-Kutta Method reduces to Simpson's Rule for numerical quadrature when applied to the simple ODE $y' = f(t)$. In Figure 5.15 we compare the results of Classical RK4, Euler's Method, and the analytical solution of $y' = -y$, $y(0) = 1$ using $N = 4$ steps of size $h = 1/4$ to approximate $y(1) = e^{-1}$. Observe that even using this fairly coarse h, the numerical solution using Classical RK4 is virtually indistinguishable from the analytical solution over the interval.

There are several other notable RK4 methods, among them Ralston's Method, Gill's Method, the 3/8 method, as well as embedded pairings of fourth- and fifth-order methods, such as the Runge-Kutta-Fehlberg pair, used for local error estimation and adaptive step-size control. In the absence of stiffness giving rise to instability in an ODE, the combination of simplicity, efficiency, and accuracy of fourth-order Runge-Kutta Methods has been responsible for their popularity as workhorse methods for numerically approximating solutions of a wide variety of IVPs ever since they were first introduced.

5.4. Paired Methods: Error, Step-size, Order Control

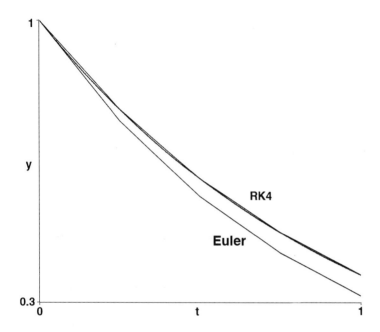

Figure 5.15. The Classical Fourth-Order Runge-Kutta Method.

5.4. Paired Methods: Error, Step-Size, Order Control

It is beyond the scope of this chapter to treat variable step-size methods and computational error estimation in any detail. However it possible and worthwhile to provide a brief taste of the concepts and principles involved using the examples we have developed. Global error estimation must necessarily involve refinement over the entire interval of computation, $[t_o, t_o + T]$. This is only done in unusual circumstances since computing multiple approximate solutions with decreasing step-sizes would dramatically extend the computational time. Furthermore, global errors reflect cumulative amplification or attenuation of local errors, depending on the absolute stability properties of a method, and stability properties may be better controlled through choice of method when a problem is stiff, not by adjusting step-sizes near an absolute stability boundary. Assuming absolute stability is taken care of, global errors should accumulate in proportion to the sum of the step-sizes. For this reason we seek to limit the local errors introduced, not simply per step, but in proportion to the length of the step itself. Even for methods with first-order accuracy, the local errors divided by the step-size decrease in proportion to the step-size.

A sort of local-global error estimation that falls under the broad heading of *extrapolation methods* can be implemented by comparing the result of one step of size h with two steps of size $h/2$. A more common strategy is to estimate the local error of a method at each step of size h by comparing the result with one step of another method using the same step-size. The companion method should be related closely to the method whose error is to be estimated. Not only is such a relationship necessary in order to imply a convenient estimate, it also means that the methods can share a substantial portion of their computation, reducing the cost of error estimation. But due to the different nature of multistep and Runge-Kutta approximations, the specific relationship most commonly used for each differs.

The local error incurred in computing one step y_{n+1} of a multistep method whose global order of accuracy is p has the form

$$E_{n+1,h} = y_{n+1} - y(t_{n+1}) = Cy^{(p+1)}(t_n)h^{p+1} + O(h^{p+2}).$$

In Appendix I we show that C can be found by Taylor expanding the approximations in the method to sufficiently high order. The *Milne device* suggests a comparison method of the same order, which computes $\hat{y}_{n+1}$ whose local error has the form

$$\hat{E}_{n+1,h} = \hat{y}_{n+1} - y(t_{n+1}) = \hat{C}y^{(p+1)}(t_n)h^{p+1} + O(h^{p+2}).$$

The value of $\hat{C}$ can also be found and must not equal C. The difference of the two approximations is equal to the difference of their local errors, i.e., $y_{n+1} - \hat{y}_{n+1} = E_{n+1,h} - \hat{E}_{n+1,h}$. This says that as long as $\hat{C} \neq C$ and if h is small, the difference of the two approximations satisfies

$$y_{n+1} - \hat{y}_{n+1} = (C - \hat{C})y^{(p+1)}(t_n)h^{p+1} + O(h^{p+2}).$$

5.4. Paired Methods: Error, Step-size, Order Control 179

The leading term of the local error of either y_{n+1} or $\hat{y}_{n+1}$ may be easily recovered from this value by multiplying by $\frac{C}{C-\hat{C}}$ or $\frac{\hat{C}}{C-\hat{C}}$, respectively. If the estimated local error satisfies the prescribed bound by an excessive proportion, it may be advantageous to increase step-size or reduce the order of the method. If it exceeds a prescribed bound on local error per step-size, the step-size can be decreased. The considerable cost of reinitialization for a multistep method is a deterrent to allowing step-sizes that require reduction and is an important consideration in the design of methods that automatically adapt their step-size and order.

The pair of approximations can also be extrapolated to cancel the leading term of the local error, giving a higher-order local error, and it is possible to vary the order of a method under appropriate circumstances. Doing so carelessly risks sacrificing potentially superior stability properties of one of the lower-order methods. In particular, a pair of multistep methods of the same order, one of which is explicit and the other implicit, can serve in dual roles, both for error estimation via the Milne device and as predictor and corrector. For instance if we extrapolated away the error of the Backward Euler Method by averaging with Euler's Method, we obtain the more accurate but less absolutely stable Heun's Method. The second-order accurate leapfrog predictor-trapezoidal corrector method suggested in the examples section is a more realistic example. Higher-order pairs of Adams Methods that fit this description are discussed in Appendix I.

The Milne approach cannot be used to estimate local error of Runge-Kutta Methods for the same reason that the accuracy model problem only determines the magnitude of local truncation error for multistep methods and the lowest-order Runge-Kutta Methods. A pair of Runge-Kutta Methods of the same order p will typically have leading order local errors that cannot be related as simply as those of multistep methods. In Appendix H we derive a rapidly proliferating set of nonlinear conditions on the coefficients of Runge-Kutta Methods that determine local error behavior. Each condition corresponds to a distinct rooted tree having $p+1$ nodes. Because of this indeterminacy, instead of a companion method of the same order, a

Runge-Kutta Methods of order p is paired with a method of higher order. Typically this will be a method of order $p+1$ that employs many of the same evaluations of the vector field as the lower-order method. The difference between the lower-order approximation whose local errors have order $p+1$ and a higher-order approximation whose local errors have order $p+2$ gives an estimate of order $p+1$ on the local error of the lower-order method. This estimate has the same order of magnitude that the Milne device yields for a multistep method, but at the cost of computing a higher order approximation. By sharing the same evaluations of the vector field $\mathbf{f}$, pairs of *embedded methods*, e.g., the Euler-trapezoidal and Runge-Kutta-Fehlberg pairs mentioned in the examples, accomplish local error estimation without degrading the overall performance excessively. Though theoretically we are only estimating the local error of the lower-order method, it is standard and obviously preferable to use the higher-order approximation to advance the solution. This has the additional benefit of making the local error estimate quite conservative and limits step-size and order adjustments to when they are truly necessary.

5.5. Behavior of Example Methods on a Model 2×2 System

The methods described above have straightforward extensions to systems of ODEs:

$$\mathbf{y}' = \mathbf{f}(t, \mathbf{y}), \quad \mathbf{y}(t_o) = \mathbf{y}_o, \quad t \in [t_o, t_o + T], \quad \mathbf{y} \in \mathbf{R}^d. \quad (5.31)$$

For example, Euler's Method for this system is defined by (5.10) along with

$$\mathbf{y}_{n+1} = \mathbf{y}_n + h\mathbf{f}(t_n, \mathbf{y}_n), \quad n = 0, 1, \ldots, \quad \mathbf{y}_0 = \mathbf{y}_o. \quad (5.32)$$

Not every method for systems arises in this simple manner as Exercise 5–9 will demonstrate.

We generalize the model problem $(M_S(\lambda))$ to the complex plane as

$$z' = \lambda z, \quad z(0) = 1, \quad t \in [0, T], \quad z, \lambda \in \mathbf{C}. \quad (5.33)$$

5.5. Example Methods on a Model 2 × 2 System

In particular, by setting $z = x + yi$ with $\lambda = a + bi$, we obtain the following model system of ODEs in the plane:

$$\begin{pmatrix} x \\ y \end{pmatrix}' = \begin{pmatrix} a & -b \\ b & a \end{pmatrix}\begin{pmatrix} x \\ y \end{pmatrix}, \quad \begin{pmatrix} x(0) \\ y(0) \end{pmatrix} = \begin{pmatrix} 1 \\ 0 \end{pmatrix}, \quad t \in [0,T]. \quad (5.34)$$

With $\lambda = 0 + bi$, this becomes the *harmonic oscillator* system,

$$\begin{pmatrix} x \\ y \end{pmatrix}' = \begin{pmatrix} -by \\ bx \end{pmatrix}, \quad \begin{pmatrix} x(0) \\ y(0) \end{pmatrix} = \begin{pmatrix} 1 \\ 0 \end{pmatrix}, \quad t \in [0,T]. \quad (5.35)$$

The analytic solution of (5.35) is $\begin{pmatrix} x(t) \\ y(t) \end{pmatrix} = \begin{pmatrix} \cos(bt) \\ \sin(bt) \end{pmatrix}$, or in complex notation, $z(t) = 1e^{ibt} = \cos(bt) + i\sin(bt)$.

▷ **Exercise 5–8.** Modify the programs implementing each of the example methods (Euler, midpoint, leapfrog, trapezoidal, Heun, Backward Euler, and y-midpoint) on the scalar model problem ($M_S(\lambda)$) to treat the model problem (5.35) in the plane, with $\lambda = 0 + bi$. In this context, Euler's Method takes the form

$$\begin{pmatrix} x_{n+1} \\ y_{n+1} \end{pmatrix} = \begin{pmatrix} x_n - bhy_n \\ y_n + bhx_n \end{pmatrix}, \quad \begin{pmatrix} x_0 \\ y_0 \end{pmatrix} = \begin{pmatrix} 1 \\ 0 \end{pmatrix}. \quad (5.36)$$

Use $T = 4\pi$ and $T = 16\pi$, with combinations of $b = \pm 2^l$, $l = 0, 1, \ldots, L$, and $h = T/N$ with $N = 2^m$, $m = 3, \ldots, 3 + M$, for $L = 5$, $M = 5$.

▷ **Exercise 5–9.** Modify Euler's Method (5.36) for the model problem in the plane as follows:

$$\begin{pmatrix} x_{n+1} \\ y_{n+1} \end{pmatrix} = \begin{pmatrix} x_n - bhy_n \\ y_n + bhx_{n+1} \end{pmatrix}, \quad \begin{pmatrix} x_0 \\ y_0 \end{pmatrix} = \begin{pmatrix} 1 \\ 0 \end{pmatrix}, \quad (5.37)$$

using the same parameters as in the previous exercise. This modification uses the updated x_{n+1} instead of x_n to compute y_{n+1} and does not directly correspond to any scalar method.

We will not illustrate any accuracy studies applying our example methods to the model problem in the plane, since the results are essentially similar to the scalar case, other than that the scalar error norm, the absolute value of the difference, is replaced by a vector norm such as the Euclidean norm of the difference. The order of the methods remains the same, and the complex value of the model parameter only affects multiplicative constants. However, the absolute stability of a method depends substantially on the location of λh in the complex plane. Rather than illustrating transitions to instability as we have above, we simply illustrate some significant differences in stability exhibited by our example methods applied to the model problem in the plane, in order to contrast them with each other and with their behavior when λ was real.

Figures 5.16 and 5.17 show some representative results of applying each of the example methods discussed above, as well as the modification (5.37), to the model problem in the plane, (5.35). In both figures, $\lambda = 0 + 1i$, $N = 128$. $T = 4\pi$ in Figure 5.16 and $T = 16\pi$ in Figure 5.17, so $h = T/N$ differs by a factor of 4. The explicit one-step methods, Euler's Method and the two second-order Runge-Kutta Methods, midpoint and Heun's, which agree on the model problem, all exhibit an exponentially growing instability. In contrast to the situation when they were applied to the scalar model problem, this behavior persists no matter how small we choose h. The leapfrog method, the explicit multistep method that exhibited similar undesirable behavior in the scalar case, is the one that, for sufficiently small h, behaves nicely in the plane.

It is straightforward to understand the behavior of several of the methods displayed in these figures simply by modifying earlier analyses to the context of the complex model problem, $z' = \lambda z$. If we set $w = \lambda h$, one step of each of the single-step methods may be written in terms of an amplification factor $c(w)$ in the form $z_{n+1} = a(w) z_n$. For the explicit first-order Euler's Method, $a_E(w) = 1 + w$. For the explicit second-order Runge-Kutta Methods (ERK2), the midpoint method and Heun's Method, $a_{ERK2}(w) = 1 + w + w^2/2$. For the implicit first-order Backward Euler Method, $a_{BE}(w) = (1-w)^{-1}$. For the implicit second-order Runge-Kutta Methods (IRK2), the trapezoidal method and the y-midpoint method, $a_{IRK2}(w) = (1+w/2)(1-w/2)^{-1}$. When $\lambda = ib$, we define $\omega = bh$ so $w = i\omega$. Then $|a_E(w)|^2 = 1 + \omega^2 > 1$, $|a_{ERK2}(w)|^2 = (1 - \omega^2/2)^2 + \omega^2 = 1 + \omega^4/4 > 1$, for any $h > 0$. These approximations will spiral exponentially outward to infinity away from the analytic solution, regardless of how small h

5.5. Example Methods on a Model 2 × 2 System

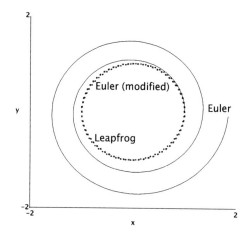

Figure 5.16. Euler, Euler (modified), and leapfrog methods in the plane.

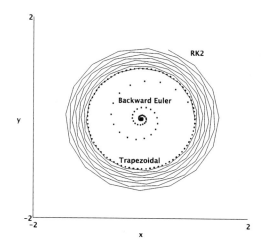

Figure 5.17. Explicit and implicit RK2 and Backward Euler Methods in the plane.

is chosen. We use $|z_1 z_2| = |z_1||z_2|$, $|z_1^{-1}| = |z_1|^{-1}$, and $|z_1| = |\bar{z}_1|$ for any $z_1, z_2 \in \mathbf{C}$ to see that $|a_{BE}(w)| = |1 - ibh|^{-1} = |1 + ibh|^{-1} = |a_E(w)|^{-1} < 1$ and $|a_{IRK2}(w)| = |(1+ibh/2)(1-ibh/2)^{-1}| = 1$ for any

h. The backward Euler approximations spiral exponentially inward to the origin away from the analytic solution, while the trapezoidal and y-midpoint solutions will always remain on a circle of constant radius.

Note that while the Backward Euler Method was qualitatively correct and the trapezoidal and y-midpoint methods were qualitatively incorrect when w was near negative infinity on the real axis, the roles are reversed when w is purely imaginary. This raises an important and subtle point. The ability to capture the qualitative behavior correctly in these regimes is not what matters in practice. Any of the standard methods can resolve the phenomena in which we are interested with reasonable values of h. In addition, the magnitudes of other modes, as well as the errors introduced in them by approximation and computation, tend to be much smaller. Because of this, the importance of absolute stability is not whether these other modes are calculated correctly in a qualitative sense, *but that they are not amplified*. To illuminate this point, we will conclude this section with exercises involving systems of two scalar and two planar model problems. But first, we will analyze the behavior of the remaining two methods in the figures, the leapfrog method and the modification (5.37) of Euler's Method.

The behavior of the leapfrog method in the real scalar case might have made one wonder if it and other multistep methods have any useful purpose. In Appendix I we discuss other implicit multistep methods—such as the Adams-Moulton Methods and backward difference formulas (BDFs) up to a certain order—that do have good stability properties on such problems. There are explicit multistep methods that behave well on the model problems on which the leapfrog method performs so poorly, i.e., those with λ real and negative. The leapfrog method has its place. It is symmetric with respect to time reversal, and this makes it more appropriate for solving equations with similar symmetry, as shown in Figure 5.16. When applied to the model problem in the complex plane, solutions of the leapfrog method are still linear combinations, $z_n = c_+ z_n^+ + c_- z_n^-$, of two basic solutions,

5.5. Example Methods on a Model 2×2 System

$z_n^+ = r_+^n$ and $z_n^- = r_-^n$, where $r_\pm$ are roots of the characteristic polynomial $p(r) = r^2 - 2wr - 1$ associated with the difference equation. For general complex w, the roots of $p(r) = r^2 - 2wr - 1$ still satisfy $r_- = -r_+^{-1}$, so if $r_+ = u + vi$, $r_- = -(u - vi)/(u^2 + v^2)$. The only case in which neither root has magnitude greater than unity is when both lie on the unit circle, $u^2 + v^2 = 1$; i.e., they are of the form $r_+ = e^{+i\theta}$, $r_- = -e^{-i\theta}$, $\theta \in \mathbf{R}$. With $\omega = bh$ as before, since $p(r) = (r - r_+)(r - r_-)$, $r_+ + r_- = 2w = 2i\omega$ and $\omega = \sin(\theta)$. So if $|\omega| > 1$, there will be an exponentially growing mode. If $\omega = \pm 1$, i.e., $w = \pm i$, then $r_0 = \pm i$ is a multiple root of the characteristic polynomial: $p(r) = r^2 - 2w - 1 = (r - r_0)^2$. In this case, the form of the general solution of the difference equation becomes $y_n = c_0 r_0^n + c_1 n r_0^n$, and the leapfrog method exhibits algebraic, not exponential, growth. This is typical of a method whose characteristic polynomial has multiple roots on the locus of marginal stability, the unit circle.

If $|\omega| < 1$, $\theta = \arcsin\omega = \omega + \omega^3/6 + \cdots$, and $r_+ = e^{+i\theta} = 1 + i\omega - \omega^2/2 + i\omega^3/6 + \cdots$ agrees with the analytical solution to order h^2, just as in the real scalar case. Initialization using Euler's Method agrees with principal root r_+ to first-order in h, so the initial magnitude of the so-called 'parasitic solution', z_-, is of order h^2. This contribution remains small, because the corresponding root, $r_- = -e^{-i\theta}$, located near $e^{i(\pi-\omega)} \approx -1$ on the unit circle, also has magnitude exactly equal to 1. What we observe for the leapfrog method in Figure 5.16 is not exactly a circular orbit, but the superposition of an orbit of radius $O(h^2)$ around the principal orbit of radius $1 + O(h)$. The small perturbation is nearly reversed as it completes about a one-half turn of its orbit for each step of the method.

To see why the behavior of modification (5.37) of Euler's Method is closer to that of the leapfrog method than Euler's Method, we write Euler's Method (5.36) in 2×2 matrix form,

$$\mathbf{u}_{n+1} = \mathbf{A}_E \mathbf{u}_n, \quad \text{where } \mathbf{u}_n = \begin{pmatrix} x_n \\ y_n \end{pmatrix} \text{ and } \mathbf{A}_E = \begin{pmatrix} 1 & -\omega \\ \omega & 1 \end{pmatrix},$$

and compare the real canonical forms of A_E and A_{ME}. Since $tr(\mathbf{A}_E) = 2$ and $\det(\mathbf{A}_E) = 1 + \omega^2$, the characteristic polynomial of $\mathbf{A}_E$ is

$r^2 - 2r + 1 + \omega^2$. Therefore, $\mathbf{A}_E$ has conjugate complex eigenvalues of magnitude > 1, $\lambda_\pm = 1 \pm i\omega$, and corresponding eigenvectors $\mathbf{z}_\pm = \begin{pmatrix} \pm i \\ 1 \end{pmatrix}$. Let λ_R and λ_I be the real and imaginary parts of the eigenvalue λ_+ and let $\mathbf{v}_R$ and $\mathbf{v}_I$ be the real and imaginary parts of the corresponding eigenvector $\mathbf{z}_+ = \mathbf{v}_R + i\mathbf{v}_I$, so $\lambda_R = 1$, $\lambda_I = \omega$, $\mathbf{v}_I = \begin{pmatrix} 1 \\ 0 \end{pmatrix}$, and $\mathbf{v}_R = \begin{pmatrix} 0 \\ 1 \end{pmatrix}$. Taking the real and imaginary parts of $\mathbf{A}_E(\mathbf{v}_R + i\mathbf{v}_I) = (\lambda_R + i\lambda_I)(\mathbf{v}_R + i\mathbf{v}_I)$ shows that $\mathbf{A}_E(\mathbf{v}_R) = \lambda_R \mathbf{v}_R - \lambda_I \mathbf{v}_I$ and $\mathbf{A}_E(\mathbf{v}_I) = \lambda_R \mathbf{v}_I + \lambda_I \mathbf{v}_R$. (Note the correspondence with the cosine and sine addition formulas.) Rearranging these formulas shows that, with respect to the properly oriented ordered basis, $\mathbf{v}_I, \mathbf{v}_R$, the action of $\mathbf{A}_E$ is given by

$$\mathbf{A}_E \mathbf{v}_I = \lambda_R \mathbf{v}_I + \lambda_I \mathbf{v}_R$$

and

$$\mathbf{A}_E \mathbf{v}_R = -\lambda_I \mathbf{v}_I + \lambda_R \mathbf{v}_R,$$

and so the matrix of $\mathbf{A}_E$ with respect to this ordered basis is $\begin{pmatrix} 1 & -\omega \\ \omega & 1 \end{pmatrix}$.

Next we write the modification in 2×2 matrix form $\mathbf{u}_{n+1} = \mathbf{A}_{ME}\mathbf{u}_n$, where

$$\mathbf{A}_{ME} = \begin{pmatrix} 1 & 0 \\ -\omega & 1 \end{pmatrix}^{-1} \begin{pmatrix} 1 & -\omega \\ 0 & 1 \end{pmatrix} = \begin{pmatrix} 1 & -\omega \\ \omega & 1 - \omega^2 \end{pmatrix},$$

and perform the parallel calculations for A_{ME}. We find $tr(\mathbf{A}_{ME}) = 2 - \omega^2$ and $\det(\mathbf{A}_{ME}) = 1$, so the characteristic polynomial of $\mathbf{A}_{ME}$ is $r^2 - (2 - \omega^2)r + 1$. For $|\omega| < 2$, $\mathbf{A}_{ME}$ has conjugate complex eigenvalues of magnitude 1,

$$\lambda_\pm = e^{\pm i\theta} = 1 - \omega^2/2 \pm i\sqrt{\omega^2 - \omega^4/4} = \cos(\theta) + i\sin(\theta),$$

and corresponding eigenvectors

$$\mathbf{z}_\pm = \begin{pmatrix} \omega/2 \pm i\sqrt{1 - \omega^2/4} \\ 1 \end{pmatrix}.$$

The real and imaginary parts of the eigenvector $\mathbf{z}_+$ are

$$\mathbf{v}_R = \begin{pmatrix} \omega/2 \\ 1 \end{pmatrix} \quad \text{and} \quad \mathbf{v}_I = \begin{pmatrix} \sqrt{1 - \omega^2/4} \\ 0 \end{pmatrix}$$

and the real and imaginary parts of the eigenvalue λ_+ are
$$\lambda_R = 1 - \omega^2/2 = \cos(\theta) \quad \text{and} \quad \lambda_I = \sqrt{\omega^2 - \omega^4/4} = \sin(\theta).$$
The matrix for $\mathbf{A}_{ME}$ with respect to the ordered basis $\{\mathbf{v}_I, \mathbf{v}_R\}$ is
$$\begin{pmatrix} \cos(\theta) & -\sin(\theta) \\ \sin(\theta) & \cos(\theta) \end{pmatrix}.$$
In this form, it is apparent that each step of the modified method (5.37) is similar to a counterclockwise rotation by an angle $\theta = \arccos(1 - \omega^2/2)$ with respect to the (nonorthogonal) basis $\{\mathbf{v}_I, \mathbf{v}_R,\}$, an $O(h)$ perturbation of the standard basis.

5.6. Stiff Systems and the Method of Lines

In this section we apply example methods to approximate solutions of sophisticated problems that arise in modern research and applications. The lessons of the previous sections regarding accuracy and absolute stability culminate here in helping us to explain the literally divergent behavior of certain methods.

We begin with a diagonal 2×2 extension of the real scalar model problem above and then generalized it to a $J \times J$ version that arises in many models in pure and applied mathematics. Let m_1 and m_2 be positive integers. The function $u(x,t) = a_1(t)\cos(m_1 x) + a_2(t)\cos(m_2 x)$ is a 2π-periodic solution of the *heat equation*,
$$\frac{\partial u}{\partial t} = \frac{\partial^2 u}{\partial x^2},$$
if $a_1(t)$ and $a_2(t)$ satisfy the diagonal 2×2 system of ODEs,
$$\begin{pmatrix} a_1 \\ a_2 \end{pmatrix}' = \begin{pmatrix} \lambda_1 & 0 \\ 0 & \lambda_2 \end{pmatrix} \begin{pmatrix} a_1 \\ a_2 \end{pmatrix}, \quad \begin{pmatrix} a_1(0) \\ a_2(0) \end{pmatrix} = \begin{pmatrix} a_{1o} \\ a_{2o} \end{pmatrix}, \quad t \in [0, T], \tag{5.38}$$
where $\lambda_1 = -m_1^2$ and $\lambda_2 = -m_2^2$.

When $m_1 = 1$, $m_2 = 10$, $a_{1o} = 1$, $a_{2o} = 0.01$, we compare the analytic solution at $t = 1$, $\mathbf{y}_A = \begin{pmatrix} 1e^{-1} \\ 0.01e^{-100} \end{pmatrix}$, with the numerical approximations for $h = 1/10$, $N = 10$: using Euler's Method, $\mathbf{y}_E = \begin{pmatrix} 0.17^{10} \\ 0.01(-9)^{10} \end{pmatrix}$; using an explicit second-order Runge-Kutta Method,

$$\mathbf{y}_{ERK2} = \begin{pmatrix} 0.1705^{10} \\ 0.01(41)^{10} \end{pmatrix};$$ and using the Backward Euler Method, $\mathbf{y}_{BE} = \begin{pmatrix} 1/1.1^{10} \\ 0.01(1/11)^{10} \end{pmatrix}$. The relative error in the first mode for $\mathbf{y}_E$ and $\mathbf{y}_{BE}$ is ≈ 0.05, and for $\mathbf{y}_{ERK2}$ it is ≈ 0.002. But it is immediately clear that in the second mode, both explicit approximations, $\mathbf{y}_E$ and $\mathbf{y}_{ERK2}$, yield absurd answers with huge absolute and relative errors for that mode individually and considered as vectors in $\mathbf{R}^2$. What might be less obvious is that though the qualitative nature of the implicit approximation $\mathbf{y}_{BE}$ is correct and the absolute error is small in both components, the relative error in the second component alone is still astronomical, on the order of 10^{33}. The point is not that, for the given step-size, the implicit method does a particularly better job at approximating the mode with the much faster time scale. The point is that in spite of doing a terrible job, this error does not disturb the approximation in the dominant mode. The other methods do as well or better on that mode, but any advantage is destroyed by their behavior on the peripheral mode. The result is that the relative error of $\mathbf{y}_{BE}$, considered as a vector in $\mathbf{R}^2$, remains ≈ 0.05. This also emphasizes that the stiffness of the system, as defined by a ratio of extreme eigenvalues, is the overriding issue, rather than the absolute magnitude of an individual eigenvalue. If, for example, the extreme eigenvalues are $\lambda_1 = -10^{-3}$, $\lambda_2 = -1$, the system is still stiff even though neither mode appears to vary rapidly. But if it is the slower behavior governed by λ_1 that we are interested in, it would make sense to rescale time so that the eigenvalues in the new variables would become $\tilde{\lambda}_1 = -1$, $\tilde{\lambda}_2 = -10^3$ and the significance of absolute stability again becomes apparent. Conversely, the condition number of any scalar problem including $y' = -10^3 y$ is 1, and it is possible to rescale time so that it becomes $y' = -y$ in the new variables. For the same reason, one might suggest that a smaller time-step of size 10^{-3} would restore the superiority of the second-order method in the example above.

Perhaps this example is a contrived exception whose modes have unrealistic decay rates and relative magnitudes that would never be observed in practice? On the contrary, our next example will demonstrate that this example is somewhat universal. We now introduce a

5.6. Stiff Systems and the Method of Lines

general approach that is used to approximate solutions of PDE initial value problems. Many important partial differential equations of pure and applied mathematics are given in the form of a well-posed initial value problem. Given $u_o(\mathbf{x})$ satisfying certain smoothness and boundary conditions on its spatial domain, there exists a unique solution $u(\mathbf{x}, t)$ that satisfies the partial differential equation for $t > 0$, approaches $u_o(\mathbf{x})$ as $t \to 0$, and depends continuously on $u_o(\mathbf{x})$.

The initial value problem for the heat equation (or diffusion equation) on the unit circle is given by

$$\frac{\partial}{\partial t} u(x,t) = \nu \frac{\partial^2}{\partial x^2} u(x,t), \quad u(x,0) = u_o(x) \tag{5.39}$$

where $\nu > 0$ is the diffusion coefficient, $x \in [0, 2\pi]$, $t \geq 0$, and $u_o(x)$ is a smooth 2π-periodic function, $u^{(j)}(0) = u^{(j)}(2\pi)$, $j = 0, 1, \ldots$. Understanding the behavior of solutions of the heat equation and its relatives is important not only in applied mathematics, but also in pure areas as diverse as Probability, Geometry, and even Topology where it has been used in proofs of index theorems and the Poincaré conjecture. In what follows, we will set $\nu = 1$ and leave the appropriate modifications for general ν to the reader.

The heat equation can be solved analytically using an expansion in terms of the orthogonal eigenfunction of the symmetric operator $\frac{\partial^2}{\partial x^2}$ as follows. For $u_k(x,t) = \hat{u}_k(t) e^{ikx}$ to be a solution of the heat equation, $\hat{u}_k(t)$ must satisfy

$$\frac{d}{dt} \hat{u}_k(t) = \lambda_k \hat{u}_k(t), \quad \lambda_k = -k^2. \tag{5.40}$$

Therefore, we represent the initial value $u_o(x)$ as a Fourier series

$$u_o(x) = \sum_{k=-\infty}^{\infty} \hat{u}_k e^{ikx}, \tag{5.41}$$

where

$$\hat{u}_k = \frac{1}{2\pi} \int_0^{2\pi} u_o(x) e^{-ikx} \, dx. \tag{5.41'}$$

Then for $t > 0$,

$$u(x,t) = \sum_{k=-\infty}^{\infty} \hat{u}_k e^{\lambda_k t} e^{ikx}. \tag{5.42}$$

A fundamental method for approximating solutions of a partial differential equation initial value problem is the *method of lines*. We discretize the spatial domain and approximate the values $u(\mathbf{x_j}, t)$ using the solution of a corresponding initial value problem for a system of ODEs, $\mathbf{u_j}(t)$. The ODEs are obtained by replacing spatial derivatives of u at $(\mathbf{x_j}, t)$ by finite difference approximations of these derivatives in terms of $\mathbf{u_j}(t)$, and the initial value $\mathbf{u}_{o,\mathbf{j}}$ is obtained by evaluating $u_o(\mathbf{x})$ at $\mathbf{x_j}$.

For the example of the heat equation on the unit circle above, we discretize the interval $[0, 2\pi]$ using J points $x_j = j\Delta x, j = 0, \ldots, J$, where the grid-size is $\Delta x = \frac{2\pi}{J}$ and $\mathbf{u}_{o,j} = u_o(x_j)$. We approximate the second spatial derivative of u at x_j using the centered second divided difference formula (see Appendix J). This is obtained by interpolating $u(x,t)$ at x_{j-1}, x_j, x_{j+1} with a polynomial $p_2(x)$ of degree ≤ 2 and evaluating

$$p''(x_j) = \frac{\frac{u_{j+1}-u_j}{\Delta x} - \frac{u_j - u_{j-1}}{\Delta x}}{\Delta x} = \frac{1}{\Delta x^2}(u_{j+1} - 2u_j + u_{j-1}),$$

for $j = 1, \ldots, J-2$. For $j = 0$ and $j = J-1$, we extend periodically and use $u_{-1}(t) = u_{J-1}(t)$ and $u_J(t) = u_0(t)$ to keep the system closed. Therefore, the method of lines approximation has the form

$$\frac{d}{dt}u_j(t) = \frac{1}{\Delta x^2}(u_{j+1}(t) - 2u_j(t) + u_{j-1}(t)), \quad u_j(0) = u_{o,j}. \quad (5.43)$$

We also write these equations in matrix form

$$\frac{d}{dt}\mathbf{u}_J(t) = \mathbf{A}_J \mathbf{u}_J(t), \quad (5.43')$$

where $\mathbf{A}_J$ is a $J \times J$ periodic tridiagonal matrix (i.e., nonzero entries in the nondiagonal corners reflect the periodicity).

Not only can we solve this system analytically using the same method of eigenfunction expansion that we used for the continuous problem, but the orthogonal eigenfunctions of the symmetric difference operator $\mathbf{A}_J$ obtained by discretizing the symmetric differential operator $\frac{\partial^2}{\partial x^2}$ are just the discretized eigenfunctions of $\frac{\partial^2}{\partial x^2}$! For $\mathbf{f}_{k,J}(t) = \hat{u}_k(t)e^{ikx_j}$ to be a solution of

$$\frac{d}{dt}\mathbf{f}_{k,J}(t) = \mathbf{A}_J \mathbf{f}_{k,J}(t), \quad (5.44)$$

5.6. Stiff Systems and the Method of Lines

$\hat{u}_k(t)$ must satisfy

$$\frac{d}{dt}\hat{u}_k(t) = \lambda_{k,J}\hat{u}_k(t), \quad \lambda_{k,J} = \frac{2(\cos(k\Delta x) - 1)}{\Delta x^2}. \tag{5.45}$$

Therefore, we represent the initial value $u_{o,j} = u_o(x_j)$ as a finite discrete Fourier series

$$u_{o,j} = \sum_{k=0}^{J-1} \hat{u}_{k,J} e^{ikx_j} \tag{5.46}$$

where

$$\hat{u}_k = \frac{1}{J}\sum_{j=0}^{J-1} u_{o,j} e^{-ikx_j}. \tag{5.46'}$$

Then for $t > 0$,

$$u_j(t) = \sum_{k=0}^{J-1} \hat{u}_{k,J} e^{\lambda_{k,J} t} e^{ikx_j}. \tag{5.47}$$

The apparent asymmetry between the discrete summation over nonnegative frequencies k, $\sum_{k=0}^{J-1}$, and the continuous summation over all integers is illusory, since the exponentials whose arguments they index are periodic. If $k \in [J/2, J]$, using $e^{iJj\Delta x} = 1$, we may write $e^{ikx_j} = e^{ik'x_j}$ with $k' = -(J-k)$ in the octave $[-J/2, 0]$. The frequencies of greatest magnitude that can be represented on this grid, $k_{\max}$, depend on whether J is even or odd. If J is even, there is a single k corresponding to $k_{\max} = J/2 = -k_{\max}(\mod J)$, but if J is odd, $k_{\max} = \frac{J-1}{2}$ and $-k_{\max}$ are distinct mod J, with $-k_{\max} = k_{\max} + 1 = J - k_{\max}(\mod J)$.

Note that as $\Delta x \to 0$, $\lambda_{k,J} \approx -k^2 + O(k^4 \Delta x^2)$. If J is even, the minimum (negative) growth rate corresponding to the highest frequencies satisfies

$$\lambda_J^* = \min_{0 \le k \le J-1} \lambda_{k,J} = \lambda_{k_{\max}} = \frac{-4}{(\Delta x)^2} = -4\frac{J^2}{(2\pi)^2}. \tag{5.48}$$

If J is odd,

$$\lambda_J^* = -\frac{2}{(\Delta x)^2}(1 + \cos(\frac{\Delta x}{2})) \approx \frac{-4}{(\Delta x)^2} = -4\frac{J^2}{(2\pi)^2}. \tag{5.48'}$$

What's more, if we approximate the solution of the spatially discretized system

$$\frac{d}{dt}\mathbf{u}_J(t) = \mathbf{A}_J \mathbf{u}_J(t)$$

using a temporally discretized numerical method, the result may be expressed by simply replacing the factor $e^{\lambda_{k,J} t}$ in the above formula, i.e., the analytical solution of the (complex) scalar differential equation describing the evolution of the coefficient of the k-eigenvector $\frac{d}{dt}\hat{u}_k(t) = \lambda_{k,J}\hat{u}_k(t)$, with the approximate solution obtained from the particular method. For example, if we approximate the solution using Euler's Method, $\mathbf{u}_{n+1,J} = \mathbf{u}_{n,J} + h\mathbf{A}_J \mathbf{u}_{n,J}$, or, in components,

$$u_{n+1,j} = u_{n,j} + \frac{h}{\Delta x^2}(u_{n,j+1} - 2u_{n,j} + u_{n,j-1}), \quad u_{0,j} = u_{o,j}, \quad (5.49)$$

the result will be

$$u_{n,j} = \sum_{k=-J}^{J} \hat{u}_{k,J}(1 + \lambda_{k,J} h)^n e^{ikx_j}. \quad (5.50)$$

If we approximate the solution using the trapezoidal method,

$$\mathbf{u}_{n+1,J} = \mathbf{u}_{n,J} + \frac{h}{2}(\mathbf{A}_J \mathbf{u}_{n,J} + \mathbf{A}_J \mathbf{u}_{n+1,J}), \quad (5.51)$$

or, in components,

$$u_{n+1,j} = u_{n,j} + \frac{h}{2\Delta x^2}((u_{n,j+1} - 2u_{n,j} + u_{n,j-1})$$
$$+ (u_{n+1,j+1} - 2u_{n+1,j} + u_{n+1,j-1})), \quad (5.52)$$

the result will be

$$u_{n,j} = \sum_{k=-J}^{J} \hat{u}_{k,J}\left(\frac{1 + \frac{\lambda_{k,J} h}{2}}{1 - \frac{\lambda_{k,J} h}{2}}\right)^n e^{ikx_j}. \quad (5.53)$$

In Figure 5.18, the Euler's Method and trapezoidal method approximations using $J = 64$ and $h = 0.005$ are depicted at $t = 2.14$, along with the analytical solution at the same t, and the initial condition $u_{o,j} = 1.0\sin(x_j)$, $j = 0, \ldots, 63$. It appears that there are only

5.6. Stiff Systems and the Method of Lines

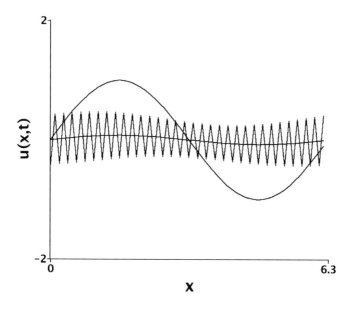

Figure 5.18. Euler's Method for the heat equation.

three curves because the trapezoidal approximation is indistinguishable from the analytical solution.

The explanation of the sawtooth error resulting from Euler's Method is contained in (5.49). Since $\lambda_J^* = \frac{-4}{\Delta x^2} < \lambda_{k,J} < 0$, unless $1 + h\frac{-4}{\Delta x^2} > -1$, i.e., $h < \frac{1}{2}\Delta x^2$, $w = \lambda_J^* h$ will be outside of the region of absolute stability of Euler's Method. The terms $\hat{u}_{\pm k_{\max},J}(1 + \lambda_J^* h)^n e^{\pm i k_{\max} x_j}$ in the numerical solution will not only fail to approximate the corresponding terms $\hat{u}_{\pm k_{\max}} e^{-k_{\max}^2 t} e^{\pm i k_{\max} x}$ in the analytical solution, they will grow exponentially and render the approximation useless. For the parameters in Figure 5.18, we check that $0.5(\frac{2\pi}{64})^2 = 0.0048\ldots$ and our step-size of 0.005 barely violates the stability requirement. Observe not only that it is the highest frequency mode that amplifies (count exactly 32 peaks on the interval) but also that the errors that are amplified only exist due to finite precision computer arithmetic! In exact arithmetic, for the initial data $u_o(x) = \sin(x)$ the FFT gives $\hat{u}_{k,J} = 0$ for $k \neq \pm 1$ (and

$\hat{u}_{\pm 1,J} = \mp 0.5i$), in which case $\hat{u}_{\pm k_{\max},J}(1 + \lambda_J^* h)^n = 0$ regardless of $|1 + \lambda_J^* h| > 1$. Even when $h < \frac{1}{2}\Delta x^2$ and the highest modes are stable, they will not be approximated at all well in the sense of relative error unless h is considerably smaller, as demonstrated in the two-mode example discussed above. But as we pointed out there, the modes with highest k are negligible compared to those with lower k, a consequence of the rapid decay of Fourier coefficients of smooth functions. These errors do not appreciably affect the relative error of the overall solution as long as they are not amplified. The restriction $h < \frac{1}{2}\Delta x^2$ is quite severe. If we set $h = \Delta x$ and send both to zero, the jointly discretized system will formally approximate the heat equation in the limit. But the above analysis demonstrates that the numerical solutions will blow up faster and faster, becoming worse and worse approximations of the analytical solution, not better. The solutions of the approximations are not approximations of the solutions! For $\Delta x = 0.01$, we would require $h < 0.00005$ and 20,000 steps to reach $T = 1$ in order to maintain stability.

In other words, *it is impossible to beat the instability with smaller step-sizes*. The highest frequencies m that occur in practice are proportional to the number of grid-points used to approximate spatial derivatives in the original PDE. The amplitudes of these frequencies typically decay rapidly with m. This example explains why, when using an explicit numerical method whose absolute stability region is bounded, increasing the spatial resolution of an approximation to a PDE such as the heat equation can actually make the temporal approximation drastically worse, unless step-sizes are reduced and computational effort is increased quadratically.

When we replace Euler's Method with an A-stable method, such as the trapezoidal method, or the Backward Euler Method, the computational effort per time-step will be increased (especially when the original equation is nonlinear). But the advantage of unconditional stability means that only accuracy considerations need to be taken into account in determining the appropriate step-size. The reduction in number of steps required to obtain the approximation within a specified tolerance will always outweigh the increase in effort per step

5.6. Stiff Systems and the Method of Lines

in situations like these, and thus we get a sense of why absolute stability considerations are paramount when we apply numerical methods to discretized PDEs.

We cannot proceed without mentioning one of the most significant numerical methods of the past half-century, the Fast Fourier Transform, or FFT, that simply performs the calculations of (5.46), (5.46′) in a very efficient manner. If symmetries are ignored, it should require $O(J^2)$ complex multiplications to compute the discrete Fourier coefficients of the vector $\mathbf{u}$ with respect to the orthogonal basis of J vectors $\mathbf{f}_k \in \mathbf{C}^J$, and the same number again to form the superposition $\mathbf{u}(t) = \hat{u}_k(t)\mathbf{f}_k$. The FFT takes advantage of symmetries of the matrices $\mathbf{F}_J$ and $\mathbf{F}_J^*$ built from these basis vectors to reduce recursively the number of multiplications required to $O(J \log J)$. In particular, if $J = 2^M$ is a power of 2, we define the $J \times J$ discrete Fourier transform (DFT) matrix $\mathbf{F}_J$ by $f_{k,j} = e^{-ikj\Delta x_J}$ where $\Delta x_J = 2\pi/J$, and we index matrices and vectors naturally, from 0 to $J-1$. The reader should confirm that $\mathbf{F}_J^*\mathbf{F}_J = J\mathbf{I}$, i.e.,

$$\sum_{p=0}^{J-1} e^{i(m-n)p\Delta x_J} = J\delta_{mn}, \qquad (5.54)$$

using geometric summation or the discrete fundamental theorem. Therefore, we may multiply $\mathbf{u}$ by $\frac{1}{J}\mathbf{F}_J$ to form the complex dot products that define the coefficients of

$$\mathbf{u} = \sum_k \hat{u}_k \mathbf{f}_k$$

with respect to the orthogonal columns $\mathbf{f}_k$ of $\bar{\mathbf{F}}$. Using this notation, the forward and inverse discrete Fourier transforms are given in vector and component form by

$$\hat{\mathbf{u}} = \frac{1}{J}\mathbf{F}_J\mathbf{u}, \quad \text{or } \hat{u}_k = \frac{1}{J}\sum_{j=0}^{J-1} e^{-ikj\Delta x_J} u_j, \qquad (5.55)$$

and

$$\mathbf{u} = \mathbf{F}_J^*\hat{\mathbf{u}}, \quad \text{or } u_j = \sum_{k=0}^{J-1} \hat{u}_k e^{ikj\Delta x_J}. \qquad (5.56)$$

The FFT is based on the observation that $\mathbf{F}_J$ is related to the matrix for the $J/2 \times J/2$ DFT, $\mathbf{F}_{J/2}$, in the following simple manner. The matrix consisting of the upper halves of the even columns of $\mathbf{F}_J$ (the first $J/2$ rows) is simply $\mathbf{F}_{J/2}$ and the same is true for the lower halves of the even columns. For $j, k = 0, J/2 - 1$,

$$f_{k,2j} = e^{-ik(2j)\Delta x_J} = e^{-ikj\Delta x_{J/2}} \tag{5.57}$$

and so

$$\begin{aligned} f_{k+J/2,2j} &= e^{-i(k+J/2)j\Delta x_{J/2}} \\ &= e^{-ikj\Delta x_{J/2}} e^{-i(J/2)\Delta x_{J/2}} = e^{-ikj\Delta x_{J/2}}. \end{aligned} \tag{5.58}$$

The corresponding matrices consisting of the odd columns of these submatrices are just plus or minus a diagonal 'twiddle-factor' matrix times $\mathbf{F}_{J/2}$, respectively, $\pm \mathbf{D}_{J/2}\mathbf{F}_{J/2}$, where $\mathbf{D}_{J/2} = \operatorname{diag}\{e^{-ij\Delta x_J}\}$. Again for $j, k = 0, J/2 - 1$,

$$f_{k,2j+1} = e^{-ik(2j+1)\Delta x_J} = e^{-ik\Delta x_J} e^{-ikj\Delta x_{J/2}} \tag{5.59}$$

and so

$$\begin{aligned} f_{k+J/2,2j+1} &= e^{-i(k+J/2)\Delta x_J} e^{-i(k+J/2)j\Delta x_{J/2}} \\ &= e^{-ik\Delta x_J} e^{-i(J/2)\Delta x_J} e^{-ikj\Delta x_{J/2}} \\ &= -e^{-ik\Delta x_J} e^{-ikj\Delta x_{J/2}}. \end{aligned} \tag{5.60}$$

The pattern at the heart of the FFT is one of two important discoveries in modern applied mathematics named for the superfamily *Papilionidae*, the butterfly. In chaotic dynamics, the butterfly represents the small change in initial conditions to which a system is sensitive, but here it refers to the crisscross nature of the calculation expressed by

$$\begin{aligned} \mathbf{F}_J^{\text{upper}} \mathbf{u} &= \mathbf{F}_{J/2} \mathbf{u}_{\text{even}} + \mathbf{D}_{J/2} \mathbf{F}_{J/2} \mathbf{u}_{\text{odd}}, \\ \mathbf{F}_J^{\text{lower}} \mathbf{u} &= \mathbf{F}_{J/2} \mathbf{u}_{\text{even}} - \mathbf{D}_{J/2} \mathbf{F}_{J/2} \mathbf{u}_{\text{odd}}, \end{aligned} \tag{5.61}$$

Applying the butterfly formula (5.61) recursively, one transform of length 2^M is reduced to computing two transforms of length 2^{M-1}; these are reduced to computing four transforms of length 2^{M-2} until 2^{M-1} transforms of length two of the form $u_e + u_o, u_e - u_o$ are performed. Since it only takes a total of J multiplications to combine

5.6. Stiff Systems and the Method of Lines

all 2^m transforms of length 2^{M-m} into 2^{m-1} transforms of length 2^{M+1-m}, the original transform can be computed with fewer than $Jm = J\log_2(J)$ complex multiplications!

See the Web Companion for additional details of implementing this algorithm. Numerous variations on the continuous, discrete, and fast Fourier transforms and ways to use them have been developed for different constant coefficient partial differential (and even integro-differential/pseudodifferential) equations, boundary conditions, numbers of dimensions, and the prime or composite nature of the order J. For excellent overviews, we recommend [BH], [FB], [SG].

Many initial value problems for PDEs can be classified as *parabolic* or *hyperbolic*. Parabolic equations model diffusive behavior, and the heat equation is a prototype and canonical form. Hyperbolic equations model waves, and there are several prototypes and canonical forms, including the wave equation on the unit circle

$$(\frac{\partial^2}{\partial t^2} - c^2 \frac{\partial^2}{\partial x^2})u(x,t) = 0, \quad u(x,0) = u_o(x), \quad (5.62)$$

where $|c|$ is the wave speed, $x \in [0, 2\pi]$, $t \geq 0$, and $u_o(x)$ is a smooth 2π-periodic function, $u^{(j)}(0) = u^{(j)}(2\pi)$, $j = 0, 1, \ldots$. The operator in parentheses can be factored and solutions written as the superposition of solutions of the *advection equation*

$$\frac{\partial}{\partial t} + c\frac{\partial}{\partial x}u(x,t) = 0, \quad u(x,0) = u_o(x) \quad (5.63)$$

with $c = \pm\sqrt{c^2}$.

The unstable behavior observed in numerical approximations of the heat equation above is not specific to equations of parabolic type. We now construct and investigate corresponding examples for wave equations. Our first example of a wave equation will be dispersive, not hyperbolic. It has two modes in the complex plane with widely separated oscillation frequencies and amplitudes. Again we will discover how methods with advantageous absolute stability properties can trump methods with higher accuracy.

Let m_1 and m_2 be positive integers. The real and imaginary parts of $u(x,t) = c_1(t)e^{im_1 x} + c_2(t)e^{im_2 x}$ are 2π-periodic solutions of the *linearized Korteweg-deVries equation,*

$$\frac{\partial u}{\partial t} = \frac{\partial^3 u}{\partial x^3}, \tag{5.64}$$

if $c_1(t)$ and $c_2(t)$ satisfy the 2×2 system of complex ODEs:

$$\begin{pmatrix} c_1 \\ c_2 \end{pmatrix}' = \begin{pmatrix} \lambda_1 & 0 \\ 0 & \lambda_2 \end{pmatrix} \begin{pmatrix} c_1 \\ c_2 \end{pmatrix}, \quad \begin{pmatrix} c_1(0) \\ c_2(0) \end{pmatrix} = \begin{pmatrix} c_{1o} \\ c_{2o} \end{pmatrix}, \quad t \in [0,T], \tag{5.65}$$

where $\lambda_1 = -im_1^3$ and $\lambda_2 = -im_2^3$. The linearized Korteweg-deVries equation arises from the Korteweg-deVries equation, or *KdV equation*, $u_t + 6uu_x + u_{xxx} = 0$ by linearizing about the solution $u(x,t) = 0$ and reversing the sense of time. These equations are prototypical of dispersive wave phenomena. We will discuss the full KdV equation in our final example below.

When $m_1 = 1$, $m_2 = 10$, $c_{1o} = 1$, $c_{2o} = 0.001$, we compare the analytic solution at $t = 1$, $\mathbf{z}_A = \begin{pmatrix} 1e^{-1i} \\ 0.001e^{-1000i} \end{pmatrix}$, with the numerical approximations for $h = 1/100$, $N = 100$ using Euler's Method: $\mathbf{z}_E = \begin{pmatrix} (1-.01i)^{100} \\ 0.001(1-10i)^{100} \end{pmatrix}$, using an explicit second-order Runge-Kutta Method, $\mathbf{z}_{ERK2} = \begin{pmatrix} (0.179995 - .01i)^{100} \\ 0.001(-49 - 10i)^{100} \end{pmatrix}$, and using the Backward Euler Method $\mathbf{z}_{BE} = \begin{pmatrix} (1+.01i)^{-100} \\ 0.001(1+10i)^{-100} \end{pmatrix}$. The relative error in the first mode for $\mathbf{z}_E$ and $\mathbf{z}_{BE}$ is $\approx .005$, and for $\mathbf{z}_{ERK2}$ it is $\approx 1.15 \times 10^{-5}$. But again both explicit approximations, $\mathbf{z}_E$ and $\mathbf{z}_{ERK2}$, yield absurd answers in the second mode with huge absolute and relative errors for that mode individually and considered as vectors in $\mathbf{C}^2$. This time, the second component of implicit first-order approximation $\mathbf{z}_{BE}$ has magnitude $\approx 10^{-103}$, instead of 10^{-3} for the analytical solution. (The trapezoidal method would capture the magnitude but not the phase correctly). But this error does not disturb the approximation in the dominant mode as it did with the other methods. Again their accuracy advantage is overwhelmed by their unstable behavior on this peripheral mode. The result is that the relative error of $\mathbf{z}_{BE}$, considered as a vector in $\mathbf{C}^2$, remains $\approx .005$.

5.6. Stiff Systems and the Method of Lines

To see the practical relevance of this example more clearly, we compare the behavior with respect to absolute stability of two spatial discretizations and two temporal discretizations of the first-order advection, or wave equation

$$u_t + cu_x = 0, \quad u(x,0) = u_o(x), \tag{5.66}$$

where subscripts indicate partial derivatives. Along the lines given by

$$\frac{d}{dt}x(t;x_o) = c, \quad x(0;x_o) = x_o, \tag{5.67}$$

known as the *characteristics* of the PDE, we find that

$$\begin{aligned}\frac{d}{dt}u(x(t;x_o),t) &= u_x(x(t;x_o),t)\frac{dx}{dt}(t;x_o) + u_t(x(t;x_o),t) \\ &= (u_t + cu_x)(x(t;x_o),t) = 0\end{aligned} \tag{5.68}$$

so $u(x(t;x_o),t) = u_o(x_o)$. In other words, the solution is constant along each characteristic, equal to the value $u_o(x_o)$ of the initial data at the point $x_o = x - ct$ where the characteristic intersects the line $t = 0$. This value is shifted spatially at speed c and the analytical solution of this problem is $u(x,t) = u_o(x - ct)$.

If we approximate the spatial derivatives using standard forward difference quotients, the method of lines approximation takes the form

$$\frac{du_j}{dt} = -c\frac{(u_{j+1} - u_j)}{\Delta x}. \tag{5.69}$$

The eigenvalues corresponding to the eigenvectors $\mathbf{f}_{k,J} = e^{ikx_j}$ of the periodic problem are

$$\lambda_{k,J}^F = -c\frac{e^{ik\Delta x} - 1}{\Delta x} = -\frac{c}{\Delta x}(\cos(k\Delta x) - 1 + i\sin(k\Delta x)) \tag{5.70}$$

on a circle of radius $\frac{c}{\Delta x}$ centered at $\frac{c}{\Delta x}$ and therefore tangent to the imaginary axis at the origin. This circle lies in the right or left half-plane depending on whether c is positive or negative, and its radius is proportional to J.

If Euler's Method is used to approximate this system temporally, with u_j^n approximating $u_j(t_n)$ (approximating in turn $u(x_j, t_n)$), the overall approximation takes the form

$$u_j^{n+1} = u_j^n - c\frac{h}{\Delta x}(u_{j+1}^n - u_j^n). \tag{5.71}$$

Because the region of absolute stability of Euler's Method is the disc of radius 1 centered at -1, when forward spatial differencing is used, absolute stability analysis predicts that Euler's Method will behave well if $c < 0$ and $-\frac{c}{\Delta x}h < 1$, i.e., $h \leq -\frac{\Delta x}{c}$, but otherwise exponential instability will result.

This restriction obtained purely from absolute stability considerations can also be obtained by a completely different perspective. From the form of (5.71), we can see that regardless of the sign of c, the value of the numerical solution u_j^n comes from values u_{j+p}^{n-q} with $0 \leq p \leq q$, corresponding to the earlier time $t_n - qh$ and the interval to its right, $[x_j, x_j + q\Delta x]$. The value of the analytical solution at (x, t) comes from the value at $(x - cqh, t - qh)$. If $c > 0$ and the value of the analytical solution comes only from the left, there is no possibility that such an approximation could converge! Even if $c < 0$, unless $-ch < \Delta x$, the location of the value on which the analytical solution depends, $(x - cqh, t - qh)$, is outside the interval $[x_j, x_j + q\Delta x]$ from which the numerical solution is computed. If the constraint we found using absolute stability analysis is violated, the analytical solution depends on values to which it has no access through the numerical method. The fact that absolute stability analysis of eigenvalues gives the same result as this spatial analysis is compelling evidence of the intrinsic significance of absolute stability.

The condition

$$-ch < \Delta x \tag{5.72}$$

is called a *CFL condition*, named for Richard Courant, Kurt Friedrichs, and Hans Lewy, the authors of an influential 1928 paper on finite difference approximations of PDEs, [CFL]. The CFL condition says that the step-size used in numerical approximation must not exceed the time required for information traveling on the characteristics of the PDE to travel between adjacent spatial grid points. If the step-size is larger, the analytical solution corresponding to that step depends on data outside of the domain from which the numerical solution is computed.

5.6. Stiff Systems and the Method of Lines

If instead of using forward differences in space, we use centered differences, the method of lines approximation takes the form

$$\frac{du_j}{dt} = -c\frac{(u_{j+1} - u_{j-1})}{2\Delta x}. \tag{5.73}$$

The eigenvalues corresponding to the eigenvectors $\mathbf{f}_{k,J}$ are

$$\lambda_{k,J}^C = -c\frac{e^{ik\Delta x} - e^{-ik\Delta x}}{2\Delta x} = \frac{-ic\sin(k\Delta x)}{\Delta x}. \tag{5.74}$$

If J is a multiple of 4, these $\lambda_{k,J}^C$ span the interval $[-\frac{|c|}{\Delta x}i, +\frac{|c|}{\Delta x}i]$ on the imaginary axis whose size is proportional to J, with small modifications for nonmultiples. If we use Euler's Method to approximate the resulting system of ODEs, it will not behave well for any h because the purely imaginary $\lambda_{k,J}^C$ are outside its region of absolute stability. The exponential growth we saw when we used Euler's Method to approximate solutions of $z' = iz$ will be reproduced in every mode. If we use backward differences, the behavior is simply reversed in comparison with forward differences, with respect to the sign of c.

Figure 5.19 (upper) shows the eigenvalues corresponding to forward, backward, and centered difference approximations of $\mathbf{A}_J = c\frac{\partial}{\partial x}$ with periodic boundary conditions, using $c = -1$ and $J = 2^5, 2^6$. The eigenvalues for the forward difference approximations are given in (5.70). They lie on circles with radius proportional to J passing through the origin and otherwise lying in the left half-plane. The eigenvalues for the corresponding backward difference approximation, or equivalently with $c = +1$, are the reflections of these circles across the y-axis, the circles in the right half-plane. The eigenvalues for the centered difference approximation are given in (5.74). They are the averages of the eigenvalues of the forward and backward difference approximations, and therefore they span intervals on the imaginary axis with length proportional to J. For $J = 2^6$, this corresponds to a range of $\pm(2\pi/64)^{-1}i \approx \pm 10.2i$ as observed in the figure.

Figure 5.19 (lower) shows the eigenvalues corresponding to centered difference approximations of $\frac{\partial^2}{\partial x^2}$ and $\frac{\partial^3}{\partial x^3}$ with periodic boundary conditions, using $J = 2^4$, along with the corresponding eigenvalues ($|J| \leq 8$) of $\frac{\partial^2}{\partial x^2}$ itself. The eigenvalues of the centered difference

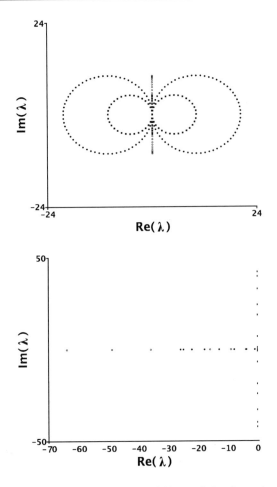

Figure 5.19. Eigenvalues of forward, backward, and centered difference operators.

approximations of $\frac{\partial^2}{\partial x^2}$ are given in (5.45) and span an interval on the negative real axis with one end at the origin and the other at λ_J^* given in (5.48). For $J = 2^4$, this $\lambda_J^* = -4J^2/(2\pi)^2 - 25.9$ overlaps the analytic double eigenvalue $\lambda_{\pm 5} = -25$. The maximum spectral eigenvalue, $\lambda^* = -8^2$, is at the far left. The approximation of $\frac{\partial^3}{\partial x^3}$ and its eigenvalues are discussed below.

5.6. Stiff Systems and the Method of Lines

In Figure 5.20, we illustrate the results of applying Euler's Method with forward, backward, and centered difference method of lines approximations of the left-moving wave solution of $u_t = u_x$ with initial data $u_o(x) = 1.0\sin(x)$, with step-size $h = 0.005$. The smooth small amplitude wave in the figure results from forward differencing after time $T = 10\pi$. For this approximation, all of the eigenvalues are in the left half-plane, and for the step-size employed, they are all strictly inside the region of absolute stability of Euler's Method. No instability is observed in any mode, but the complex amplitude of the single mode initially present decays at a slow exponential rate. The jagged wave whose peak is near the ends of the interval results from backward differencing after little more than a quarter-period, $T = 1.85$. For this approximation, all of the eigenvalues are in the right half-plane, outside the region of absolute stability of Euler's Method for this and any step-size. The highest frequency mode is the most unstable, and round-off errors have already been amplified to be clearly visible. If the computation were carried any further, these errors would overwhelm the desired content of the figure and eventually cause overflow. The incipient jaggedness of the wave whose peak is near the center of the interval results from centered differencing after approximately 20 periods, $T = 130$. For this approximation, all of the eigenvalues are on the imaginary axis, outside the region of absolute stability of Euler's Method for this and any step-size. Due to the small step-size, however, the most unstable, highest frequency mode has a relatively slow growth rate, and its growth is only initiated by round-off errors. This explains the longer time required for the instability to manifest itself.

If, instead of Euler's Method, we use the leapfrog method to approximate the system of ODEs obtained from the advection equation using centered differencing, it will behave well as long as the time-step satisfies the CFL condition $h < \frac{\Delta x}{|c|}$, regardless of the sign of c, because its region of absolute stability is the interval $(-i, +i)$. However, this method behaves poorly when forward or backward spatial differencing is used, since the $\lambda^F_{k,J}$ are nonimaginary regardless of the sign of c.

5. Numerical Methods

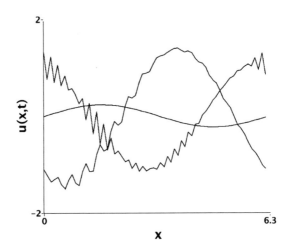

Figure 5.20. Euler's Method for the advection equation.

In Figure 5.21, we illustrate the results of applying the leapfrog method with forward, backward, and centered difference method of lines approximations of the left-moving wave solution of $u_t = u_x$ with initial data $u_o(x) = 1.0\sin(x)$, with step-size $h = 0.005$. The smooth wave with amplitude 1 in the figure results from centered differencing after fifty periods, $T = 100\pi$. For this approximation, all of the eigenvalues are on the imaginary axis, and for the step-size employed, they are all strictly inside the region of absolute stability of the leapfrog method. No instability is observed in any mode, and the complex amplitude of the single mode initially present is preserved exactly. The speed of the wave is not exactly 1, but even after 50 periods, it has returned very nearly to its original position. The jagged waves whose peaks are near the ends of the interval result from forward and backward differencing after approximately a quarter-period, $T = 1.75$. For these approximation, all of the eigenvalues are in the left and half-planes, respectively, outside the region of absolute stability of the leapfrog method for this and any step-size. The highest frequency mode is the most unstable, and round-off errors have already amplified to be clearly visible. The wave with slightly higher amplitude corresponds to backward differencing. If the computation

5.6. Stiff Systems and the Method of Lines 205

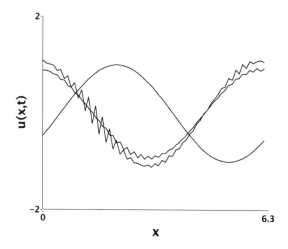

Figure 5.21. Leapfrog method for the advection equation.

were carried any further, the instability would cause overflow and would overwhelm the desired content of the figure.

The central lesson of our development is now summarized in just a few figures. The vastly superior performance of temporal discretization methods with lower-order accuracy over methods with higher-order accuracy in approximating solutions of the systems of ODEs shown in Figures 5.18, 5.20, and 5.21 can be completely and simply understood in terms of the correlation of the absolute stability regions shown in Figure 5.14 and the location of the eigenvalues of chosen spatial discretization of the differential operator as shown in Figure 5.19.

Our concluding example involves a PDE related to the two-mode example above, just as the heat equation example was related to the first two-mode example. The Korteweg-DeVries equation, or KdV equation,

$$\frac{\partial u}{\partial t} + 6u\frac{\partial u}{\partial x} + \frac{\partial^3 u}{\partial x^3} = 0 \qquad (5.75)$$

describes the nonlinear evolution of waves in a shallow channel. The initial value problem for the KdV equation on the unit circle is given

by
$$\frac{\partial}{\partial t}u(x,t) = -6u(x,t)\frac{\partial}{\partial x}u(x,t) - \frac{\partial^3}{\partial x^3}u(x,t), \quad u(x,0) = u_o(x)$$
(5.76)

where $x \in [0, 2\pi]$, $t \geq 0$, and $u_o(x)$ is a smooth 2π-periodic function, $u^{(j)}(0) = u^{(j)}(2\pi)$, $j = 0, 1, \ldots$. The KdV equation and its discrete cousin, the Fermi-Pasta-Ulam system, are charter members of a family of nonlinear PDEs and related lattice models that have many properties that are quite unusual and notable among such equations. Like the heat equation, the KdV equation can be solved analytically, though with a nonlinear 'inverse scattering transform' instead of the linear Fourier transform. After nonlinear interaction, certain pulse-like solutions called solitons pass through each other with their shape asymptotically unchanged. Solutions possess an infinite number of independent integral functionals that are conserved under the evolution. It can be viewed as an infinite-dimensional completely integrable Hamiltonian system, as well as an isospectral flow arising from a Lax-pair formulation. Solitons are often significant in algebraic and differential geometry. For example, solitons of the Sine-Gordon equation correspond to surfaces of constant negative curvature. Geometric Backlund transformations that generate new surfaces of constant negative curvature from old ones also generate new solitons from old ones. See [GBL], [LG] for further background and references on this beautiful subject.

We will perform a method of lines approximation for the KdV equation, using centered first and third divided differences. The result is the periodic pentadiagonal system

$$\frac{du_j}{dt} = -6u_j \frac{(u_{j+1} - u_{j-1})}{2\Delta x} - \frac{(u_{j+2} - u_{j+1} + 2u_j - u_{j-1} + u_{j-1})}{\Delta x^3},$$
$$u_j(0) = u_{o,j}.$$
(5.77)

For $j = 0, 1$ and $j = 2J - 1, 2J$, we extend periodically and use $u_{-2}(t) = u_{2J-1}(t)$ $u_{-1}(t) = u_{2J}(t)$, $u_{2J+1}(t) = u_0(t)$, and $u_{2J+2}(t) = u_1(t)$ to keep the system closed.

When the classical explicit fourth-order Runge-Kutta Method is applied to the resulting system of equations, the time-steps have to be

5.6. Stiff Systems and the Method of Lines

tiny to maintain absolute stability; otherwise the computation quickly yields spurious results and terminates due to overflow as shown in Figures 5.22 and 5.23. Due to the presence of the nonlinear term, the explanation for this behavior is not as simple as for the heat equation. The discussion of operator splitting below provides some justification for analyzing the linear terms in both the PDE and the approximating system, individually. When we analyze individually, the reason is apparent. The eigenvalues of the continuous third derivative operator are $\lambda_k = -ik^3$. The eigenvalues of the discretized operator span an interval on the imaginary axis $[-i\lambda_{\max}(J), i\lambda_{\max}(J)]$ where $\lambda_{\max}(J)$ grows like J^3. So unless h shrinks like Δx^3, $\lambda_j^* h$ will escape the portion of the region of absolute stability of RK4 on the imaginary axis, and the exponentially growing sawtooth solutions of the figure are inevitable. The consequences of instability appear much sooner in the calculation than they did for the heat equation. As soon as $t > 0$, the nonlinearity causes every mode (including the highest) to become nonzero, for both the PDE and the approximating system of ODEs.

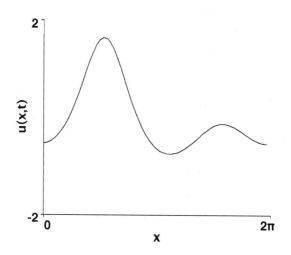

Figure 5.22. RK4 method for the KdV equation (stable).

If a lower-order but unconditionally stable method such as the A-stable trapezoidal method (second-order) or the Backward Euler

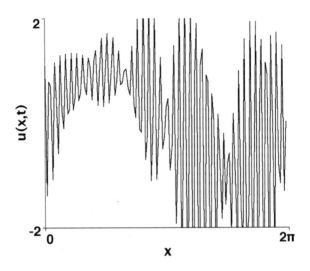

Figure 5.23. RK4 method for the KdV equation (unstable).

Method (first-order) is used, the step-size dictated by accuracy is still far less restrictive than the step-size required for RK4 to be stable. The additional computation needed to solve nonlinear systems at each step of the implicit methods is more than rewarded by the far fewer steps that are needed to obtain a satisfactory approximation.

We conclude this section with a brief description of an elegant *splitting method* for approximating solutions of the KdV equation numerically, suggested by Tappert [TF]. Splitting is a technique used to approximate solutions of an evolution equation governed by a sum of terms

$$\mathbf{u}' = \mathbf{A}(\mathbf{u}) + \mathbf{B}(\mathbf{u}), \quad \mathbf{u}(0) = \mathbf{u}_o \tag{5.78}$$

using various combinations of approximate solutions of

$$\mathbf{v}' = \mathbf{A}(\mathbf{v}), \; \mathbf{v}(0) = \mathbf{v}_o \quad \text{and} \quad \mathbf{w}' = \mathbf{B}(\mathbf{w}), \; \mathbf{w}(0) = \mathbf{w}_o. \tag{5.79}$$

A formal justification of this when $\mathbf{A}$ and $\mathbf{B}$ are linear is based on expanding $\mathbf{u}(h) \approx \exp((\mathbf{A} + \mathbf{B})h)\mathbf{u}_o$, where

$$\exp((\mathbf{A} + \mathbf{B})h) \approx \mathbf{I} + h(\mathbf{A} + \mathbf{B}) + \frac{h^2}{2}(\mathbf{A} + \mathbf{B})^2 + O(h^3), \tag{5.80}$$

5.6. Stiff Systems and the Method of Lines

and performing the corresponding expansions on $\mathbf{v}(h) = \exp(\mathbf{A})\mathbf{v}_o$ and $\mathbf{w}(h) = \exp(\mathbf{B})\mathbf{w}_o$:

$$\exp(\mathbf{A}h) \approx \mathbf{I} + h\mathbf{A} + \frac{h^2}{2}\mathbf{A}^2 + O(h^3), \qquad (5.81)$$

$$\exp(\mathbf{B}h) \approx \mathbf{I} + h\mathbf{B} + \frac{h^2}{2}\mathbf{B}^2 + O(h^3).$$

Using these expressions, we find

$$\exp((\mathbf{A} + \mathbf{B})h) = \exp(\mathbf{B}h)\exp(\mathbf{A}h) + O(h^2), \qquad (5.82)$$
$$\exp((\mathbf{A} + \mathbf{B})h) = \exp(\mathbf{A}h/2)\exp(\mathbf{B}h)\exp(\mathbf{A}h/2) + O(h^3).$$

We denote by $S_{\mathbf{f},p,h}$ any pth-order accurate numerical solution operator that advances an approximation of $\mathbf{y}' = \mathbf{f}(t, \mathbf{y})$ one time-step of size h. Then the $O(h^2)$ approximation in (5.82) suggests that if we alternate steps of size h of methods corresponding to the individual equations (5.79) that are at least first-order accurate, $S_{\mathbf{A}_o,1,h}, S_{\mathbf{B}_o,1,h}$, we will obtain a first-order accurate method $S_{\mathbf{A}_o+\mathbf{B}_o,1,h}$ for approximating solutions of (5.78). The $O(h^3)$ approximation in (5.82), $S_{\mathbf{A}_o,2,h/2}S_{\mathbf{B}_o,2,h}S_{\mathbf{A}_o,2,h/2}$, is used to obtain a second-order accurate method, $S_{\mathbf{A}_o+\mathbf{B}_o,2,h}$ for (5.78). This becomes even more notable when we observe that as long as no intermediate output is required, multiple steps can be combined in the form

$$\begin{aligned}(S_{\mathbf{A}_o,2,h/2}S_{\mathbf{B}_o,2,h}S_{\mathbf{A}_o,2,h/2})^n \\ = S_{\mathbf{A}_o,2,h/2}(S_{\mathbf{B}_o,2,h}S_{\mathbf{A}_o,2,h})^{n-1}S_{\mathbf{B}_o,2,h}S_{\mathbf{A}_o,2,h/2}.\end{aligned} \qquad (5.83)$$

This says that when the individual methods are at least second-order accurate, we only need to shift half of the initial step of n steps of first-order accurate splitting, $(S_{\mathbf{B}_o,2,h}S_{\mathbf{A}_o,2,h})^n$, to the end of the computation in order to turn it into second-order accurate splitting! In the numerical methods literature, this second-order splitting method is commonly known as *Strang splitting*. One of the earliest and best known uses of this technique was Strang's method for approximating solutions of an advection PDE in two spatial dimensions. Again using subscripts to indicate partial derivatives, the equation

$$u_t = c_1 u_x + c_2 u_y$$

was split into two equations,
$$v_t = c_1 v_x$$
and
$$w_t = c_2 w_y,$$
and each was approximated with second-order accurate time-stepping in one-space dimension. If implicit methods are called for, approximate solutions of the resulting one-dimensional problems can be computed more efficiently because the the corresponding linear systems have substantially smaller bandwidth.

For the KdV equation with periodic boundary conditions, Tappert split
$$u_t + 6uu_x + u_{xxx} = 0$$
into linear and nonlinear pieces. Except for the minus sign,
$$w_t = -w_{xxx}, \quad w(x,0) = w_o(x),$$
the linear part is exactly the equation whose two-mode solution we analyzed above for the effects of absolute stability. Solutions were efficiently approximated using the spectrally ('infinite order') accurate and A-stable Fourier Method, in which the evolution of each Fourier coefficient is determined from the analytical equation, not the discretized equation. In other words, we use $\frac{d\hat{w}_k}{dt} = \lambda_k \hat{w}_k$ where $\lambda_k = -(ik)^3$ from the third derivative operator, not the corresponding $\lambda_{k,J}$ of the third difference operator. So to advance the solution by one step of size h, we let $\hat{w}_k(t+h) = \hat{w}_k(t) e^{-(ik)^3 t}$. There are only two sources of error in the numerical solution. The first is due to the absence of any terms with $|k| > J$ in the solution. The second is due to the fact that these missing modes reappear in terms with $|k| \leq J$ through *aliasing*. Aliasing is used to describe the fact that since $e^{imJ\Delta x} = 1$ for any integer m, when such a component is sampled on the grid $x_j = j\Delta x$, then $\hat{w}_{k+mJ} e^{i(k+mJ)x_j} = \hat{w}_{k+mJ} e^{ikx_j}$ and it is treated as a mode in the range $|k| \leq J$. For smooth functions of x, the magnitudes of the coefficients of these terms decay faster than any power of J, so aliasing effects go to zero rapidly.

5.6. Stiff Systems and the Method of Lines

If we ignore the conventional factor of 6 from the KdV equation, the nonlinear part,

$$v_t = -6vv_x, \quad v(x,0) = v_o(x), \tag{5.84}$$

is known as the inviscid Burgers equation. This equation may be analyzed in a manner similar to the first-order advection equation above. Along the curves given by

$$\frac{d}{dt}x(t;x_o) = 6v(x,t), \quad x(0;x_o) = x_o, \tag{5.85}$$

known as the *characteristics* of the PDE, we find that

$$\begin{aligned}\frac{d}{dt}v(x(t;x_o),t) &= v_x(x(t;x_o),t)\frac{dx}{dt}(t;x_o) + v_t(x(t;x_o),t) \\ &= (v_t + 6vv_x)(x(t;x_o),t) = 0\end{aligned} \tag{5.86}$$

so $v(x(t;x_o),t) = v_o(x_o)$. In other words, the solution is constant along each characteristic, which is therefore a straight line. That constant is equal to the value $v_o(x_0)$ of the initial data at the point where the characteristic intersects the line $t = 0$. This value is shifted spatially at speed $6v_o$, so that higher points on the graph of $v(x,t)$ travel faster. The characteristics emanating from any interval on which the initial data is decreasing inevitably collide for some $t > 0$. Because of this, the inviscid Burgers equation is the fundamental elementary model for the breaking of nonlinear waves and shock formation in compressible fluids.

Tappert used centered differencing in conservation form,

$$-6uu_x = -3(u^2)_x \approx \frac{-3}{2\Delta x}(u_{j+1}^2 - u_{j-1}^2), \tag{5.87}$$

to perform a method of lines approximation of this equation, using the same grid that was used to discretize the linear term. If the resulting equations are approximated numerically using the leapfrog method, linearized eigenvalue and absolute stability analyses similar to those we performed for the heat equation require that the time-step satisfy a CFL requirement of the form

$$h \leq C\frac{\Delta x}{\max|6v_o(x)|}.$$

Although this restriction is not as prohibitive as those we have found for equations involving higher-order spatial derivatives, in order to obtain unconditional absolute stability, solutions of this system were approximated using the A-stable and second-order accurate trapezoidal method. When the trapezoidal method is applied to the method of lines approximations of PDEs, especially those of diffusion (parabolic) type, it is known as the Crank-Nicholson Method. Further details on the split-step pseudospectral method for numerical approximation of solutions of the KdV equation, including links to an implementation in the Virtual Math Museum can be found in the Web Companion.

Operator splitting makes it possible to analyze both accuracy and absolute stability properties term by term, and different splittings may be used to obtain methods with superior characteristics. For example, even analytically, we might expect the behavior of solutions of a nonlinear diffusion equation,

$$u_t = ((m+u)u_x)_x - u_{xxxx}, \quad u(x,0) = u_o(x), \qquad (5.88)$$

and its relatives that arise in various applications to be extremely unstable if the diffusion coefficient $m + u(x)$ were ever negative. The fourth-order term does not imply a maximum-minimum principle but behaves somewhat like the diffusion term in the heat equation, guaranteeing local well-posedness regardless of the sign of $m + u_o$. Numerically, in cases involving negative diffusion, splitting (5.88) into its nonlinear second-order and linear fourth-order terms would invite disaster. But if before we split, we borrow some diffusion from the higher-order term and lend it to the lower-order nonlinear term, both of the resulting problems are numerically well-behaved. We split (5.88) as

$$v_t = ((m+v+c)v_x)_x \quad \text{and} \quad w_t = -cw_{xx} - w_{xxxx}, \qquad (5.89)$$

where $c \geq 0$ is adaptively chosen so that $m + u + c > 0$ for all (x,t). We then solve the individual problems on the interval $[0, 2\pi]$ with periodic boundary conditions using the same methods as above. We use the A-stable second-order Crank-Nicholson Method to approximate solutions of the nonlinear second-order portion, tamed to become a relatively safe forward diffusion equation that obeys a maximum principle. We use the spectral method for the constant coefficient linear portion, and with $\lambda_k = -k^4 + ck^2$, all exponential growth is restricted to a finite band of frequencies and solved as exactly as possible with the analytical expression $e^{\lambda_k h}$. This elegant extension of Tappert's Method was proposed by Andy Majda for a nonlocal version of (5.89) arising in vortex dynamics. It led to the discovery of a Fourier-based method for rigorously proving finite time blowup of solutions of a large class of evolution equations including (5.89). This work is reported in [BP], and more recent applications of the method to fourth-order equations such as (5.89) and the complexified three-dimensional Euler equations are found in [BB] and [FP].

5.7. Convergence Analysis: Euler's Method

In this section we will prove that Euler's Method (5.11) is convergent for the scalar IVP (5.1). The proof relies on identifying and analyzing two factors responsible for the growth in error bounds as the number of steps taken to advance the solution over a fixed time interval is increased. One, the *local truncation error*, describes the magnitude of new local errors being introduced at each step. The other, the *amplification factor*, describes how each step transforms prior errors and characterizes the stability of the method for a particular equation and time-step. We will derive a recurrence inequality satisfied by errors, containing an additive term corresponding to the first effect and a multiplicative factor corresponding to the second. We will focus on the scalar case and leave the relatively straightforward generalization to systems of ODEs to the reader.

Though many treatments begin with this proof, its signficance may be better understood after observing the behavior of different examples, especially with regard to these two interacting contributions to global errors. The reader is encouraged to consider how the analysis may be generalized to other Runge-Kutta and multistep methods in order to improve these estimates. Rigorous theory helps us design methods that are advantageous for certain problems and explains their failure for other problems and why other methods can have the opposite behavior. We will briefly survey two alternative methods for quantifying the errors introduced at each step, *residual error*, used in backward error analysis, and *local error* (sans 'truncation'), used in automatic step-size control.

To begin the proof, we assume that $f(t,y)$ is continuous on $[t_o, t_o+b)$ and satisfies the Lipschitz condition (5.2), so that the existence of a unique solution $y(t)$ is guaranteed on some maximal subinterval $[t_o, t_o + T_*)$ with $0 < T_* \leq b \leq \infty$. Under these conditions, we will now show that numerical solutions obtained from Euler's Method (5.11) converge to solutions of (5.1) with first-order accuracy. It is possible to analyze convergence by shrinking the step-size $h \to 0$ continuously, but to do so, we must continuously adjust the times at which we compare the numerical and analytical solutions. It is more convenient to keep a time $t_o + T$ fixed and to restrict the step-size h to discrete values satisfying $Nh = T$ as we increase the number of steps $N \to \infty$. In doing so, we must adjust our notation to avoid any ambiguous meaning of the quantity y_j. For example, y_4 could mean y_N with $N = 4$, the fourth step of four, or it could mean $y_{N/2}$ with $N = 8$, the fourth step of eight. We will assume that a specific value of T has been set, and we will index the intermediate approximations by both their step number, n, and step-size, $h = T/N$, in the form $y_{n,h}$. Extending this, we define $e_{n,h} = y(t_n) - y_{n,h}$, where $t_n = t_o + nh$ as in (5.10). Thus, $e_{N,h}$ signifies

$$y(t_N) - y_{N,h} = y(t_o + T) - y_{N,T/N},$$

the global error of approximating $y(t_o + T)$ using N steps of Euler's Method. We now derive the estimate that shows $|e_{N,h}| \to 0$ as $N \to \infty$.

We begin by defining two local numerical and analytical solutions and their corresponding errors. The local analytical solution (initialized using the numerical solution) $\hat{y}_{n,h}(t)$ is defined as the analytical solution of the ODE whose initial value is the point (t_n, y_n) on the numerical solution,

$$\hat{y}'_{n,h} = f(t, \hat{y}_{n,h}), \quad \hat{y}_{n,h}(t_n) = y_n. \quad (5.90)$$

The corresponding local error is the difference between the result of advancing the numerical solution and the local analytical solution by

5.7. Convergence Analysis: Euler's Method

the same time-step, h,

$$le_{n,h} = y_{n+1} - \hat{y}_{n,h}.$$

These are the local errors that we have estimated using paired methods for step-size control. We can relate the change in global errors to this kind of local error as follows:

$$\begin{aligned} e_{n+1,h} &= y(t_{n+1}) - y_{n+1,h} \\ &= (y(t_{n+1}) - \hat{y}_{n,h}(t_{n+1})) + (\hat{y}_{n,h}(t_{n+1}) - y_{n+1,h}) \quad (5.91) \\ &= y(t_{n+1}) - \hat{y}_{n,h}(t_{n+1}) + le_{n,h}. \end{aligned}$$

The term $y(t_{n+1}) - \hat{y}_{n,h}(t_{n+1})$ is the difference at t_{n+1} of two solutions of the ODE whose values at t_n differ by $y(t_n) - y_n = e_{n,h}$. This behavior of this term depends on the stability of the ODE that is to be solved.

The local numerical solution (initialized using the analytical solution) $\hat{y}_{n+1,h}$ is the result of one step of the numerical method initialized with the analytical solution through $(t_n, y_{n,h})$. The *local truncation error*, $\epsilon_{n,h}$, is the difference between the result of advancing the analytical solution and the local numerical solution by the same time-step, h, $y(t_{n+1}) - \hat{y}_{n+1,h}$. The local truncation error measures by how much the solution of the differential equation fails to satisfy the numerical method, while the local error measures by how much the solution of the numerical method fails to satisfy the differential equation.

We can relate the change in global errors to local truncation error as follows:

$$\begin{aligned} e_{n+1} &= y(t_{n+1}) - y_{n+1,h} \\ &= y(t_{n+1}) - \hat{y}_{n+1,h} + \hat{y}_{n+1,h} - y_{n+1,h} \quad (5.92) \\ &= \epsilon_{n,h} + \hat{y}_{n+1,h} - y_{n+1,h}. \end{aligned}$$

The term $\hat{y}_{n+1,h} - y_{n+1,h}$ is the difference at t_{n+1} of two solutions of the numerical approximation method whose values at t_n differ by $y(t_n) - y_{n,h} = e_{n,h}$. This behavior of this term depends on the stability of the numerical method as applied to the particular ODE using the time-step h.

We will bound the growth of the magnitude of global errors for Euler's Method by estimating the stability factor and bounding the local truncation error. The two types of local solution and local error are depicted in Figure 5.24. We have used the same example as in Figure 5.1, i.e., Euler's Method applied to the IVP $y' = -y$, $y(0) = 1$ with $T = 1$, $N = 3$, $h = 1/3$. A third type of local error, the *residual*, measures by how much a certain *continuous extension* $S_{n,h}(t)$ of the discrete numerical approximation over each time-step interval fails to satisfy the differential equation. The continuous extension for a step of Euler's Method is also shown in the figure and will be defined and analyzed later when we discuss the residual error.

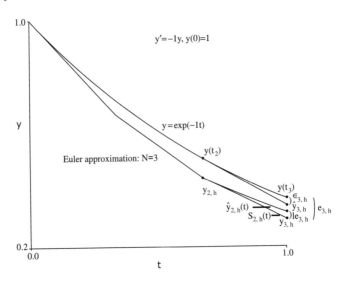

Figure 5.24. Local solutions and truncation errors for error analysis.

We proceed with different estimates of the local truncation error, depending on whether or not additional assumptions on the analytical solution $y(t)$ are satisfied. Henrici [HP] gives a proof with no additional assumptions on f beyond continuity in t and Lipschitz continuity in y. In [GCW], a proof is presented using an additional assumption that $f(t, y)$ is also Lipschitz in its first argument with

5.7. Convergence Analysis: Euler's Method

Lipschitz constant K. In this case, the mean value theorem guarantees

$$|\epsilon_{n,h}| = |y(t_{n+1}) - (y(t_n) - hy'(t_n))|$$
$$\leq Kh^2 + Lh^2 \max_{t \in [t_o, t_o+T]} |y'(t)|. \quad (5.93)$$

Here, for simplicity of exposition, we will assume $y \in C^2[t_o, t_o + T]$. In this situation, Taylor's Theorem with Remainder guarantees that

$$\epsilon_{n,h} = y(t_{n+1}) - (y(t_n) + hy'(t_n)) = \frac{h^2}{2} y''(\xi_n) \quad (5.94)$$

for some $\xi_n \in [t_n, t_{n+1}]$. Since we can bound $\frac{|y''|}{2} \leq M$ on $[t_o, t_o + T]$, we can bound the local truncation error of Euler's Method by

$$|\epsilon_{n,h}| \leq Mh^2. \quad (5.95)$$

Substituting $y'(t_n) = f(t_n, y(t_n))$ in the formula for the change in global error in terms of local truncation error, we collect terms of order h to obtain

$$e_{n+1} = e_n + h(f(t_n, y(t_n)) - f(t_n, y_n)) + \epsilon_{n,h}. \quad (5.96)$$

The Lipschitz condition (5.2) allows us to estimate

$$|f(t_n, y(t_n)) - f(t_n, y_n)| \leq L|y(t_n) - y_n| = L|e_n|$$

and the bound on y'' above shows that

$$|e_{n+1}| \leq (1 + hL)|e_n| + Mh^2. \quad (5.97)$$

For any fixed $h > 0$, (5.97) is a linear constant coefficient inhomogeneous first-order difference inequality satisfied by the sequence $e_n, n = 0, 1, \ldots$. If we let $a = 1 + hL$, $b = Mh^2$, and $u_0 = |e_0|$ and if u_n is defined by the corresponding difference equation,

$$u_{n+1} = au_n + b, \quad n = 0, 1, \ldots, \quad (5.98)$$

then $|e_n| \leq u_n$.

▷ **Exercise 5–10.** Use induction to verify that $|e_n| \leq u_n$, $n = 0, 1, \ldots$.

There are several ways to obtain the closed form solution of (5.98). If we write out the first few terms concretely, u_0, $au_0 + b$, $a^2 u_0 +$

$b(1+a)$, $a^3 u_0 + b(1+a+a^2)$, ..., we can recognize the expressions multiplying b as partial sums of a geometric series and replace it with a familiar formula to get

$$u_n = a^n u_o + \frac{b(1-a^n)}{(1-a)}. \tag{5.99}$$

▷ **Exercise 5–11.** Verify (5.99) in three ways. First use (5.98) and mathematical induction. Second, write (5.99) as the sum $u_n = u_n^H + u_n^P$, where $u_n^H = ka^n$ is a solution of the corresponding homogeneous equation and $u_{n+1}^H = b/(1-a)$ is the constant particular solution of (5.98), Finally, obtain (5.99) by finding a 'summation factor' by which we can multiply the difference equation $u_{n+1} - au_n = b$ to make the left-hand side a perfect difference of another sequence. Then apply the 'discrete fundamental theorem of calculus',

$$f_N - f_0 = \sum_{n=0}^{N-1} (f_{n+1} - f_n),$$

to solve the resulting equation. This is the discrete analogue of finding an integrating factor by which we can multiply the differential equation $y' - ay = b$ to make the left-hand side a perfect derivative. The discrete and continuous factors are solutions of corresponding homogeneous equations.

Now we substitute the definitions of a, b, and u_0 in (5.99). The result is

$$|e_n| \leq u_n = (1+hL)^n |e_o| + \frac{Mh^2(1-(1+hL)^n)}{-hL}.$$

After further simplification,

$$|e_n| \leq (1+hL)^n |e_o| + \frac{Mh((1+hL)^n - 1)}{L}. \tag{5.100}$$

Then to complete our original goal of estimating e_N, we use the fact that for $x \geq -1$,

$$0 \leq 1 + x \leq e^x, \tag{5.101}$$

and by induction,

$$0 \leq (1+x)^n \leq e^{nx}. \tag{5.102}$$

5.7. Convergence Analysis: Euler's Method

Setting $x = hL > 0$ and $Nh = T$ in (5.102), we get

$$1 \leq (1 + hL)^N \leq e^{LT} \qquad (5.103)$$

and finally, (5.100) becomes

$$|e_{N,h}| \leq e^{LT}|e_o| + h\frac{M}{L}(e^{LT} - 1). \qquad (5.104)$$

We refer to (5.104) as the *global error* estimate for Euler's Method.

▷ **Exercise 5–12.** Confirm (5.101) using two arguments. First, use the fact that $f(x) = e^x$ is everywhere concave up and therefore is above its tangent line at $x = 0$. Alternatively, prove (5.101) by showing that the remainder for the first degree Maclaurin expansion of $f(x)$ at $x = 0$ is positive.

Finally, to complete the proof of convergence, we observe that as long as $y_0 \to y_o$ as $N \to +\infty$ and in particular if $y_0 = y_o$, then $e_o \to 0$ as $N \to +\infty$. Furthermore, for any $T \in [0, b]$, we have $h = T/N \to 0^+$, and so using (5.104),

$$\max_{0 \leq T \leq b} |e_{N,h}(T)| \to 0$$

uniformly as $N \to +\infty$. We conclude that the sequence $y_{N,h}$ obtained using Euler's Method (5.11) converges to the analytical solution $y(t)$ of (5.1) for all $t \in [t_o, t_o + b]$.

Three quantities, a, b, and u_o, contribute to the form (5.99) of the global error bound (5.104). We now examine how they might be modified for more rapid convergence. The first-order dependence of the second term arises from the interaction of a and b. If the inhomogeneous forcing term $b = Mh^2$ in (5.97) were modified to become $b = Mh^{p+1}$, the corresponding term of the estimate (5.104) would behave like h^p.

In Appendix H we will present a method based on rooted trees that can be used to determine the conditions that must be satisfied by the $r(r+1)/2$ coefficients of an explicit r-stage Runge-Kutta Method in order for the local truncation error to satisfy

$$|\epsilon_{n,h}| \leq Ch^{p+1}. \qquad (5.105)$$

In Appendix I we will show that a linear multistep method has formal accuracy of order P if and only if its appropriately defined local truncation error also satisfies (5.105). Other equivalent conditions involve the coefficients of the methods and the asymptotic behavior of the characteristic polynomial of the method.

The factor $\frac{a^n-1}{a-1} = \frac{e^{LT}-1}{hL}$ that multiplies Mh^2 and leads to first-order convergence bounds the cumulative amplification of all local truncation errors. We call a the *amplification factor* of the method, and it governs the stability of the method. (We might distinguish a as the local amplification factor and $\frac{a^n-1}{a-1}$ as a global amplification factor, though this is not standard terminology.) The

$$au_n = (1+hL)e_n$$

term in (5.98) arose from estimating

$$y(t_n) - y_n + h(f(t_n, y(t_n)) - f(t_n, y_n)) \qquad (5.106)$$

in (5.96). The amplification factor satisfies $a \geq 1$ and the only situation in which $a = 1$ is when f is independent of y. Amplification of errors is due to the separation of different solutions of the numerical method from each other and does not require or forbid separation of solutions of the differential equation. The phenomenon is present whether different integral curves are spreading apart or approaching each other. The third contribution, the initial error, is reflected in the global error with its order unchanged.

Together, a bound on the initial error, $|e_0|, \ldots, |e_{m-1}| \leq Ch^P$, a 0-stability bound on the amplification factor, and the bound (5.105) on the local truncation error are necessary and sufficient to prove convergence with order of accuracy P. The conditions that determine the accuracy of Runge-Kutta Methods are somewhat involved, but the basic stability required for convergence is essentially built into the definition of these methods. In contrast, the accuracy of linear multistep methods is settled rather simply by the notion of formal accuracy, equivalent to a single condition on the coefficients of the method for each degree of accuracy. It is stability that is more subtle for these methods. A linear multistep method whose formal accuracy is P is convergent and has order of accuracy P if and only if the

5.7. Convergence Analysis: Euler's Method

method is 0-stable, i.e., the characteristic polynomial of the method at $h = 0$, $\rho(r)$, satisfies the root condition (R). A complete proof of this theorem in even greater generality, but based on the same ideas used in our proof of convergence for Euler's Method, is given in [IK].

Depending on the type of numerical method used, the numerical solution y_n can be associated in different ways to a *continuous extension*. This is a function $S_{n,h}(t)$ defined over each interval $[t_n, t_{n+1}]$ that is equal to the numerical solution at the endpoints of the interval. To extend a one-step method such as a Runge-Kutta Method, let $h' \in (0, h)$ and for $t = t_n + h'$, define $S_{n,h}(t)$ using the same method used to obtain y_{n+1} from y_n, only replacing h with h'. Note that the discretization parameter h and the corresponding discrete numerical approximation remain fixed. Only in between these values do we use the distinct continuous independent variable h' in place of h in the formulas (5.3) that define steps of the underlying Runge-Kutta Method. To extend a multistep method for $t = t_n + h'$, define $S_{n,h}(t)$ using the same interpolation on which the underlying multistep method has been derived. The values of $S_{n,h}(t)$ are sometimes known as the *dense output* of a method. Using the continuous extension, we can define the *residual*,

$$r_{n,h}(t) = S'_{n,h}(t) - f(t, S_{n,h}), \qquad (5.107)$$

using a term from backward error analysis. (By this distinction, the local truncation error estimates correspond to forward error analysis.) The residual is another quantity that, like the local error, measures by how much the numerical solution fails to satisfy the differential equation on each interval. Because of the continuous nature of its argument, the residual can measure the difference at the level of derivatives. In the work of Shampine [SLF], as well as that of Enright [EWH] and Higham [HDJ] who refer to the residual as the *defect*, continuous extension and residual estimation bridge the notions of local and global error. They have developed numerical schemes with automatic step-size control based on estimates of the residual using integral norms. Hermann Karcher [KH] has used this approach along with careful applications of Gronwall's Inequality to obtain convergence proofs that are valid in a broader setting due to weaker differentiability requirements on error norms.

In its simplest form, residual analysis can be used to provide a direct derivation of a useful form of the local truncation error estimate, in which the amplification term already appears as a continuous rather than discrete exponential. In analogy with the discrete error estimates (5.96), (5.97), we can write

$$\frac{d}{dt}(y(t) - S_{n,h}(t)) = y'(t) - S'_{n,h}(t) = f(t, y(t)) - S'_{n,h}(t)$$
$$= (f(t, y(t)) - f(t, S_{n,h}(t)))$$
$$+ (f(t, S_{n,h}(t)) - S'_{n,h}(t)), \qquad (5.108)$$

so

$$\frac{d}{dt}|y(t) - S_{n,h}(t)| \qquad (5.109)$$
$$\leq L|y(t) - S_{n,h}(t)| + |f(t, S_{n,h}(t)) - S'_{n,h}(t)|.$$

The first term on the right plays a similar role to the stability or amplification factor in the discrete approach. But in this approach the inhomogeneous term that measures by how much the approximation fails to satisfy the differential equation is the residual rather than the local truncation error. In general, a local truncation estimate of order h^{p+1} corresponds to a residual estimate of order h^p, since the former compares approximate solutions and the latter compares their derivatives. Again we use Euler's Method as an example since it is both a Runge-Kutta Method and a multistep method and the respective continuous extensions coincide. The Runge-Kutta approach extends y_n and y_{n+1} to $S_{n,h}(t)$ by letting $S_{n,h}(t) = y_n + (h')y_n'$, where $y_n' = f(t_n, y_n)$. The multistep approach integrates the constant left endpoint approximation of $y(t)$, $S_{n,h}(t) = y_n + \int_0^{h'} y_n' \, ds$, yielding the same result.

Note that $S_{n,h}(t)$ is continuous with $S'_{n,h}(t_n) = y_n$ for all $n = 0, 1, \ldots, N$. Also $S'_{n,h}(t)$ is piecewise constant:

$$S'_{n,h}(t) = y_n' = f(t_n, y_n) = f(t_n, S_{n,h}(t_n))$$

on (t_n, t_{n+1}). If we define $g(t) = f(t, S_{n,h}(t))$, then $S_{n,h}(t) = g(t_n)$ is the degree zero (constant) polynomial that interpolates $g(t)$ on $[t_n, t_n + h]$. By the standard error of interpolation estimate that in

5.7. Convergence Analysis: Euler's Method

this simplest case reduces to the mean value theorem, the magnitude of the residual satisfies

$$|f(t, S_{n,h}(t)) - S'_{n,h}(t)| = |g(t) - g(t_n)|$$
$$= |g'(\xi)||t - t_n| \leq Ch \quad (5.110)$$

where $C = \max_{[t_n, t_n+h]} |g'(t)|$. If we set $e(t) = |y(t) - S_{n,h}(t)|$, we have $e'(t) \leq Le(t) + Ch$ so Gronwall's Inequality tells us

$$e(t) \leq e(t_n)e^{(t-t_n)L} + C(e^{(t-t_n)L} - (1 + (t-t_n)L))$$
$$\leq e(t_n)e^{(t-t_n)L} + C'((t-t_n)L)^2). \quad (5.111)$$

This may be viewed as a local truncation error estimate for Euler's Method, derived without Taylor expansion, blended with an exponential stability estimate over the interval $[t_n, t_n + h]$. In the discrete approach above, the exponential stability estimate only appeared after iterating the difference inequality.

For Runge-Kutta Methods, residual estimates for the continuous extension correspond directly to local truncation error estimates of one order higher, and the stability analysis carries over just as easily. For multistep methods with more steps, 0-stability and the root condition govern the instantaneous growth rate of perturbations. To obtain higher-order residual estimates, we make use of the degree $p-1$ interpolation of the derivative of the solution, $S'_{t_n,\ldots,t_{n+1-m},h}(t)$. In Appendix I, these very interpolants are even used to derive many families of higher-order multistep methods. This allows us to replace (5.110) with higher-order interpolation error estimates

$$g(t) - S'_{t_n,\ldots,t_{n+1-p}}(t) = \frac{g^{(p)}(\xi)}{p!}(t - t_n)\cdots(t - t_{n+1-p}), \quad (5.112)$$

and for $t \in [t_{n+1-p}, t_n + h]$,

$$|g(t) - S'_{t_n,\ldots,t_{n+1-p}}(t)| \leq \max_{t \in [t_{n+1-p}, t_n+h]} K|g^{(p)}(t)|h^p, \quad (5.113)$$

where K depends only on the number p of intervals of length h involved in the interpolation. The interpolation estimates above are derived in Appendix J. They are also useful in deriving multistep methods in Appendix I.

In either case, once we have $e'(t) \leq Le(t) + Ch^p$ with $e(t)$ defined as above, Gronwall's Inequality tells us

$$e(t) \leq e(t_n)e^{(t-t_n)L} + C(e^{(t-t_n)L} - T_p((t-t_n)L)), \quad (5.114)$$

where $T_p(x)$ is the Taylor polynomial for e^x at $x = 0$, and so

$$e(t) \leq e(t_n)e^{(t-t_n)L} + C'((t-t_n)L)^{p+1}. \quad (5.115)$$

This single interval estimate can then be iterated just as in the discrete case to obtain global error estimates of the same order.

Our discussion of numerical methods for initial value problems has invoked many major topics from numerical analysis, including interpolation, root finding, quadrature, and linear algebra. The reader will find that pursuing further study of numerical methods will be rewarded with many more suprising and beautiful connections among an ever expanding network of mathematical subjects. A brief guide to recommend further reading appears in the Web Companion.

Appendix A

Linear Algebra and Analysis

This short review is not intended as an introduction or tutorial. On the contrary, it is assumed that the reader is already familiar with multi-variable calculus, the basic facts concerning metric spaces, and the elementary theory of finite-dimensional real and complex vector spaces. The goal of this appendix is rather to clarify just what material is assumed, to develop a consistent notation and point of view towards these subjects for use in the rest of the book, and to formulate some of their concepts and propositions in ways that will be convenient for applications elsewhere in the text.

A.1. Metric and Normed Spaces

A metric space is just a set X with a *distance* $\rho(x_1, x_2)$ defined between any two of its points x_1 and x_2. The distance should be a non-negative real number that is zero if and only if $x_1 = x_2$, and it should be symmetric in x_1 and x_2. Aside from these obvious properties of anything we would consider calling a distance function, the only other property we demand of the function ρ (which is also called the *metric* of X) is that the "triangle inequality" hold for any three points x_1, x_2, and x_3 of X. This just means that $\rho(x_1, x_3) \leq \rho(x_1, x_2) + \rho(x_2, x_3)$, and what it says in words is that "things close to the same thing are close to each other".

If $\{x_n\}$ is a sequence of points in X, then we say this sequence converges to a point x in X if $\lim_{n \to \infty} \rho(x_n, x) = 0$. It follows from the triangle inequality that if the sequence also converges to x', then

$\rho(x, x') = 0$, so $x = x'$, i.e., the limit of $\{x_n\}$ is unique if it exists, and we write $\lim_{n \to \infty} x_n = x$. The sequence $\{x_n\}$ is called a *Cauchy sequence* if $\rho(x_n, x_m)$ converges to zero as both m and n tend to infinity. It is easy to check that a convergent sequence is Cauchy, but the reverse need not be true, and if every Cauchy sequence in X does in fact converge, then we call X a *complete* metric space.

If X and Y are metric spaces and $f : X \to Y$ is a function, then we call f continuous if and only if $f(x_n)$ converges to $f(x)$ whenever x_n converges to x. An equivalent definition is that given any positive ϵ and an x in X, there exists a $\delta > 0$ such that if $\rho_X(x, x') < \delta$, then $\rho_Y(f(x), f(x')) < \epsilon$, and if we can choose this δ independent of x, then we call f *uniformly continuous*. A positive constant K is called a Lipschitz constant for f if for all x_1, x_2 in X, $\rho_Y(f(x_1), f(x_2)) < K\rho_X(x_1, x_2)$, and we call f a *contraction* if it has a Lipschitz constant K satisfying $K < 1$. Note that if f has a Lipschitz constant (in particular, if it is a contraction), then f is automatically uniformly continuous (take $\delta = \epsilon/K$).

▷ **Exercise A–1.** Show that if K is a Lipschitz constant for $f : X \to Y$ and L is a Lipschitz constant for $g : Y \to Z$, then KL is a Lipschitz constant for $g \circ f : X \to Z$.

The classic example of a metric space is **R** with $\rho(x, y) = |x - y|$, and this has an important generalization. Namely, if V is any (real or complex) vector space, then a nonnegative real-valued function $v \mapsto \|v\|$ on V is called a *norm* for V if it shares three basic properties of the absolute value of a real number, namely i) positve homogeneity, i.e., $\|\alpha v\| = |\alpha| \|v\|$ for a scalar α; ii) $\|v\| = 0$ only if $v = 0$; and iii) the triangle inequality $\|v_1 + v_2\| \leq \|v_1\| + \|v_2\|$ for all $v_1, v_2 \in V$. A vector space V together with a choice of norm for V is called a normed vector space, and on such a V we get a metric by defining $\rho(v_1, v_2) = \|v_1 - v_2\|$. If this makes V a complete metric space, then the normed space V is called a Banach space. In particular, $\mathbf{R}^n$ and $\mathbf{C}^n$ with their usual norms are Banach spaces.

If A is a subset of a metric space X, then the metric for X restricted to $A \times A$ defines a metric for A, and the resulting metric space is called a subspace of X.

A.2. Inner-Product Spaces

If x is a point of the metric space X and $\epsilon > 0$, then we denote $\{y \in X \mid \rho(x,y) < \epsilon\}$, the so-called "open ball of radius ϵ about x in X", by $B_\epsilon(x, X)$. A subset A of X is a neighborhood of x if it includes $B_\epsilon(x, X)$ for some positive ϵ, and A is called an open subset of X if it is a neighborhood of each of its points. On the other hand A is called a closed subset of X if the limit of any sequence of points in A is itself in A. It is easily proved that A is closed in X if and only if its complement in X is open and that $f : X \to Y$ is continuous if and only if the inverse image, $f^{-1}(O)$, of every open subset O of Y is open in X.

If X is a metric space, then X itself is both open and closed in X, and hence the same is true of its complement, the empty set. If there are no other subsets of X that are both open and closed (or equivalently if X cannot be partitioned into two complementary nonempty open sets), then X is called a *connected* space. We say that a subset A of X is connected if the corresponding subspace is a connected metric space. There is an important characterization of the connected subsets of $\mathbf{R}$, namely $A \subseteq \mathbf{R}$ is connected if and only if, it is an interval, i.e., if and only if, whenever it contains two points, it also contains all points in between.

A subset A of a metric space is called *compact* if every sequence in A has a subsequence that converges to a point of A. The Bolzano-Weierstrass Theorem characterizes the compact subsets of $\mathbf{R}^n$ (and $\mathbf{C}^n$). Namely they are precisely those sets that are both closed and bounded. (A subset A of a metric space is bounded if and only if the distances between points of A are bounded above.)

A.2. Inner-Product Spaces

An *inner product* on a real vector space V is a real-valued function on $V \times V$, $(x, y) \mapsto \langle x, y \rangle$ having the following three properties:

1) Symmetry: $\langle x, y \rangle = \langle y, x \rangle$ for all $x, y \in V$.
2) Positive definiteness: $\langle x, x \rangle \geq 0$, with equality if and only if $x = 0$.
3) Bilinearity: $\langle \alpha x + \beta y, z \rangle = \alpha \langle x, z \rangle + \beta \langle y, z \rangle$, for all $x, y, z \in V$ and all $\alpha, \beta \in \mathbf{R}$.

An *inner-product space* is a pair $(V, \langle\ ,\ \rangle)$ consisting of a real vector space V and a choice of inner product for V, but it is customary to suppress reference to the inner product. The motivating example of an inner-product space is of course $\mathbf{R}^n$ with the usual "dot-product" $\langle x, y \rangle := \sum_{i=1}^{n} x_i y_i$.

In what follows, V will denote an arbitrary inner-product space, and we define $\|x\|$, the norm of an element x of V, by $\|x\| := \sqrt{\langle x, x \rangle}$.

By bilinearity, if $x, y \in V$ and $t \in \mathbf{R}$, then $\|tx + y\|^2$ is a quadratic polynomial function of t, namely,

$$\|tx + y\|^2 = \langle tx + y, tx + y \rangle = \|x\|^2 t^2 + 2 \langle x, y \rangle t + \|y\|^2,$$

and note the important special case

$$\|x + y\|^2 = \|x\|^2 + 2 \langle x, y \rangle + \|y\|^2.$$

Finally, for reasons we shall see a little later, the two vectors x and y are called *orthogonal* if $\langle x, y \rangle = 0$, so in this case we have

A.2.1. Pythagorean Identity. *If x and y are orthogonal vectors in an inner product space, then $\|x + y\|^2 = \|x\|^2 + \|y\|^2$.*

Recall some basic high-school mathematics concerning a quadratic polynomial $P(t) = at^2 + bt + c$. (For simplicity, we assume a is positive.) The *discriminant* of $P(t)$ is the quantity $b^2 - 4ac$, and it distinguishes what kind of roots the polynomial has. In fact, the so-called "quadratic formula" says that the two (possibly complex) roots of $P(t)$ are $(-b \pm \sqrt{b^2 - 4ac})/2a$. So there are three cases:

Case 1: $b^2 - 4ac > 0$. Then $P(t)$ has two real roots. Between these roots $P(t)$ is negative, and outside of the interval between the roots it is positive.

Case 2: $b^2 - 4ac = 0$. Then $P(t)$ has only the single real root $-b/2a$, and elsewhere $P(t) > 0$.

Case 3: $b^2 - 4ac < 0$. Then $P(t)$ has no real roots, and $P(t)$ is positive for all real t.

For the polynomial $\|tx + y\|^2$ we see that $a = \|x\|^2$, $c = \|y\|^2$, and $b = 2 \langle x, y \rangle$, so the discriminant is $4(|\langle x, y \rangle|^2 - \|x\|^2 \|y\|^2)$. Case 1 is ruled out by positive definiteness. In Case 2, we have $|\langle x, y \rangle| = \|x\| \|y\|$,

A.2. Inner-Product Spaces

so if t is the root of the polynomial, then $\|x+ty\| = 0$, so $x = -ty$, and we see that in this case x and y are linearly dependent. Finally, in Case 3, $|\langle x,y \rangle| < \|x\|\|y\|$, and since $x+ty$ is never zero, x and y are linearly independent. This proves one of the most important inequalities in all of mathematics.

A.2.2. Schwartz Inequality. *For all $x,y \in V$, $|\langle x,y\rangle| \leq \|x\|\|y\|$, with equality if and only if x and y are linearly dependent.*

▷ **Exercise A–2.** Use the Schwartz Inequality to deduce the triangle inequality:
$$\|x+y\| \leq \|x\| + \|y\|.$$
(Hint: Square both sides.)

This shows that an inner-product space is a normed space.

• **Example A–1.** Let $C([a,b])$ denote the vector space of continuous real-valued functions on the interval $[a,b]$. For $f,g \in C([a,b])$ define $\langle f,g \rangle = \int_a^b f(x)g(x)\,dx$. It is easy to check that this satisfies our three conditions for an inner product.

In what follows, we assume that V is an inner-product space. If $v \in V$ is a nonzero vector, we define a unit vector e with the same direction as V by $e := v/\|v\|$. This is called *normalizing* v, and if v already has unit length, then we say that v is *normalized*. We say that k vectors $e_1,\ldots,e_k$ in V are *orthonormal* if each e_i is normalized and if the e_i are mutually orthogonal. Note that these conditions can be written succinctly as $\langle e_i, e_j \rangle = \delta_j^i$, where δ_j^i is the so-called Kronecker delta symbol and is defined to be zero if i and j are different and 1 if they are equal.

▷ **Exercise A–3.** Show that if $e_1,\ldots,e_k$ are orthonormal and v is a linear combination of the e_i, say $v = \alpha_1 v_1 + \cdots + \alpha_k v_k$, then the α_i are uniquely determined by the formulas $\alpha_i = \langle v, e_i \rangle$. Deduce from this that orthonormal vectors are automatically linearly independent.

Orthonormal bases are also referred to as *frames* and they play an very important role in all explicit computation in inner-product spaces. Note that if $e_1,\ldots,e_n$ is an orthonormal basis for V, then every element of V is a linear combination of the e_i, so that by the exercise each $v \in V$ has the expansion $v = \sum_{i=1}^n \langle v, e_i \rangle e_i$.

• **Example A–2.** The "standard basis" for $\mathbf{R}^n$ is $\delta^1, \ldots, \delta^n$, where $\delta^i = (\delta_1^i, \ldots, \delta_n^i)$. It is clearly orthonormal.

Let V be an inner product space and W a linear subspace of V. We recall that the *orthogonal complement* of W, denoted by $W^\perp$, is the set of those v in V that are orthogonal to every w in W.

▷ **Exercise A–4.** Show that $W^\perp$ is a linear subspace of V and that $W \cap W^\perp = 0$.

If $v \in V$, we will say that a vector w in W is its orthogonal projection on W if $u = v - w$ is in $W^\perp$.

▷ **Exercise A–5.** Show that there can be at most one such w. (Hint: If w' is another, so $u' = v - u \in W^\perp$, then $u - u' = w' - w$ is in both W and $W^\perp$.)

A.2.3. Remark. Suppose $\omega \in W$. Then since $v - \omega = (v - w) + (w - \omega)$ and $v - w \in W^\perp$ while $(w - \omega) \in W$, it follows from the Pythagorean identity that $\|v - \omega\|^2 = \|v - w\|^2 + \|w - \omega\|^2$. Thus, $\|v - \omega\|$ is strictly greater than $\|v - w\|$ unless $\omega = w$. In other words, **the orthogonal projection of v on w is the unique point of W that has minimum distance from v.**

We call a map $P : V \to W$ the *orthogonal projection of V onto W* if $v - Pv$ is in $W^\perp$ for all $v \in V$. By the previous exercise this mapping is uniquely determined if it exists (and we will see below that it always does exist).

▷ **Exercise A–6.** Show that if $P : V \to W$ is the orthogonal projection onto W, then P is a linear map. Show also that if $v \in W$, then $Pv = v$ and hence $P^2 = P$.

▷ **Exercise A–7.** Show that if $e_1, \ldots, e_n$ is an orthonormal basis for W and if for each $v \in V$ we define $Pv := \sum_{i=1}^n \langle v, e_i \rangle e_i$, then P is orthogonal projection onto W. In particular, orthogonal projection onto W exists for any subspace W of V that has some orthonormal basis. Since, as we now will show, any W has an orthonormal basis, orthogonal projection on a subspace is always defined.

A.2. Inner-Product Spaces

There is a beautiful algorithm, called the **Gram-Schmidt Procedure**, for starting with an arbitrary sequence $w_1, w_2, \ldots, w_k$ of linearly independent vectors in an inner-product space V and manufacturing an orthonormal sequence $e_1, \ldots, e_k$ out of them. Moreover it has the nice property that for all $j \leq k$, the sequence $e_1, \ldots, e_j$ spans the same subspace W_j of V as is spanned by $w_1, \ldots, w_j$.

In case $k = 1$ this is easy. To say that w_1 is linearly independent just means that it is nonzero, and we take e_1 to be its normalization: $e_1 := w_1 / \|w_1\|$. Surprisingly, this trivial special case is the crucial first step in an inductive procedure.

In fact, suppose that we have constructed orthonormal vectors $e_1, \ldots, e_m$ (where $m < k$) and that they span the same subspace W_m that is spanned by $w_1, \ldots, w_m$. How can we take the next step and construct e_{m+1} so that $e_1, \ldots, e_{m+1}$ is orthonormal and spans the same subspace as $w_1, \ldots, w_{m+1}$?

First note that since the $e_1, \ldots, e_m$ are linearly independent and span W_m, they are an orthonormal basis for W_m, and hence we can find the orthogonal projection ω_{m+1} of w_{m+1} onto W_m using the formula $\omega_{m+1} = \sum_{i=1}^{m} \langle w_{m+1}, e_i \rangle e_i$. Recall that this means that $\epsilon_{m+1} = w_{m+1} - \omega_{m+1}$ is orthogonal to W_m, and in particular to $e_1, \ldots, e_m$. Now ϵ_{m+1} **cannot be zero!** Why? Because if it were, then we would have $w_{m+1} = \omega_{m+1} \in W_m$, so w_{m+1} would be a linear combination of $w_1, \ldots, w_m$, contradicting the assumption that $w_1, \ldots, w_k$ are linearly independent. But then we can define e_{m+1} to be the normalization of ϵ_{m+1}, i.e., $e_{m+1} := \epsilon_{m+1} / \|\epsilon_{m+1}\|$, and it follows that e_{m+1} is also orthogonal to $e_1, \ldots, e_m$, so that $e_1, \ldots, e_{m+1}$ is orthonormal. Finally, it is immediate from its definition that e_{m+1} is a linear combination of $e_1, \ldots, e_m$ and w_{m+1} and hence of $w_1, \ldots, w_{m+1}$, completing the induction. Let's write the first few steps in the Gram-Schmidt Process explicitly.

(1) $e_1 := w_1 / \|w_1\|$ % normalize w_1 to get e_1.
(2a) $\omega_2 := \langle w_2, e_1 \rangle e_1$ % get projection ω_2 of w_2 on W_1.
(2b) $\epsilon_2 := w_2 - \omega_2$ % subtract ω_2 from w_2 to get $W_1^\perp$ component ϵ_2 of w_2.

(2c) $e_2 := \epsilon_2 / \|\epsilon_2\|$　　　% and normalize it to get e_2.

(3a) $\omega_3 := \langle w_3, e_1 \rangle e_1$　　　% get projection ω_3 of w_3 on W_2.
　　　　$+ \langle w_3, e_2 \rangle e_2$

(3b) $\epsilon_3 := w_3 - \omega_3$　　　% subtract ω_3 from w_3 to get $W_2^\perp$
　　　　　　　　　　　　component ϵ_3 of w_3.

(3c) $e_3 := \epsilon_3 / \|\epsilon_3\|$　　　% and normalize it to get e_3.

...

If W is a k-dimensional subspace of an n-dimensional inner-product space V, then we can start with a basis for W and extend it to a basis for V. If we now apply the Gram-Schmidt Procedure to this basis, we end up with an orthonormal basis for V with the first k elements in W and with the remaining $n-k$ in $W^\perp$. This tells us several things:

- $W^\perp$ has dimension $n-k$.
- V is the direct sum of W and $W^\perp$. This just means that every element of V can be written uniquely as the sum $w+u$ where $w \in W$ and $u \in W^\perp$.
- $(W^\perp)^\perp = W$.
- If P is the orthogonal projection of V on W and I denotes the identity map of V, then $I-P$ is orthogonal projection of V on $W^\perp$.

Appendix B

The Magic of Iteration

The subject of this appendix is one of our favorites in all of mathematics, and it's not hard to explain why. As you will see, the basic theorem, the Banach Contraction Principle, has a simple and elegant statement and a proof to match. And yet, at the same time, it is extremely powerful, having as easy consequences two of the most important foundations of advanced analysis, the Implicit Function Theorem and the Local Existence and Uniqueness Theorem for systems of ODE.

But there is another aspect that we find very appealing, and that is that the basic technique that goes into the contraction principle, namely iteration of a mapping, leads to remarkably simple and effective algorithms for solving equations. Indeed what the Banach Contraction Principle teaches us is that if we have a good algorithm for evaluating a function $f(x)$, then we can often turn it into an algorithm for inverting f, i.e., for solving $f(x) = y$!

B.1. The Banach Contraction Principle

In what follows we will assume that X is a metric space and that $f : X \to X$ is a continuous mapping of X to itself. Since f maps X to itself, we can compose f with itself any number of times, so we can define $f^0(x) = x$, $f^1(x) = f(x)$, $f^2(x) = f(f(x))$, and inductively $f^{n+1}(x) = f(f^n(x))$. The sequence $f^n(x)$ is called the sequence of iterates of x under f, or the orbit of x under f. By associativity of composition, $f^n(f^m(x)) = f^{n+m}(x)$, and by Exercise A-1 of Appendix A, if K is a Lipschitz constant for f, then K^n is a Lipschitz constant for f^n. We shall use both of these facts below without further mention.

A point x of X is called a fixed point of f if $f(x) = x$. Notice that finding a fixed point amounts to solving a special kind of equation. What may not be obvious is that solving many other types of equations can often be reduced to solving a fixed-point equation. We will give other examples later, but here is a typical reduction. Assume that V is a vector space and that we want to solve the equation $g(x) = y$ for some (usually nonlinear) map $g : V \to V$. Define a new map $f : V \to V$ by $f(x) = x - g(x) + y$. Then clearly x is a fixed point of f if and only if it solves $g(x) = y$. This is in fact the trick used to reduce the Inverse Function Theorem to the Banach Contraction Principle.

The Banach Contraction Principle is a very general technique for finding fixed points. First notice the following: if x is a point of X such that the sequence $f^n(x)$ of iterates of x converges to some point p, then p is a fixed point of f. In fact, by the continuity of f, $f(p) = f(\lim_{n \to \infty} f^n(x)) = \lim_{n \to \infty} f(f^n(x)) = \lim_{n \to \infty} f^{n+1}(x) = p$. We will see that if f is a contraction, then for any point x of X the sequence of iterates of x is in any case a Cauchy sequence, so if X is complete, then it converges to a fixed point p of f. In fact, we will see that a contraction can have at most one fixed point p, and so to locate this p when X is complete, we can start at any point x and "follow the iterates of x to their limit". This in essence is the Banach Contraction Principle. Here are the details.

B.1.1. Fundamental Contraction Inequality. If $f : X \to X$ is a contraction mapping and if $K < 1$ is a Lipschitz constant for f, then for all x_1 and x_2 in X,

$$\rho(x_1, x_2) \leq \frac{1}{1 - K}(\rho(x_1, f(x_1)) + \rho(x_2, f(x_2))).$$

Proof. The triangle inequality,

$$\rho(x_1, x_2) \leq \rho(x_1, f(x_1)) + \rho(f(x_1), f(x_2)) + \rho(f(x_2), x_2),$$

B.1. The Banach Contraction Principle

together with $\rho(f(x_1), f(x_2)) \leq K\rho(x_1, x_2)$ gives

$$\rho(x_1, x_2) - K\rho(x_1, x_2) \leq \rho(x_1, f(x_1)) + \rho(f(x_2), x_2).$$

Since $1 - K > 0$, the desired inequality follows. ∎

This is a very strange inequality: it says that we can estimate how far apart any two points x_1 and x_2 are just from knowing how far x_1 is from its image $f(x_1)$ and how far x_2 is from its image $f(x_2)$. As a first application we have

B.1.2. Corollary. *A contraction can have at most one fixed point.*

Proof. If x_1 and x_2 are both fixed points, then $\rho(x_1, f(x_1))$ and $\rho(x_2, f(x_2))$ are zero, so by the Fundamental Inequality $\rho(x_1, x_2)$ is also zero. ∎

B.1.3. Proposition. *If $f : X \to X$ is a contraction mapping, then, for any x in X, the sequence $f^n(x)$ of iterates of x under f is a Cauchy sequence.*

Proof. Taking $x_1 = f^n(x)$ and $x_2 = f^m(x)$ in the Fundamental Inequality gives

$$\rho(f^n(x), f^m(x)) \leq \frac{1}{1-K}(\rho(f^n(x), f^n(f(x))) + \rho(f^m(x), f^m(f(x)))).$$

Since K^n is a Lipschitz constant for f^n,

$$\rho(f^n(x), f^m(x)) \leq \frac{K^n + K^m}{1-K}\rho(x, f(x)),$$

and since $0 \leq K < 1$, $K^n \to 0$, so $\rho(f^n(x), f^m(x)) \to 0$ as n and m tend to infinity. ∎

B.1.4. Banach Contraction Principle. *If X is a complete metric space and $f : X \to X$ is a contraction mapping, then f has a unique fixed point p, and for any x in X the sequence $f^n(x)$ converges to p.*

Proof. The proof is immediate from the above. ∎

▷ **Exercise B–1.** Use the mean value theorem of differential calculus to show that if $X = [a, b]$ is a closed interval and $f : X \to R$ is a continuously differentiable real-valued function on X, then the maximum value of $|f'|$ is the smallest possible Lipschitz constant for f. In particular $\sin(1)$ (which is less than 1) is a Lipschitz constant for the cosine function on the interval $X = [-1, 1]$. Note that for any x in R the iterates of x under cosine are all in X. Deduce that no matter where you start, the successive iterates of cosine will always converge to the same limit. Put your calculator in radian mode, enter a random real number, and keep hitting the cos button. What do the iterates converge to?

As the above exercise suggests, if we can reinterpret the solution of an equation as the fixed point of a contraction mapping, then it is an easy matter to write an algorithm to find it. Well, almost— something important is still missing, namely, when should we stop iterating and take the current value as the "answer"? One possibility is to just keep iterating until the distance between two successive iterates is smaller than some predetermined "tolerance" (perhaps the machine precision). But this seems a little unsatisfactory, and there is actually a much neater "stopping rule".

Suppose we are willing to accept an "error" of ϵ in our solution; i.e., instead of the actual fixed point p of f we will be happy with any point p' of X satisfying $\rho(p, p') < \epsilon$. Suppose also that we start our iteration at some point x in X. It turns out that it is easy to specify an integer N so that $p' = f^N(x)$ will be a satisfactory answer. The key, not surprisingly, lies in the Fundamental Inequality, which we apply now with $x_1 = f^N(x)$ and $x_2 = p$. It tells us that $\rho(f^N(x), p) \leq \frac{1}{1-K} \rho(f^N(x), f^N(f(x))) \leq \frac{K^N}{1-K} \rho(x, f(x))$. Since we want $\rho(f^N(x), p) \leq \epsilon$, we just have to pick N so large that $\frac{K^N}{1-K} \rho(x, f(x)) < \epsilon$. Now the quantity $d = \rho(x, f(x))$ is something that we can compute after the first iteration and we can then compute how large N has to be by taking the log of the above inequality and solving for N (remembering that $\log(K)$ is negative). We can express our result as

B.1. The Banach Contraction Principle

B.1.5. Stopping Rule. If $d = \rho(x, f(x))$ and
$$N > \frac{\log(\epsilon) + \log(1-K) - \log(d)}{\log(K)},$$
then $\rho(f^N(x), p) < \epsilon$.

From a practical programming point of view, this allows us to express our iterative algorithm with a "for loop" rather than a "while loop", but this inequality has another interesting interpretation. Suppose we take $\epsilon = 10^{-m}$ in our stopping rule inequality. What we see is that the growth of N with m is a constant plus $m/|\log(K)|$, or in other words, to get one more decimal digit of precision we have to do (roughly) $1/|\log(K)|$ more iteration steps. Stated a little differently, if we need N iterative steps to get m decimal digits of precision, then we need another N to double the precision to $2m$ digits.

We say a numerical algorithm has linear convergence if it exhibits this kind of error behavior, and if you did the exercise above for locating the fixed point of the cosine function, you would have noticed it was indeed linear. Linear convergence is usually considered somewhat unsatisfactory. A *much* better kind of convergence is quadratic, which means that each iteration should (roughly) double the number of correct decimal digits. Notice that the actual linear rate of convergence predicted by the above stopping rule is $1/|\log(K)|$. So one obvious trick to get better convergence is to see to it that the best Lipschitz constant for our iterating function f in a neighborhood of the fixed point p actually approaches zero as the diameter of the neighborhood goes to zero. If this happens at a fast enough rate, we may even achieve quadratic convergence, and that is what actually occurs in "Newton's Method", which we study next.

▷ **Exercise B–2.** Newton's Method for finding $\sqrt{2}$ gives the iteration $x_{n+1} = x_n/2 + 1/x_n$. Start with $x_0 = 1$, and carry out a few steps to see the impressive effects of quadratic convergence.

B.1.6. Remark. Suppose V and W are orthogonal vector spaces, U is a convex open set in V, and $f : U \to W$ is a continuously differentiable map. Let's try to generalize the exercise above to find a

Lipschitz constant for f. If p is in U, then recall that Df_p, the differential of f at p, is a linear map of V to W defined by $Df_p(v) = (d/dt)_{t=0} f(p+tv)$, and it then follows that if $\sigma(t)$ is any smooth path in U, then $d/dt f(\sigma(t)) = Df_{\sigma(t)}(\sigma'(t))$. If p and q are any two points of U and if $\sigma(t) = p+t(q-p)$ is the line joining them, then integrating the latter derivative from 0 to 1 gives the so-called "finite difference formula": $f(q) - f(p) = \int_0^1 Df_{\sigma(t)}(q-p)\,dt$. Now recall that if T is any linear map of V to W, then its norm $\|T\|$ is the smallest nonnegative real number r so that $\|Tv\| \leq r\|v\|$ for all v in V. Since $\left\|\int_a^b g(t)\,dt\right\| \leq \int_a^b \|g(t)\|\,dt$, $\|f(q) - f(p)\| \leq (\int_0^1 \|Df_{\sigma(t)}\|\,dt)\|(q-p)\|$, and it follows that the supremum of $\|Df_p\|$ for p in U is a Lipschitz constant for f. (In fact, it is the smallest one.)

B.2. Newton's Method

The algorithm called "Newton's Method" has proved to be an extremely valuable tool with countless interesting generalizations, but the first time one sees the basic idea explained, it seems so utterly obvious that it is hard to be very impressed.

Suppose $g : R \to R$ is a continuously differentiable real-valued function of a real variable and x_0 is an "approximate root" of g, in the sense that there is an actual root p of g close to x_0. Newton's Method says that to get an even better approximation x_1 to p, we should take the point where the tangent line to the graph of g at x_0 meets the x-axis, namely $x_1 = x_0 - g(x_0)/g'(x_0)$. Recursively, we can then define $x_{n+1} = x_n - g(x_n)/g'(x_n)$ and get the root p as the limit of the resulting sequence $\{x_n\}$.

Typically one illustrates this with some function like $g(x) = x^2 - 2$ and $x_0 = 1$ (see the exercise above). But the simple picture in this case hides vast difficulties that could arise in other situations. The $g'(x_0)$ in the denominator is a tip-off that things are not going to be simple. Even if $g'(x_0)$ is different from zero, g' could still vanish several times (even infinitely often) between x_0 and p. In fact, determining the exact conditions under which Newton's Method "works" is a subject in itself, and generalizations of this problem constitute an interesting and lively branch of discrete dynamical systems theory.

B.2. Newton's Method

We will not go into any of these interesting but complicated questions, but rather content ourselves with showing that under certain simple circumstances we can derive the correctness of Newton's Method from the Banach Contraction Principle.

It is obvious that the right function f to use in order to make the Contraction Principle iteration reduce to Newton's Method is $f(x) = x - g(x)/g'(x)$ and that a fixed point of this f is indeed a root of g. On the other hand it is clear that this cannot work if $g'(p) = 0$, so we will assume that p is a "simple root" of g, i.e., that $g'(p) \neq 0$. Given $\delta > 0$, let $N_\delta(p) = \{x \in R \mid |x - p| \leq \delta\}$. We will show that if g is C^2 and δ is sufficiently small, then f maps $X = N_\delta(p)$ into itself and is a contraction on X. Of course we choose δ so small that g' does not vanish on X, so f is well-defined on X. It will suffice to show that f has a Lipschitz constant $K < 1$ on X, for then if $x \in X$, then

$$|f(x) - p| = |f(x) - f(p)| \leq K|x - p| < \delta,$$

so $f(x)$ is also in X.

But, by one of the exercises, to prove that K is a Lipschitz bound for f in X, we only have to show that $|f'(x)| \leq K$ in X. Now an easy calculation shows that $f'(x) = g(x)g''(x)/g'(x)^2$. Since $g(p) = 0$, it follows that $f'(p) = 0$ so, by the evident continuity of f', given any $K > 0$, $|f'(x)| \leq K$ in X if δ is sufficiently small.

The fact that the best Lipschitz bound goes to zero as we approach the fixed point is a clue that we should have better than linear convergence with Newton's Method, but quadratic convergence is not quite a consequence. Here is the proof of that.

Let C denote the maximum of $|f''(x)|$ for x in X. Since $f(p) = p$ and $f'(p) = 0$, Taylor's Theorem with Remainder gives $|f(x) - p| \leq C|x - p|^2$. This just says that the error after $n + 1$ iterations is essentially the square of the error after n iterations.

Generalizing Newton's Method to find zeros of a C^2 map $G : \mathbf{R}^n \to \mathbf{R}^n$ is relatively straightforward. Let $x_0 \in \mathbf{R}^n$ be an approximate zero of G, again in the sense that there is a p close to x with $G(p) = 0$. Let's assume now that DG_p, the differential of G at p, is nonsingular and hence that DG_x is nonsingular for x near

p. The natural analogue of Newton's Method is to define $x_{n+1} = x_n - DG_{x_n}^{-1}(G(x_n))$, or in other words to consider the sequence of iterates of the map $F : N_\delta(p) \to \mathbf{R}^n$ given by $F(x) = x - DG_x^{-1}(G(x))$. Again it is clear that a fixed point of F is a zero of G, and an argument analogous to the one-dimensional case shows that for δ sufficiently small $F : N_\delta(p) \to N_\delta(p)$ is a contraction.

B.3. The Inverse Function Theorem

Let V and W be orthogonal vector spaces and $g : V \to W$ a C^k map, $k > 0$. Suppose that for some v_0 in V the differential Dg_{v_0} of g at v_0 is a linear isomorphism of V with W. Then the Inverse Function Theorem says that g maps a neighborhood of v_0 in V one-to-one onto a neighborhood U of $g(v_0)$ in W and that the inverse map from U into V is also C^k.

It is easy to reduce to the case that v_0 and $g(v_0)$ are the respective origins of V and W, by replacing g by $v \mapsto g(v + v_0) - g(v_0)$. We can then further reduce to the case that $W = V$ and Dg_0 is the identity mapping I of V by replacing this new g by $(Dg_0)^{-1} \circ g$.

Given y in V, define $f = f_y : V \to V$ by $f(v) = v - g(v) + y$. Note that a solution of the equation $g(x) = y$ is the same thing as a fixed point of f. We will show that if δ is sufficiently small, then f restricted to
$$X = N_\delta = \{v \in V | \ \|v\| \leq \delta\}$$
is a contraction mapping of N_δ to itself provided $\|y\| < \delta/2$. By the Banach Contraction Principle it then follows that g maps N_δ one-to-one into V and that the image covers the neighborhood of the origin $U = \{v \in V | \ \|v\| < \delta/2\}$. This proves the Inverse Function Theorem except for the fact that the inverse mapping of U into V is C^k, which we will not prove.

The first thing to notice is that since $Dg_0 = I$, $Df_0 = 0$ and hence, by the continuity of Df, $\|Df_v\| < 1/2$ for v in N_δ provided δ is sufficiently small. Since N_δ is convex, by a remark above, this proves that $1/2$ is a Lipschitz bound for f in N_δ and in particular that f restricted to N_δ is a contraction. Thus it only remains to show that f maps N_δ into itself provided $\|y\| < \delta/2$. That is, we must

B.4. The Existence and Uniqueness Theorem for ODE

show that if $\|x\| \leq \delta$, then also $\|f(x)\| \leq \delta$. But since $f(0) = y$,

$$\|f(x)\| \leq \|f(x) - f(0)\| + \|f(0)\|$$
$$\leq \frac{1}{2}\|x\| + \|y\|$$
$$\leq \delta/2 + \delta/2 \leq \delta. \quad \blacksquare$$

▷ **Exercise B–3.** The first (and main) step in proving that the inverse function $h : U \to V$ is C^k is to prove that h is Lipschitz. That is, we want to find a $K > 0$ so that given y_1 and y_2 with $\|y_i\| < \delta/2$ and x_1 and x_2 with $\|x_i\| < \delta$, if $h(y_i) = x_i$, then $\|x_1 - x_2\| \leq K \|y_1 - y_2\|$. Prove this with $K = 2$, using the facts that $h(y_i) = x_i$ is equivalent to $f_{y_i}(x_i) = x_i$ and $1/2$ is a Lipschitz constant for $h = I - g$.

B.4. The Existence and Uniqueness Theorem for ODE

Let $V : \mathbf{R}^n \times \mathbf{R} \to \mathbf{R}^n$ be a C^1 time-dependent vector field on $\mathbf{R}^n$. In the following $I = [a, b]$ will be a closed interval that contains t_0 and we will denote by $C(I, \mathbf{R}^n)$ the vector space of continuous maps $\sigma : I \to \mathbf{R}^n$ and define a distance function on $C(I, \mathbf{R}^n)$ by

$$\rho(\sigma_1, \sigma_2) = \max_{t \in I} \|\sigma_1(t) - \sigma_2(t)\|.$$

It is not hard to show that $C(I, \mathbf{R}^n)$ is a complete metric space. In fact, this just amounts to the theorem that a uniform limit of continuous functions is continuous.

Define for each v_0 in $\mathbf{R}^n$ a map $F = F^{V,v_0} : C(I, \mathbf{R}^n) \to C(I, \mathbf{R}^n)$ by $F(\sigma)(t) := v_0 + \int_{t_0}^{t} V(\sigma(s), s) \, ds$. The Fundamental Theorem of Calculus gives $\frac{d}{dt}(F(\sigma)(t)) = V(\sigma(t), t)$, and clearly $F(\sigma)(t_0) = v_0$. It follows that if σ is a fixed point of F, then it is a solution of the ODE $\sigma'(t) = V(\sigma(t), t)$ with initial condition v_0, and the converse is equally obvious. Thus it is natural to try to find a solution of this differential equation with initial condition v_0 by starting with the constant path $\sigma_0(t) \equiv v_0$ and applying successive approximations using the function F. We will now see that this idea works and leads to the following result, called the Local Existence and Uniqueness Theorem for C^1 ODE.

B.4.1. Theorem. Let $V : \mathbf{R}^n \times \mathbf{R} \to \mathbf{R}^n$ be a C^1 time-dependent vector field on $\mathbf{R}^n$, $p \in \mathbf{R}^n$, and $t_0 \in \mathbf{R}$. There are positive constants ϵ and δ depending on V, p, and t_0 such that if $I = [t_0 - \delta, t_0 + \delta]$, then for each $v_0 \in V$ with $\|v_0 - p\| < \epsilon$ the differential equation $\sigma'(t) = V(\sigma(t), t)$ has a unique solution $\sigma : I \to \mathbf{R}^n$ with $\sigma(t_0) = v_0$.

Proof. If $\epsilon > 0$, then, using the technique explained earlier, we can find a Lipschitz constant M for V restricted to the set of $(x,t) \in \mathbf{R}^n \times \mathbf{R}$ such that $\|x - p\| \leq 2\epsilon$ and $|t - t_0| \leq \epsilon$. Let B be the maximum value of $F(x,t)$ on this same set, and choose $\delta > 0$ so that $K = M\delta < 1$ and $B\delta < \epsilon$, and define X to be the set of σ in $C(I, V)$ such that $\|\sigma(t) - p\| \leq 2\epsilon$ for all $|t| \leq \delta$. It is easy to see that X is closed in $C(I, V)$ and hence a complete metric space. The theorem will follow from the Banach Contraction Principle if we can show that for $\|v_0\| < \epsilon$, F^{V,v_0} maps X into itself and has K as a Lipschitz bound.

If $\sigma \in X$, then $\|F(\sigma)(t) - p\| \leq \|v_0 - p\| + \int_0^t \|V(\sigma(s), s)\| \, ds \leq \epsilon + \delta B \leq 2\epsilon$, so F maps X to itself. And if $\sigma_1, \sigma_2 \in X$, then $\|V(\sigma_1(t), t) - V(\sigma_2(t), t)\| \leq M \|\sigma_1(t) - \sigma_2(t)\|$, so

$$\|F(\sigma_1)(t) - F(\sigma_2)(t)\| \leq \int_0^t \|V(\sigma_1(s), s) - V(\sigma_2(s), s)\| \, ds$$
$$\leq \int_0^t M \|\sigma_1(s) - \sigma_2(s)\| \, ds$$
$$\leq \int_0^t M \rho(\sigma_1, \sigma_2) \, ds$$
$$\leq \delta M \rho(\sigma_1, \sigma_2) \leq K \rho(\sigma_1, \sigma_2)$$

and it follows that $\rho(F(\sigma_1), F(\sigma_2)) \leq K \rho(\sigma_1, \sigma_2)$. ∎

Appendix C

Vector Fields as Differential Operators

Let $V = (p, v)$ be a point of $\mathbf{R}^n \times \mathbf{R}^n$. We are going to regard such a pair asymmetrically as a "vector v based at the point p", and as such we will refer to it as a tangent vector at p. If $\sigma : I \to \mathbf{R}^n$ is a C^1 curve, then for each t_0 in I, we get such a pair, $(\sigma(t_0), \sigma'(t_0))$, which we will denote by $\dot\sigma(t_0)$ and call the *tangent vector to σ at time t_0*. Let $C^\infty(\mathbf{R}^n)$ denote the algebra of smooth real-valued functions on $\mathbf{R}^n$. If $f \in C^\infty(\mathbf{R}^n)$, then the directional derivative of f at $p = \sigma(t_0)$ in the direction $v = \sigma'(t_0)$ is by definition $\left(\frac{d}{dt}\right)_{t=t_0} f(\sigma(t))$, which by the chain rule is equal to $\sum_{i=1}^n v_i \frac{\partial f}{\partial x_i}(p)$. An important consequence of the latter formula is that the directional derivative depends only on $\dot\sigma(t_0) = (p, v)$ and not on the choice of curve σ. (So we can for example take σ to be the straight line $\sigma(t) = p + tv$.)

This justifies using Vf to denote the directional derivative and regarding V as a (clearly linear) map $V : C^\infty(\mathbf{R}^n) \to \mathbf{R}$. Moreover, since $Vx_i = v_i$, this map determines V, and it has become customary to identify the tangent vector V with this linear map and denote V alternatively by $\sum_{i=1}^n v_i \left(\frac{\partial}{\partial x_i}\right)_p$. In particular, taking $v_i = 1$ for $i = k$ and $v_i = 0$ for $i \neq k$ gives the tangent vector at p in the direction of the x_k coordinate curve, which we denote by $\left(\frac{\partial}{\partial x_k}\right)_p$.

It is an immediate consequence of the product rule of differentiation that the mapping V satisfies the so-called Liebniz Identity:

$$V(fg) = (Vf)g(p) + f(p)(Vg).$$

C. Vector Fields as Differential Operators

Any linear map $L : C^\infty(\mathbf{R}^n) \to \mathbf{R}$ that satisfies this Leibniz Identity is called a *derivation at p*. Note that such an L vanishes on a product fg if both f and g vanish at p (and hence also on any linear combination of such products).

▷ **Exercise C–1.** Show that if L is a derivation at p, then $Lf = 0$ for any constant function. (Hint: It is enough to prove this for $f \equiv 1$ [why?], but then $f^2 = f$.)

▷ **Exercise C–2.** Show that if L is a derivation at p, then it is the directional derivative operator defined by some tangent vector at p. (Hint: Use Taylor's Theorem with Integral Remainder to write any $f \in C^\infty(\mathbf{R}^n)$ as

$$f = f(p) + \sum_{i=1}^n \frac{\partial f}{\partial x_i}(p)(x_i - p_i) + R,$$

where R is a linear combination of products of functions vanishing at p.)

Now let O be open in $\mathbf{R}^n$. A vector field in O is a function that assigns to each p in O a tangent vector at p, $(p, V(p))$. Usually one simplifies the notation by dropping the redundant first component, p, and identifies the vector field with the mapping $V : O \to \mathbf{R}^n$. If $f : O \to \mathbf{R}$ is a smooth function on O, then $Vf : O \to \mathbf{R}$ is the function whose value at p is $V(p)f$, the directional derivative of f at p in the direction $V(p)$. If both V and f are C^∞, then clearly so is Vf, so that we may regard V as a linear operator on the vector space $C^\infty(O)$ of smooth real-valued functions on O.

▷ **Exercise C–3.** Suppose that $V : O \to \mathbf{R}^n$ is a C^∞ vector field in O. Show that $V : C^\infty(O) \to C^\infty(O)$ is a derivation of the algebra $C^\infty(O)$, i.e., a linear map satisfying $V(fg) = (Vf)g + f(Vg)$, and show also that every derivation of $C^\infty(O)$ arises in this way.

A vector field V is often identified with (and denoted by) the differential operator $\sum_{i=1}^n V_i \frac{\partial}{\partial x_i}$.

There is an important special vector field R in $\mathbf{R}^n$ called the *radial vector field*, or the Euler vector field. As a mapping $R : \mathbf{R}^n \to \mathbf{R}^n$,

C. Vector Fields as Differential Operators

it is just the identity map, while as a differential operator it is given by $R := \sum_{i=1}^{n} x_i \frac{\partial}{\partial x_i}$. Recall that a function $f : \mathbf{R}^k \to \mathbf{R}$ is said to be positively homogeneous of degree k if $f(tx) = t^k f(x)$ for all $t > 0$ and $x \neq 0$.

▷ **Exercise C–4.** Prove Euler's Formula $\sum_{i=1}^{n} x_i \frac{\partial f}{\partial x_i} = kf$ for a C^1 function $f : \mathbf{R}^n \to \mathbf{R}$ that is positively homogeneous of degree k.

Appendix D

Coordinate Systems and Canonical Forms

D.1. Local Coordinates

Let O be an open set in $\mathbf{R}^n$. We say that an n-tuple of smooth real-valued functions defined in O, $(\phi_1, \ldots, \phi_n)$, forms a *local coordinate system* for O if the map $\phi : p \mapsto (\phi_1(p), \ldots, \phi_n(p))$ is a *diffeomorphism*, that is, if it is a one-to-one map of O onto some other open set U of $\mathbf{R}^n$ and if the inverse map $\psi := \phi^{-1} : U \to \mathbf{R}^n$ is also smooth. The relation of ψ to ϕ is clearly completely symmetrical, and in particular ψ defines a coordinate system $(\psi_1, \ldots, \psi_n)$ in U.

D.1.1. Remark. By the Inverse Function Theorem, the necessary and sufficient condition for ϕ to have a smooth inverse in some neighborhood of a point p is that the $D\phi_p$ is an invertible linear map, or equivalently that the differentials $(d\phi_i)_p$ are linearly independent and hence a basis for $(\mathbf{R}^n)^*$. In other words, given n smooth real-valued functions $(\phi_1, \ldots, \phi_n)$ defined near p and having linearly independent differentials at p, they always form a coordinate system in some neighborhood O of p.

The most obvious coordinates are the "standard coordinates", $\phi_i(p) = p_i$, with O all of $\mathbf{R}^n$ (so ϕ is just the identity map). If $e_1, \ldots, e_n$ is the standard basis for $\mathbf{R}^n$, then $\phi_1, \ldots, \phi_n$ is just the dual basis for $(\mathbf{R}^n)^*$. We will usually denote these standard coordinates by $(x^1, \ldots, x^n)$. More generally, given any basis $f_1, \ldots, f_n$ for $\mathbf{R}^n$, we can let $(\phi_1, \ldots, \phi_n)$ be the corresponding dual basis. Such coordinates are called *Cartesian*. In this case, ϕ is the linear isomorphism of $\mathbf{R}^n$ that maps e_i to f_i. If the f_i are orthonormal, then these are called orthogonal Cartesian coordinates and ϕ is an orthogonal transformation.

Why not just always stick with the standard coordinates? One reason is that once we understand a concept in $\mathbf{R}^n$ in terms of arbitrary coordinates, it is easy to make sense of that concept on a general "differentiable manifold". But there is another important reason. Namely, it is often possible to simplify the analysis of a problem considerably by choosing a well-adapted coordinate system. In more detail, various kinds of geometric and analytic objects have numerical descriptions in terms of a coordinate system. This observation by Descartes is of course the basis of the powerful "analytic geometry" approach to studying geometric questions. Now, the precise numerical description of an object is usually highly dependent on the choice of coordinate system, and it can be more or less complicated depending on that choice. Frequently, there will be certain special "adapted" coordinates with respect to which the numerical description of the object has a particularly simple so-called "canonical form", and facts that are difficult to deduce from the description with respect to general coordinates can be obvious from the canonical form.

Here is a well-known simple example. An ellipse in the plane, $\mathbf{R}^2$, is given by an implicit equation of the form $ax^2 + by^2 + cxy + dx + ey + f = 0$, but if we choose the diffeomorphism that translates the origin to the center of the ellipse and rotates the coordinate axes to be the axes of the ellipse, then in the resulting coordinates ξ, η the implicit equation for the ellipse will have the simpler form $\alpha^2 \xi^2 + \beta^2 \eta^2 = 1$. Notice that this diffeomorphism is actually a Euclidean motion, so ξ and η are orthogonal Cartesian coordinates and even the metric properties of the ellipse are preserved by this change of coordinates. If that is not important in some context, we could instead use $u := \alpha\xi$ and $v := \beta\eta$ as our coordinates and work with the even simpler equation $u^2 + v^2 = 1$.

This example illustrates the general idea behind choosing coordinates adapted to a particular object Ω in some open set O. Namely,

D.1. Local Coordinates

you should think intuitively of finding a diffeomorphism ϕ that moves, bends, and twists Ω into an object Ω^ϕ in $U = \phi(O)$, one that is in a "canonical configuration" having a particularly simple description with respect to standard coordinates. While one can apply this technique to all kinds of geometric and analytic objects, here we will concentrate on three of the objects of greatest interest to us, namely real-valued functions, $f : O \to \mathbf{R}$; smooth curves, $\sigma : I \to O$; and vector fields V defined in O.

First let us consider how to define f^ϕ, σ^ϕ, and V^ϕ. For functions and curves the definition is almost obvious; namely $f^\phi : U \to \mathbf{R}$ is defined by $f^\phi := f \circ \phi^{-1}$ (so that $f^\phi(\phi_1(x), \ldots \phi_n(x)) = f(x_1, \ldots, x_n)$ for all $x \in O$) and $\sigma^\phi : I \to U$ is defined by $\sigma^\phi := \phi \circ \sigma$.

Defining the vector field V^ϕ in U from the vector field V in O is slightly more tricky. At p in O, V defines a tangent vector, $(p, V(p))$, that the differential of ϕ at p maps to a tangent vector at $q = \phi(p)$, $V^\phi(q) := (q, D\phi_p(V(p)))$. Written explicitly in terms of q (and dropping the first component) gives the somewhat ugly formula $V^\phi(q) := D\phi_{\phi^{-1}(q)} V(\phi^{-1}(q))$, but note that if the diffeomorphism ϕ is linear (i.e., if the coordinates $(\phi_1, \ldots, \phi_n)$ are Cartesian), then $D\phi_p = \phi$, so the formula simplifies to $V^\phi = \phi V \phi^{-1}$.

▷ **Exercise D–1.** Recall that the radial (or Euler) vector field R on $\mathbf{R}^n$ is defined by $R(x) = x$, or equivalently, written as a differential operator using standard coordinates, $R = \sum_{i=1}^n x^i \frac{\partial}{\partial x^i}$. Show that if ϕ is any linear diffeomorphism of $\mathbf{R}^n$, then $R^\phi = R$, or equivalently, if $(y_1, \ldots, y_n)$ is any Cartesian coordinate system, then $R = \sum_{i=1}^n y_i \frac{\partial}{\partial y_i}$. That is, the radial vector field has the remakable property that it "looks the same" in all Cartesian coordinate systems. Show that any linear vector field L on $\mathbf{R}^n$ with this property must be a constant multiple of the radial field. (Hint: The only linear transformations that commute with all linear isomorphisms of $\mathbf{R}^n$ are constant multiples of the identity.)

▷ **Exercise D–2.** Let f, σ, and V be as above.

a) Show that if σ is a solution curve of V, that is, if $\sigma'(t) = V(\sigma(t))$, then σ^ϕ is a solution of V^ϕ.

b) Show that if f is a constant of the motion for V (i.e., $Vf \equiv 0$), then f^ϕ is a constant of the motion for V^ϕ.

A common reason for using a particular coordinate system is that these coordinates reflect the symmetry properties of a geometrical problem under consideration. While Cartesian coordinates are good for problems with translational symmetry, they are not well adapted to problems with rotational symmetry, and the analysis of such problems can often be simplified by using some sort of polar coordinates. In $\mathbf{R}^2$ we have the standard polar coordinates $\phi(x, y) = (r(x, y), \theta(x, y))$ defined in $O = $ the complement of the negative x-axis by $r := \sqrt{x^2 + y^2}$ and $\theta := $ the branch of $\arctan(\frac{y}{x})$ taking values in $-\pi$ to π. In this case U is the infinite rectangle $(0, \infty) \times (-\pi, \pi)$ and the inverse diffeomorphism is given by $\psi(r, \theta) = (r\cos\theta, r\sin\theta)$. Similarly, in $\mathbf{R}^3$ we can use polar cylindrical coordinates r, θ, z to deal with problems that are symmetric under rotations about the z-axis or polar spherical coordinates r, θ, φ for problems with symmetry under all rotations about the origin.

▷ **Exercise D–3.** If $f(x, y)$ is a real-valued function, then $f^\phi(r, \theta) = f \circ \phi^{-1}(r, \theta) = f(r\cos\theta, r\sin\theta)$, so if f is C^1, then, by the chain rule, $\frac{\partial f^\phi}{\partial r}(r, \theta) = \frac{\partial f}{\partial x}\cos\theta + \frac{\partial f}{\partial y}\sin\theta$, and similarly, $\frac{\partial f^\phi}{\partial \theta}(r, \theta) = -\frac{\partial f}{\partial x}r\sin\theta + \frac{\partial f}{\partial y}r\cos\theta = -y\frac{\partial f}{\partial x} + x\frac{\partial f}{\partial y}$. Generalize this to find formulas for $\frac{\partial f^\phi}{\partial \phi_i}$ in terms of the $\frac{\partial f}{\partial x^i}$ for a general coordinate system ϕ.

▷ **Exercise D–4.** Show that a C^1 real-valued function in the plane, $f : \mathbf{R}^2 \to \mathbf{R}$, is invariant under rotation if and only if $y\frac{\partial f}{\partial x} = x\frac{\partial f}{\partial y}$. (Hint: The condition for f to be invariant under rotation is that $f^\phi(r, \theta)$ should be a function of r only, i.e., $\frac{\partial f^\phi}{\partial \theta} \equiv 0$.)

D.2. Some Canonical Forms

Now let us look at the standard canonical form theorems for functions, curves, and vector fields.

Recall that if f is a C^1 real-valued function defined in some open set G of $\mathbf{R}^n$, then a point $p \in G$ is called a critical point of f if $df_p = 0$ and otherwise it is called a regular point of f. If p is a regular

D.2. Some Canonical Forms

point, then we can choose a basis $\ell_1, \ldots, \ell_n$ of $(\mathbf{R}^n)^*$ with $\ell_1 = df_p$, and by the remark at the beginning of this appendix it follows that $(f, \ell_2, \ldots, \ell_n)$ is a coordinate system in some neighborhood O of p. This proves the following canonical form theorem for smooth real-valued functions:

D.2.1. Proposition. *Let f be a smooth real-valued function defined in an open set G of $\mathbf{R}^n$ and let $p \in G$ be a regular point of f. Then there exists a coordinate system $(\phi_1, \ldots, \phi_n)$ defined in some neighborhood of p such that $f^\phi = \phi_1$.*

Informally speaking, we can say that near any regular point a smooth function looks linear in a suitable coordinate system.

Next we will see that a straight line is the canonical form for a smooth curve $\sigma : I \to \mathbf{R}^n$ at a regular point, i.e., a point $t_0 \in I$ such that $\sigma'(t_0) \neq 0$.

D.2.2. Proposition. *If t_0 is a regular point of the smooth curve $\sigma : I \to \mathbf{R}^n$, then there is a diffeomorphism ϕ of a neighborhood of $\sigma(t_0)$ into $\mathbf{R}^n$ such that $\sigma^\phi(t) = \gamma(t)$, where $\gamma : \mathbf{R} \to \mathbf{R}^n$ is the straight line $t \mapsto (t, 0, \ldots, 0)$.*

Proof. Without loss of generality we can assume that $t_0 = 0$. Also, since we can anyway translate $\sigma(t_0)$ to the origin and apply a linear isomorphism mapping $\sigma'(t_0)$ to $e_1 = (1, 0, \ldots, 0)$, we will assume that $\sigma(0) = 0$ and $\sigma'(0) = e_1$. Then if we define a map ψ near the origin of $\mathbf{R}^n$ by $\psi(x_1, \ldots, x_n) = \sigma(x_1) + (0, x_2, \ldots, x_n)$, it is clear that $D\psi_0$ is the identity, so by the inverse mapping theorem, ψ maps some neighborhood U of the origin diffeomorphically onto another neighborhood O. By definition, $\psi \circ \gamma(t) = \sigma(t)$, so if $\phi = \psi^{-1}$, then $\sigma^\phi(t) = \phi \circ \sigma(t) = \gamma(t)$. ∎

Notice a pattern in the canonical form theorems for functions and curves. If we keep away from "singularities", then locally a function or curve looks like the simplest example. This pattern is repeated for vector fields. Recall that a singularity of a vector field V is a point p where $V(p) = 0$, and the simplest vector fields are the constant vector fields, such as $\frac{\partial}{\partial x^1}$. The canonical form theorem for vector fields, often called the "Straightening Theorem", just says that near a nonsingular point a smooth vector field looks like a constant vector field. Let's try to make this more precise.

D.2.3. Definition. Let V be a vector field defined in an open set O of $\mathbf{R}^n$, and let $\phi = (\phi_1, \ldots, \phi_n)$ be local coordinates in O. We call ϕ the *flow-box coordinates* for V in O if $V^\phi = \frac{\partial}{\partial \phi_1}$.

D.2.4. Remark. Let $\phi(x, y) = (r(x, y), \theta(x, y))$ denote polar coordinates in $\mathbf{R}^2$. We saw above that if $V = x\frac{\partial}{\partial y} - y\frac{\partial}{\partial x}$, then $V^\phi = \frac{\partial}{\partial \theta}$, so that polar coordinates are flow-box coordinates for V.

D.2.5. The Straightening Theorem. *If V is a smooth vector field defined in an open set O of $\mathbf{R}^n$ and $p_0 \in O$ is not a singularity of V, then there exist flow-box coordinates for V in some neighborhood of p_0.*

Proof. This is a very strong result; it easily implies both local existence and uniqueness of solutions and smooth dependence on initial conditions. And as we shall now see, these conversely quickly give the Straightening Theorem. Without loss of generality, we can assume that p_0 is the origin and $V(0) = e_1 = (1, 0, \ldots, 0)$. Choose $\epsilon > 0$ so that for $\|p\| < \epsilon$ there is a unique solution curve of V, $t \mapsto \sigma(t, p)$, defined for $|t| < \epsilon$ and satisfying $\sigma(0, p) = p$. The existence of ϵ follows from the local existence and uniqueness theorem for solutions of ODE (Appendix B). Let U denote the disk of radius ϵ in $\mathbf{R}^n$ and define $\psi : U \to \mathbf{R}^n$ by $\psi(x) = \sigma(x_1, (0, x_2, \ldots, x_n))$. It follows from smooth dependence on initial conditions (Appendix G) that ψ is a smooth map.

▷ **Exercise D–5.** Complete the proof of the Straightening Theorem by first showing that $D\psi_0$ is the identity map (so ψ does define a local coordinate system near the origin) and secondly showing that $V^\phi = \frac{\partial}{\partial \phi_1}$, where $\phi := \psi^{-1}$. (Hint: Note that $\sigma(0, (0, x_2, \ldots, x_n)) = (0, x_2, \ldots, x_n)$, while $\frac{\partial}{\partial x_1} \sigma(x_1, (0, \ldots, 0)) = e_1$. The fact that $D\psi_0$ is the identity is an easy consequence. Since $t \mapsto \sigma(t, p)$ is a solution curve of V, it follows that $V(\psi(x)) = \frac{\partial}{\partial x_1} \psi(x)$, and it follows that $D\psi_p$ maps $(\frac{\partial}{\partial x_1})_p$ to $V(\psi(p))$. Use this to deduce $V^\phi = \frac{\partial}{\partial \phi_1}$. ∎

D.2. Some Canonical Forms

D.2.6. Definition. Let $V : \mathbf{R}^n \to \mathbf{R}^n$ be a smooth vector field. If O is an open set in $\mathbf{R}^n$, then a smooth real-valued function $f : O \to \mathbf{R}$ is called a *local constant of the motion* for V if $Vf \equiv 0$ in O.

Notice that $x^2, x^3, \ldots, x^n$ are constants of the motion for the vector field $\frac{\partial}{\partial x_1}$. Hence,

D.2.7. Corollary of the Straightening Theorem. *If V is a smooth vector field on $\mathbf{R}^n$ and p is any nonsingular point of V, then there exist $n-1$ functionally independent local constants of the motion for V defined in some neighborhood O of p.*

(Functionally independent means that they are part of a coordinate system.)

D.2.8. CAUTION. There are numerous places in the mathematics and physics literature where one can find a statement to the effect that every vector field on $\mathbf{R}^n$ has $n-1$ constants of the motion. What is presumably meant is something like the above corollary, but it is important to realize that such statements should **not** be taken literally—there are examples of vector fields with no global constants of the motion except constants. A local constant of the motion is very different from a global one. If V is a vector field in $\mathbf{R}^n$ and f is a local constant of the motion for V, defined in some open set O, then if σ is a solution of V, $f(\sigma(t))$ will be constant on any interval I such that $\sigma(I) \subseteq O$; however it will in general have different constant values on different such intervals.

Appendix E

Parametrized Curves and Arclength

For many purposes, the precise parametrization of a curve σ is not important, in the sense that some property of the curve that we are interested in is unchanged if we "reparametrize" the curve. Let us look at just what reparametrization means. Suppose that t is a C^1 function with a strictly positive derivative on a closed interval $[\alpha, \beta]$. Then t is strictly monotonic, and hence it maps $[\alpha, \beta]$ one-to-one onto some other closed interval $[a, b]$. Thus if $\sigma : [a, b] \to \mathbf{R}^n$ is a C^1 parametrized curve, then $\tilde{\sigma} = \sigma \circ t : [\alpha, \beta] \to \mathbf{R}^n$ is another C^1 parametrized curve which clearly has the same image as σ and is called the *reparametrization* of σ defined by the parameter transformation t. (If you like, you can think of t as a "variable that parameterizes the points of the interval $[\alpha, \beta]$ by points of the interval $[a, b]$" and with this interpretation σ and $\tilde{\sigma}$ become "the same".) In particular, given any interval $[\alpha, \beta]$, we always find an affine map $t(\tau) = c\tau + k$ that maps it onto $[a, b]$, so reparametrization allows us to adjust a parameter interval as convenient in situations where parametrization is not relevant.

A reparametrization of $\sigma : [a, b] \to \mathbf{R}$ can always be thought of as arising by starting from a positive, continuous function $\rho : [a, b] \to \mathbf{R}$ and letting t be the inverse function of its indefinite integral, τ. In fact $\tau(t) := \int_a^t \rho(\xi) \, d\xi$ is a smooth C^1 function with a positive derivative, so it does indeed map $[a, b]$ one-to-one onto some interval $[\alpha, \beta]$, and by the inverse function theorem $t := \tau^{-1} : [\alpha, \beta] \to \mathbf{R}$ is C^1 with a positive derivative. A very important special case of this is reparametrization by arclength. Suppose that σ is nonsingular, i.e.,

σ' never vanishes. Define $s : [a,b] \to \mathbf{R}$ by $s(t) := \int_a^t \|\sigma'(\xi)\| \, d\xi$, and recall that by definition it gives the arclength along σ from a to t. This is a smooth map with positive derivative $\|\sigma'(t)\|$ mapping $[a,b]$ onto $[0,L]$, where L is the length of σ. The inverse function, $t(s)$, mapping $[0,L]$ to $[a,b]$, gives the point of $[a,b]$ where the arclength of σ measured from its left endpoint is s, and the curve $s \mapsto \sigma(t(s))$ is a reparametrization of σ called its reparametrization by arclength. More generally, we say that a curve $\sigma : [a,b] \to \mathbf{R}^n$ is *parameterized by arclength* if the length of σ between $\sigma(a)$ and $\sigma(t)$ is equal to $t-a$, and we say that σ is *parametrized proportionally to arclength* if that length is proportional to $t-a$.

▷ **Exercise E–1.** Show that the length of a curve is unchanged by reparametrization. (Hint: This follows from a combination of the chain rule and the change of variables formula for an integral.)

▷ **Exercise E–2.** Show that a curve σ is parametrized proportionally to arclength if and only if $\|\sigma'(t)\|$ is a constant, and it is parametrized by arclength if and only if that constant equals one.

▷ **Exercise E–3.** Prove the old saying, "A straight line is the shortest distance between two points." That is, if $\sigma : [a,b] \to \mathbf{R}^n$ is a C^1 path of length L, then $\|\sigma(b) - \sigma(a)\| \leq L$, with equality if and only if σ is a straight line from $\sigma(a)$ to $\sigma(b)$. (Hint: As we have just seen, we can assume without loss of generality that σ is parametrized proportionally to arclength, i.e., that $\|\sigma'(t)\|$ is a constant. Let $v := \sigma(b) - \sigma(a)$, so that what we must show is that $\|v\| \leq L$ with equality if and only if σ' is a constant. If $v = 0$, i.e., if $\sigma(b) = \sigma(a)$, the result is trivial so we can assume $v \neq 0$ and define a unit vector $e = \frac{v}{\|v\|}$, so that $\|v\| = \langle v, e \rangle$. Now $v = \int_a^b \sigma'(t) \, dt$, and since e is a constant vector, $\|v\| = \langle v, e \rangle = \int_a^b \langle \sigma'(t), e \rangle \, dt$. Finally note that by the Schwarz Inequality, $\langle \sigma'(t), e \rangle \leq \|\sigma'(t)\|$ and equality holds for all t if and only if $\sigma'(t)$ is a multiple of e for each t, and this multiple must be a constant since $\|\sigma'(t)\|$ is a constant.)

Appendix F

Smoothness with Respect to Initial Conditions

Suppose that V is a C^1 vector field on $\mathbf{R}^n$ and assume that the maximal solution σ_p of $\frac{dx}{dt} = V(x)$ is defined on $I = [a, b]$. For each $x \in \mathbf{R}^n$, the differential of V at x is a linear map $DV_x : \mathbf{R}^n \to \mathbf{R}^n$, and it is continuous in x since V is C^1. Thus $A(t) = DV_{\sigma_p(t)}$ defines a continuous map $A : I \to \mathbf{L}(\mathbf{R}^n)$. The differential equation $\frac{dx}{dt} = A(t)x$ is an example of the nonautonomous linear ODEs studied in Section 2.2. It is called the *variational equation* associated to the solution σ. By the general theory of such equations developed in Chapter 2, we know that for each ξ in $\mathbf{R}^n$, the variational equation will have a unique solution $u(t, \xi)$ defined for $t \in I$ and satisfying the initial condition $u(t_0, \xi) = \xi$. For each t in I, the map $\xi \mapsto u(t, \xi)$ is a linear map of $\mathbf{R}^n$ to itself that we will denote by $\delta\sigma_p(t)$. What we are going to see next is that the map $(t, p) \mapsto \sigma_p(t)$ is C^1 and that $\delta\sigma_p(t)$ is the differential at p of the map $q \mapsto \sigma_q(t)$ of $\mathbf{R}^n$ to itself. (Note that the derivative of $\sigma_p(t)$ with respect to t obviously exists and is continuous since $\sigma_p(t)$ satisfies $\sigma'_p(t) = V(\sigma_p(t))$.)

▷ **Exercise F–1.** Check that *if* $q \mapsto \sigma_q(t)$ is indeed differentiable at p, then its differential must in fact be $\delta\sigma_p(t)$. Hint: Calculate the differential of both sides of the differential equation with respect to p to see that $D\sigma_p(t)(\xi)$ satisfies the variational equation. On the right side of the equation use the chain rule and on the left side interchange the order of differentiation.

F. Smoothness with Respect to Initial Conditions

Recall that (by definition of the differential of a mapping) in order to prove that $q \mapsto \sigma_q(t)$ is differentiable at p and that $u(t, \xi) = \delta\sigma_p(t)(\xi)$ is its differential at p in the direction ξ, what we need to show is that if $g(t) := \|(\sigma_{p+\xi}(t) - \sigma_p(t)) - u(t, \xi)\|$, then $\frac{1}{\|\xi\|} g(t)$ goes to zero with $\|\xi\|$. What we will show is that there are fixed positive constants C and M such that for any positive ϵ there exists a δ so that $g(t) < C\epsilon \|\xi\| e^{Mt}$ provided $\|\xi\| < \delta$, which clearly implies that $\frac{1}{\|\xi\|} g(t)$ goes to zero with $\|\xi\|$, uniformly in t. To prove the latter estimate, it will suffice by Gronwall's Inequality to show that $g(t) < C\epsilon \|\xi\| + M \int_0^t g(s)\, ds$.

▷ **Exercise F–2.** Derive this integral estimate. Hint: $\sigma_{p+\xi}(t) = p + \xi + \int_0^t V(\sigma_{p+\xi}(s))\, ds$ and $\sigma_p(t) = p + \int_0^t V(\sigma_p(s))\, ds$, while $u(t, \xi) = \xi + \int_0^t DV_{\sigma_p(s)} u(s, \xi)\, ds$. Taylor's Theorem with Remainder gives $V(q+x) - V(q) = DV_q(x) + \|x\| r(q, x)$ where $\|r(q, x)\|$ goes to zero with x, uniformly for q in some compact set. Take $q = \sigma_p(s)$ and $x = \sigma_{p+\xi}(s) - \sigma_p(s)$ and verify that
$$g(t) = \|\xi\| \int_0^t \rho(\sigma_p(s), \sigma_{p+\xi}(s) - \sigma_p(s))\, ds + \int_0^t DV_{\sigma_p(s)} g(s)\, ds.$$
Now choose $M = \sup_{s \in I} \left\| DV_{\sigma_p(s)} \right\|$ and recall that from the theorem on continuity with respect to initial conditions we know that $\|\sigma_{p+\xi}(s) - \sigma_p(s)\| < \|\xi\| e^{Ks}$. The rest is easy, and we have now proved the case $r = 1$ of the following theorem.

F.0.1. Theorem on Smoothness w.r.t. Initial Conditions. Let V be a C^r vector field on $\mathbf{R}^n$, $r \geq 1$, and let $\sigma_p(t)$ denote the maximal solution curve of $\frac{dx}{dt} = V(x)$ with initial condition p. Then the map $(p, t) \mapsto \sigma_p(t)$ is C^r.

▷ **Exercise F–3.** Prove the general case by induction on r. Hint: As we saw, the first-order partial derivatives are solutions of an ODE whose right-hand side is of class C^{r-1}.

Appendix G

Canonical Form for Linear Operators

G.1. The Spectral Theorem

If V is an orthogonal vector space, then each element v of V defines a linear functional $f_v : V \to R$, namely $u \mapsto \langle u, v \rangle$, and since $f_v(u) = \langle u, v \rangle$ is clearly linear in v as well as u, we have a linear map $v \mapsto f_v$ of V into its dual space V^*. Moreover the kernel of this map is clearly 0 (since, if v is in the kernel, then $\|v\|^2 = \langle v, v \rangle = f_v(v) = 0$), and since V^* has the same dimension as V, it follows by basic linear algebra that this map is in fact a linear isomorphism of V with V^*. We say that v is *dual* to f_v and vice versa.

Now let $A : V \to V$ be a linear map, and for each v in V let A^*v in V be the element dual to the linear functional $u \mapsto \langle Au, v \rangle$; that is, A^*v is defined by the identity $\langle Au, v \rangle = \langle u, A^*v \rangle$. It is clear that $v \mapsto A^*v$ is linear, and we call this linear map $A^* : V \to V$ the *adjoint* of A. If $A^* = A$, then we say that A is *self-adjoint*.

▷ **Exercise G–1.** Let $L(V, V)$ denote the space of linear operators on V. Show that $A \mapsto A^*$ is a linear map of $L(V, V)$ to itself and that it is its own inverse (i.e., $A^{**} = A$). Show also that $(AB)^* = B^*A^*$.

G.1.1. Proposition. *Let A be a self-adjoint linear operator on V and let W be a linear subspace of V. If W is invariant under A, then so is $W^\perp$.*

Proof. If $u \in W^\perp$, we must show that Au is also in $W^\perp$, i.e., that $\langle w, Au \rangle = 0$ for any $w \in W$. Since $Aw \in W$ by assumption, $\langle w, Au \rangle = \langle Aw, u \rangle = 0$ follows from $u \in W^\perp$. ∎

In what follows, A will denote a self-adjoint linear operator on V. If λ is any scalar, than we denote by $E_\lambda(A)$ the set of v in V such that $Av = \lambda v$. It is clear that $E_\lambda(A)$ is a linear subspace of V, called the λ-*eigenspace of* A. If $E_\lambda(A)$ is not the 0 subspace of V, then we call λ an *eigenvalue* of A, and every nonzero element of $E_\lambda(A)$ is called an *eigenvector* corresponding to the eigenvalue λ.

G.1.2. Proposition. *If $\lambda \neq \mu$, then $E_\lambda(A)$ and $E_\mu(A)$ are orthogonal subspaces of V.*

▷ **Exercise G–2.** Prove this. (Hint: Let $u \in E_\lambda(A)$ and $v \in E_\mu(A)$. You must show $\langle u, v \rangle = 0$. But $\lambda \langle u, v \rangle = \langle Au, v \rangle = \langle u, Av \rangle = \mu \langle u, v \rangle$.)

Note that it follows that a self-adjoint operator on an N-dimensional orthogonal vector space can have at most N distinct eigenvalues.

G.1.3. Spectral Theorem for Self-Adjoint Operators. *If A is a self-adjoint operator on an orthogonal vector space V, then V is the orthogonal direct sum of the eigenspaces $E_\lambda(A)$ corresponding to the eigenvalues λ of A. Equivalently, we can find an orthonormal basis for V consisting of eigenvectors of A.*

▷ **Exercise G–3.** Prove the equivalence of the two formulations.

We will base the proof of the Spectral Theorem on the following lemma.

G.1.4. Spectral Lemma. *A self-adjoint operator $A : V \to V$ always has at least one eigenvalue unless $V = 0$.*

Here is the proof of the Spectral Theorem. Let W be the direct sum of the eigenspaces $E_\lambda(A)$ corresponding to the eigenvalues λ of A. We must show that $W = V$, or equivalently that $W^\perp = 0$. Now W is clearly invariant under A, so by the first proposition of this section, so is $W^\perp$. Since the restriction of a self-adjoint operator

G.1. The Spectral Theorem

to an invariant subspace is clearly still self-adjoint, by the Spectral Lemma, if $W^\perp \neq 0$, then there would be an eigenvector of A in $W^\perp$, contradicting the fact that all eigenvectors of A are in W.

The proof of the Spectral Lemma involves a rather pretty geometric idea. Recall that we have seen that A is derivable from the potential function $U(v) = \frac{1}{2}\langle Av, v\rangle$, i.e., $Av = (\nabla U)_v$ for all v in V. So what we must do is find a unit vector v where $(\nabla U)_v$ is proportional to v provided $V \neq 0$, i.e., provided the unit sphere in V is not empty. In fact, something more general is true.

G.1.5. Lagrange Multiplier Theorem (Special Case). *Let V be an orthogonal vector space and $f : V \to R$ a smooth real-valued function on V. Let v denote a unit vector in V where f assumes its maximum value on the unit sphere S of V. Then $(\nabla f)_v$ is a scalar multiple of v.*

Proof. The scalar multiples of v are exactly the vectors normal to S at v, i.e., orthogonal to all vectors tangent to S at v. So we have to show that if u is tangent to S at v, then $(\nabla f)_v$ is orthogonal to u, i.e., that $\langle u, (\nabla f)_v\rangle = df_v(u) = 0$. Choose a smooth curve $\sigma(t)$ on S with $\sigma(0) = v$ and $\sigma'(0) = u$ (for example, normalize $v + tu$). Then since $f(\sigma(t))$ has a maximum at $t = 0$, it follows that $(d/dt)_{t=0}f(\sigma(t)) = 0$. But by definition of df, $(d/dt)_{t=0}f(\sigma(t)) = df_v(u)$. ∎

G.1.6. Definition. An operator A on an orthogonal vector space V is *positive* if it is self-adjoint and if $\langle Av, v\rangle > 0$ for all $v \neq 0$ in V.

▷ **Exercise G–4.** Show that a self-adjoint operator is positive if and only if all of its eigenvalues are positive.

▷ **Exercise G–5.** Verify the intuitive fact that a unit vector v is orthogonal to all vectors tangent to the unit sphere at v. (Hint: Choose σ as above and differentiate the identity $\langle \sigma(t), \sigma(t)\rangle = 1$.)

▷ **Exercise G–6.** Show that another equivalent formulation of the Spectral Theorem is that a linear operator on an orthogonal vector space is self-adjoint if and only if it has a diagonal matrix in some orthonormal basis.

Appendix H

Runge-Kutta Methods

In this appendix we will analyze the conditions on the coefficients of an explicit Runge-Kutta Method that are necessary and sufficient to guarantee convergence with accuracy of order P. In particular, we will establish the connection between these conditions and the set of rooted trees with no more than P nodes. As a consequence, we will be able to show that there are r-stage methods of order r for $r \leq 4$ but not for $r > 4$.

We begin by briefly considering more general one-step methods, $y_{n+1} = F(t_n, y_n, f, h)$, for approximating solutions of the scalar ODE $y' = f(t, y(t))$.

The local truncation error at t_n is the quantity ϵ_n defined by

$$y(t_{n+1}) := F(t_n, y(t_n), f, h) + \epsilon_n.$$

From our discussion following the convergence analysis of Euler's Method in the body of the text, we can show that a 0-stable one-step method will converge to a solution of the ODE $y \in C^{P+1}[t_o, t_o + T]$ with global order of accuracy P if $|\epsilon_n| \leq Ch^{P+1}$ for some $C > 0$ depending only on $\max_{t \in (t_o, t_o+T)} |y^{(P+1)}(t)|$.

One approach to constructing methods satisfying such an estimate is to define them using Taylor's Theorem with Remainder by letting $F(t_n, y_n, f, h) = \sum_{p=0}^{P} y^{(p)}(t_n)/p! \, h^p$ be the Taylor polynomial of degree p for $y(t)$ centered at t_n and evaluated at t_{n+1}. To implement this idea, we must be able to express $y^{(k)}(t_n)$ in terms of f and its derivatives evaluated at (t_n, y_n). The resulting one-step methods are known as *Taylor Methods*. Taylor Methods are an option if the vector field that defines the ODE is given in a form that can be differentiated symbolically, which is not always the case.

To demonstrate how this would be carried out, and for later use, we examine the expressions for the first few derivatives of y in terms of f and its derivatives. We ignore the differential equation at first and differentiate $g(t) = f(t, y(t))$ and use multi-index notation for mixed partial derivatives,

$$f^{k,l} = \partial^{k,l} f(t,y) = \frac{\partial^{k+l} f}{\partial t^k \partial y^l}.$$

In this form, we can distinguish terms arising from differentiating f from those that arise by differentiating factors of y coming from the chain rule—terms that we will eventually also write in terms of f. Because of equality of mixed partial derivatives, these terms exhibit a binomial pattern,

$$\begin{aligned} y' &= f, \\ y'' &= f^{1,0} + f f^{0,1}, \\ y''' &= [f^{2,0} + 2f f^{1,1} + f^2 f^{0,2}] + [(f^{1,0} + f f^{0,1}) f^{0,1}]. \end{aligned} \quad \text{(H.1)}$$

Even when this procedure is possible, by hand or with automatic symbolic differentiation, the number of terms required to carry the expansion to high order can yield diminishing returns with the growing cost of evaluation.

An alternate approach originally proposed and developed by Runge and Kutta only requires evaluation of f at arbitrary (t, y) values to match the terms of Taylor polynomial above to order p. Runge-Kutta Methods approximate $(\mathbf{y}(t_{n+1}) - \mathbf{y}(t_n))/h$ using a weighted average of samples of the vector field $f(t, y)$ that defines an ODE. For the method to be explicit, locations of the samples must be chosen based upon information obtained in previous samples. Because of this, the general form of an explicit one-stage Runge-Kutta Method is $y_{n+1} = y_n + h\gamma_0 f(t_n, y_n)$. For the right-hand side to match the first-order terms of the Taylor expansion above, we must have $\gamma_0 = 1$. This tells us that Euler's Method is the unique explicit one-stage Runge-Kutta Method that is convergent. No higher-order terms occur when a one-stage method is used.

H. Runge-Kutta Methods

The general form of an explicit two-stage Runge-Kutta Method is

$$y'_{n,1} = f(t_n, y_n),$$
$$y'_{n,2} = f(t_n + \beta_{21}h, y_n + h\beta_{21}y'_{n,1}), \quad \text{(H.2)}$$
$$y_{n+1} = y_n + h(\gamma_1 y'_{n,1} + \gamma_2 y'_{n,2}).$$

Two of the example methods in the text fit this pattern, the midpoint method ($\beta_{12} = 1/2$, $\gamma_1 = 0$, $\gamma_2 = 1$) and Heun's Method ($\beta_{12} = 1$, $\gamma_1 = \gamma_2 = 1/2$). Both solved the second-order accuracy model problem exactly and also appeared to converge to the solution of the absolute stability model problem with second-order accuracy.

To estimate the local truncation error of these methods, we perform Taylor expansions of the terms of the general explicit 2-stage Runge-Kutta Methods. Substituting $y'_{n,1}$ in the definition of $y'_{n,2}$,

$$y'_{n,2} = f(t_n + \beta_{21}h, y_n + h\beta_{21}f).$$

Then by Taylor expanding in powers of the perturbations (to first order to obtain hy' terms to second order),

$$y'_{n,2} = f + \beta_{21}h(f^{1,0} + ff^{0,1}) + O(h^2).$$

When this is inserted in the expression for y_{n+1}, we find

$$y_{n+1} = y_n + h(\gamma_1 + \gamma_2)f + \frac{h^2}{2}2\gamma_2\beta_{21}(f^{1,0} + ff^{0,1}) + O(h^3).$$

Comparing this with (H.1), the conditions for this expansion to match the first two terms of the Taylor series

$$y(t_{n+1}) = y(t_n) + hy'_n + \frac{h^2}{2}y''_n + O(h^3)$$

are

$$\gamma_1 + \gamma_2 = 1,$$
$$2\gamma_2\beta_{21} = 1.$$

We may use γ_2 to parametrize a family, $\gamma_1 = 1 - \gamma_2$, $\beta_{21} = 1/(2\gamma_2)$, of solutions of these equations. It is straightforward to check that the midpoint method and Heun's Method satisfy these conditions.

The parameters of a Runge-Kutta Method are often displayed in the form of a so-called *Butcher tableau*:

$$\begin{array}{c|cccc} 0 & & & & \\ \alpha_2 & \beta_{21} & & & \\ \vdots & \vdots & \ddots & & \\ \alpha_r & \beta_{r1} & \cdots & \beta_{r(r-1)} & \\ \hline & \gamma_1 & \gamma_2 & \cdots & \gamma_r \end{array}$$

The Butcher tableau for the midpoint method is

$$\begin{array}{c|cc} 0 & & \\ \frac{1}{2} & \frac{1}{2} & \\ \hline & 0 & 1 \end{array}$$

The modified trapezoidal method is displayed in this format as

$$\begin{array}{c|cc} 0 & & \\ 1 & 1 & \\ \hline & \frac{1}{2} & \frac{1}{2} \end{array}$$

If we expanded $y'_{n,2}$ to higher order, we would discover that three parameters do not provide enough freedom to obtain a method of order 3. In order to satisfy the two additional h^3 conditions appearing in square brackets in the expression (H.1) for y''', another stage is needed.

The form of an explicit Runge-Kutta Method with $r = 3$ stages is

$$y'_{n,1} = f(t_n, y_n),$$
$$y'_{n,2} = f(t_n + \beta_{21}h, y_n + h\beta_{21}y'_{n,1}),$$
$$y'_{n,3} = f(t_n + (\beta_{31} + \beta_{32})h, y_n + h(\beta_{31}y'_{n,1} + \beta_{32}y'_{n,2})),$$
$$y_{n+1} = y_n + h(\gamma_1 y'_{n,1} + \gamma_2 y'_{n,2} + \gamma_3 y'_{n,3}).$$

H. Runge-Kutta Methods

The coefficient of $\frac{h^4}{4!}$ in the Taylor expansion of $y(t+h)$ in terms of f and its derivatives is

$$y^{(4)} = [f^{3,0} + 3ff^{2,1} + 3f^2 f^{1,2} + f^3 f^{0,3}]$$
$$+ [3(f^{1,0} + ff^{0,1})(f^{1,1} + ff^{0,2})]$$
$$+ [(f^{2,0} + 2ff^{1,1} + f^2 f^{0,2} + (f^{1,0} + ff^{0,1})f^{0,1})f^{0,1}].$$

It clearly becomes worthwhile to find a framework to simplify the development and comparison of the Taylor and Runge-Kutta sides of these expansions to higher orders. The autonomous scalar case is exceptional, as can be seen by setting all t-derivatives to zero in the expressions above. For greater generality, we shift our setting and notation and now consider an $\mathbf{R}^D$ vector-valued $\mathbf{f}(\mathbf{y})$ and $\mathbf{y}(t)$ that is a solution of $\mathbf{y}' = \mathbf{f}(\mathbf{y})$. The nonautonomous case can be put into this form using the standard device of replacing t by additional dependent variables y_{D+1} satisfying $y'_{D+1} = 1$. In this setting, the general r-stage explicit Runge-Kutta Method takes the form

$$\mathbf{y}_{n+1} = \mathbf{y}_n + h \sum_{i=1}^{r} \gamma_i \mathbf{y}'_{n,i}, \tag{H.3}$$

where

$$\mathbf{y}'_{n,i} = \mathbf{f}(\mathbf{y}_{n,i}), \text{ with } \mathbf{y}_{n,i} = \mathbf{y}_n + h \sum_{j=1}^{i-1} \beta_{i,j} \mathbf{y}'_{n,j}. \tag{H.4}$$

An elegant formalism for organizing, visualizing, and understanding both the Taylor expansion of the solution $\mathbf{y}(t_n+h)$ and the *Runge-Kutta expansion* of $\mathbf{y}_{n+1}$ obtained by Taylor expanding the terms in (H.3) and (H.4) has been developed and advocated by Butcher [BJ], following on the work of Gill [GS] and Merson [MRH]. This approach associates terms in both expansions with *rooted trees*. To motivate it, we begin by reviewing the formal Taylor expansion to degree 5 for a function $\mathbf{y}(t) : \mathbf{R} \to \mathbf{R}^n$ satisfying $\mathbf{y}' = \mathbf{f}(\mathbf{y})$, where $\mathbf{f}$ is a smooth function from $\mathbf{R}^n$ to $\mathbf{R}^n$,

$$\mathbf{y}(t+h) = \mathbf{y}(t) + \mathbf{y}'(t)h + \mathbf{y}''(t)\frac{h^2}{2!} + \cdots + \mathbf{y}^{(k)}(t)\frac{h^k}{k!} + \cdots,$$

and then give the representation of its terms using rooted trees. We wish to represent the derivatives $\mathbf{y}^{(k)}(t)$ in terms of $\mathbf{f}$ and its derivatives. Here, in the column on the left, we list successive derivatives of the function $\mathbf{g}(t) = \mathbf{f}(\mathbf{y}(t))$ where again at first we ignore the differential equation:

$$
\begin{aligned}
\mathbf{g}(t) &= \mathbf{f}(\mathbf{y}(t)) = \mathbf{f}, & & \mathbf{y}', 1, \\
\mathbf{g}'(t) &= \mathbf{f_y}\mathbf{y}', & & \mathbf{y}'', 1, \\
\mathbf{g}''(t) &= \mathbf{f_{yy}}\mathbf{y}'^2 + \mathbf{f_y}\mathbf{y}'', & & \mathbf{y}''', 1+1 = 2, \\
\mathbf{g}'''(t) &= \mathbf{f_{yyy}}\mathbf{y}'^3 + 3\mathbf{f_{yy}}\mathbf{y}'\mathbf{y}'' + \mathbf{f_y}\mathbf{y}''', & & \mathbf{y}^{(4)}, 1+1+2 = 4, \\
\mathbf{g}^{(4)}(t) &= \mathbf{f_{yyyy}}\mathbf{y}'^4 + 6\mathbf{f_{yyy}}\mathbf{y}'^2\mathbf{y}'' & & \\
&\quad + 3\mathbf{f_{yy}}\mathbf{y}''\mathbf{y}'' & & \\
&\quad + 4\mathbf{f_{yy}}\mathbf{y}'\mathbf{y}''' + \mathbf{f_y}\mathbf{y}^{(4)}, & & \mathbf{y}^{(5)}, 1+1+1+2+4 = 9.
\end{aligned}
$$

In the column on the right, we list the correspondence between each derivative of $\mathbf{g}$ and the next higher derivative of y. We also list the number of terms that each row represents as a sum. The terms of the sum refer recursively to terms from previous rows that appear in subsequent rows and the numbers of terms they represent. Recall that the kth derivative of $\mathbf{f}$ with respect to $\mathbf{y}$ is a symmetric k-linear function from $(\mathbf{R}^n)^k \to \mathbf{R}^n$. For $k > 1$, the symmetry is nontrivial and decreases the number of its independent coefficients with respect to a basis from n^{k+1} accordingly. For example, when $k = 2$ there are $n(n(n+1))/2$ independent components.

Next we expand the rows recursively to write the Taylor expansion as a linear combination of *elementary differentials*. These are multilinear operator compositions that express $\mathbf{y}^{(k)}$ in terms of $\mathbf{f}$ and its derivatives evaluated at t. Since $\mathbf{f_{y\cdots y}}$ is an operator with k arguments, we will use a naturally related notation for lists that will be familiar to those who have encountered the artificial intelligence programming language LISP. For our purposes, a list begins with an open parentheses and the first element, the kth partial derivative of f with respect to y for some $k \geq 0$, followed by k sublists, then a close parentheses. If a sublist has zero sublists, we omit its parentheses, and we can close all open parentheses with a right square bracket. (Our convention will be to do so when more than two are open.)

H. Runge-Kutta Methods

The internal representation of such a list in a LISP interpreter is in the form of a rooted tree data structure, the same algebraic structure that has been used to visualize and organize the terms of Taylor and Runge-Kutta expansions and their relationship. This suggests that LISP may be convenient for performing calculations involved in the derivation and analysis of Runge-Kutta Methods. A rooted tree is a set of nodes connected by edges oriented away from a distinguished node called the root, so it is a connected simple graph that contains no cycles, i.e., a tree. Graphically, we represent the lists associated with an elementary differential by a root node labeled by $\mathbf{f_{y...y}}$ (k partial derivatives) attached to k subtrees corresponding to those sublists. Lists with no sublists are leaves—terminal nodes with no edges leaving them:

$\mathbf{y}'(t) = (\mathbf{f}) = \mathbf{f},$

$\mathbf{y}''(t) = (\mathbf{f_y}\ \mathbf{f}),$

$\mathbf{y}'''(t) = (\mathbf{f_{yy}}\ \mathbf{f}\ \mathbf{f}) + (\mathbf{f_y}\ (\mathbf{f_y}\ \mathbf{f})),$

$\mathbf{y}^{(4)}(t) = (\mathbf{f_{yyy}}\ \mathbf{f}\ \mathbf{f}\ \mathbf{f}) + 3(\mathbf{f_{yy}}(\mathbf{f_y}\ \mathbf{f})\ \mathbf{f}) + (\mathbf{f_y}(\mathbf{f_{yy}}\ \mathbf{f}\ \mathbf{f})) + (\mathbf{f_y}(\mathbf{f_y}(\mathbf{f_y}\ \mathbf{f}],$

$\mathbf{y}^{(5)}(t) = (\mathbf{f_{yyyy}}\ \mathbf{f}\ \mathbf{f}\ \mathbf{f}\ \mathbf{f}) + 6(\mathbf{f_{yyy}}(\mathbf{f_y}\ \mathbf{f})\ \mathbf{f}\ \mathbf{f}) + 3(\mathbf{f_{yy}}(\mathbf{f_y}\ \mathbf{f})(\mathbf{f_y}\ \mathbf{f}))$
$\quad + 4(\mathbf{f_{yy}}(\mathbf{f_{yy}}\ \mathbf{f}\ \mathbf{f})\ \mathbf{f}) + 4(\mathbf{f_{yy}}(\mathbf{f_y}(\mathbf{f_y}\ \mathbf{f}))\ \mathbf{f}) + (\mathbf{f_y}(\mathbf{f_{yyy}}\ \mathbf{f}\ \mathbf{f}\ \mathbf{f}))$
$\quad + 3(\mathbf{f_y}(\mathbf{f_{yy}}(\mathbf{f_y}\ \mathbf{f})\ \mathbf{f})) + (\mathbf{f_y}(\mathbf{f_y}(\mathbf{f_{yy}}\ \mathbf{f}\ \mathbf{f}] + (\mathbf{f_y}(\mathbf{f_y}(\mathbf{f_y}(\mathbf{f_y}\ \mathbf{f}].$

We implicitly evaluate $\mathbf{f}$ and its derivatives at $\mathbf{y}(t)$. We will refer to the lth term of the kth-order (row) formal Taylor expansion above as T_l^k. Observe that each different term of a particular order arises from terms of the previous order from the vector-valued Leibniz rule and chain rule, through the addition of one derivative to each operator factor (we consider arguments $\mathbf{f}$ as preceded by a 0th-order identity operator) and adding a corresponding argument $\mathbf{y}' = \mathbf{f}$. In terms of rooted trees, this corresponds to the process of constructing all rooted trees with k nodes by attaching a new edge and leaf to each node (one at a time) of each rooted tree with $k-1$ nodes. The coefficients in the equations above represent the number of distinct ways each such tree can be built in this manner. Instead of expanding existing trees with new leaves, we will see that the new rooted trees that occur at the rth stage of a Runge-Kutta expansion are built by joining any number of trees built at the $(r-1)$st stage to a new root node. By considering the multiplicities of ways the trees are built in both models and the coefficients that arise from the Runge-Kutta weighting coefficients, we will obtain the matching conditions that are necessary to achieve a certain order of accuracy.

Below, we exhibit the rooted trees corresponding to each term T_l^k in the Taylor expansion above, along with their associated coefficients in that expansion. The coefficients in the numerators that represent multiplicities of the various terms in the expansion can be interpreted and determined directly in terms of the number of ways the corresponding trees can be constructed by repeatedly attaching an edge and leaf to any node of smaller trees, starting from an initial root node. For example, the factor 3 associated with T_2^4 corresponds to the fact that it can be obtained by attaching an edge and leaf to either of the two leaves of T_1^3 or by attaching an edge and leaf to the root node of T_2^3. Similarly, the factor 6 associated with T_2^5 corresponds to the fact that it can be obtained either by attaching an edge and leaf to any of the three leaves of T_1^4 or by attaching an edge and leaf to the root node of T_2^4 that itself has multiplicity 3.

$T_1^1 : (\mathbf{f}), \frac{1}{1!}$ $\qquad T_1^2 : (\mathbf{f_y}\ \mathbf{f}), \frac{1}{2!}$ $\qquad T_1^3 : (\mathbf{f_{yy}}\ \mathbf{f}\ \mathbf{f}), \frac{1}{3!}$ $\qquad T_2^3 : (\mathbf{f_y}\ (\mathbf{f_y}\ \mathbf{f})), \frac{1}{3!}$

$T_1^4 : (\mathbf{f_{yyy}}\ \mathbf{f}\ \mathbf{f}\ \mathbf{f}), \frac{1}{4!}$ $\qquad T_2^4 : (\mathbf{f_{yy}}\ (\mathbf{f_y}\ \mathbf{f})\ \mathbf{f}), \frac{3}{4!}$ $\qquad T_3^4 : (\mathbf{f_y}\ (\mathbf{f_{yy}}\ \mathbf{f}\ \mathbf{f})), \frac{1}{4}$

$T_4^4 : (\mathbf{f_y}\ (\mathbf{f_y}\ (\mathbf{f_y}\ \mathbf{f}]), \frac{1}{4!}$ $\qquad T_1^5 : (\mathbf{f_{yyyy}}\ \mathbf{f}\ \mathbf{f}\ \mathbf{f}\ \mathbf{f}), \frac{1}{5!}$ $\qquad T_2^5 : (\mathbf{f_{yyy}}\ (\mathbf{f_y}\ \mathbf{f})\ \mathbf{f}\ \mathbf{f}), \frac{6}{5!}$

H. Runge-Kutta Methods

$f_{yy} \begin{cases} f_y \text{---} f \\ f_y \text{---} f \end{cases}$

$f_{yy} \begin{cases} f_{yy} \begin{cases} f \\ f \end{cases} \end{cases}$

$f_{yy} \begin{cases} f_y \text{---} f_y \text{---} f \\ f \end{cases}$

$T_3^5 : (f_{yy} \ (f_y \ f)(f_y \ f)), \ \frac{3}{5!}$ $T_4^5 : (f_{yy} \ (f_{yy} \ f \ f) \ f), \ \frac{4}{5!}$ $T_5^5 : (f_{yy} \ (f_y \ (f_y \ f)) \ f), \ \frac{4}{5!}$

$f_y \text{---} f_{yyy} \begin{cases} f \\ f \\ f \end{cases}$

$f_y \text{---} f_{yy} \begin{cases} f_y \text{---} f \\ f \end{cases}$

$T_6^5 : (f_y \ (f_{yyy} \ f \ f \ f)), \ \frac{1}{5!}$ $T_7^5 : (f_y \ (f_{yy} \ (f_y \ f) \ f)), \ \frac{3}{5!}$

$f_y \text{---} f_y \text{---} f_{yy} \begin{cases} f \\ f \end{cases}$

$f_y \text{---} f_y \text{---} f_y \text{---} f_y \text{---} f$

$T_8^5 : (f_y \ (f_y \ (f_{yy} \ f \ f]), \ \frac{1}{5!}$ $T_9^5 : (f_y \ (f_y \ (f_y \ (f_y \ f]), \ \frac{1}{5!}$

Below, we will perform Taylor expansions of the terms in the Runge-Kutta samples (H.4) for $r = 4$ stages through order h^3. When we form their weighted sum (H.3), this yields terms up to order h^4. These Runge-Kutta expansions very quickly become horrendous, but when they are interpreted in terms of rooted trees, another surprisingly simple pattern describing the terms present at each stage and their coefficients quickly emerges, just as we saw for Taylor expansion of $\mathbf{y}(t+h)$. Before we wade through the formulas, we preview the algebraic and analytical basis for this pattern and its consequences for determining the order of an r-stage Runge-Kutta Method. In (H.4) we have used $\mathbf{y}_{n,i}$ to denote the arguments of the sample of $\mathbf{f}$ that defines the $\mathbf{y}'_{n,i}$. The first stage of any explicit Runge-Kutta Method simply samples the vector field at the current time-step, $\mathbf{y}_{n,1} = \mathbf{y}_n$. Then for any method other than Euler's Method, another stage samples $\mathbf{f}$ at $\mathbf{y}_{n,2} = \mathbf{y}_n + h(\beta_{21}\mathbf{y}'_{n,1})$, and we can formally expand $\mathbf{f}(\mathbf{y}_{n,2})$ about $\mathbf{y}_n$ in a Taylor series of the form $\sum_{l=0}^{\infty} (\frac{\mathbf{f}_{\mathbf{y}\cdots\mathbf{y}}}{l!}(h\beta_{21}\mathbf{f})^l)$. If a third stage is used, it samples $\mathbf{f}$ at $\mathbf{y}_{n,3} = \mathbf{y}_n + h(\beta_{31}\mathbf{y}'_{n,1} + \beta_{32}\mathbf{y}'_{n,2})$. Expanding $\mathbf{f}(\mathbf{y}_{n,3})$ in powers of h involves substituting the prior expansion of $\mathbf{y}'_{n,2}$, combining like terms with $\mathbf{y}'_{n,1}$ in the perturbation of $\mathbf{y}_n$ in the argument (here just $\mathbf{f}$), and then expanding in powers of the resulting power series. The lth power of this series results in terms of the form $\frac{\mathbf{f}_{\mathbf{y}\cdots\mathbf{y}}}{l!}$ operating on l-fold products of its terms. Any l terms whose orders in h sum to a particular order contribute to the overall result at that order, much like the convolution of coefficients that gives the coefficient of a certain order in a polynomial product.

Any subsequent stage can be expanded in the same manner. The description of this process in terms of rooted trees is simply that new trees are built by attaching any number of trees obtained at the prior stage to a new root node. In the ith stage we expand the evaluation of $\mathbf{f}$ at $\mathbf{y}_{n,i} = y + h(\beta_{i1}\mathbf{y}'_{n,1} + \cdots + \beta_{i(i-1)}\mathbf{y}'_{n,i-1})$. To do so, we first collect like terms in the expansions we have already obtained for $\mathbf{y}'_{n,1}, \ldots, \mathbf{y}'_{n,i-1}$ to obtain a single expansion $\mathbf{y}_{n,i} = \mathbf{y} + \sum(T_m h^m)^l$. This simply involves summing the parameters for the current stage times the corresponding coefficients obtained at the previous stage. Then the $(\frac{\mathbf{f}_{\mathbf{y}\cdots\mathbf{y}}}{l!}(\sum T_m h^m)^l)$ term of the Taylor expansion of $\mathbf{f}(\mathbf{y}_{n,i})$ is comprised of terms of the form $(\mathbf{f}_{\mathbf{y}\cdots\mathbf{y}} \, T_{i_1} \cdots T_{i_m})$. This is represented as a rooted tree whose root node is attached to the m trees corresponding to the terms $T_{i_1}, \ldots, T_{i_m}$ in the prior stage of the expansion. Now here is the expansion for $r = 4$.

$$\mathbf{y}'_{n,1} = \mathbf{f}(\mathbf{y}_n) = \mathbf{f},$$

$$\begin{aligned}\mathbf{y}'_{n,2} &= \mathbf{f}(\mathbf{y}_n + h(\beta_{21}\mathbf{y}'_{n,1})) = \mathbf{f}(\mathbf{y}_n + h(\beta_{21}\mathbf{f})) \\ &= \mathbf{f} + (\mathbf{f}_{\mathbf{y}} \ (h\beta_{21}\mathbf{f})) \\ &\quad + (\frac{\mathbf{f}_{\mathbf{yy}}}{2!} \ (h\beta_{21}\mathbf{f})^2) + (\frac{\mathbf{f}_{\mathbf{yyy}}}{3!} \ (h\beta_{21}\mathbf{f})^3) + \cdots \\ &= \mathbf{f} + h\beta_{21}(\mathbf{f}_{\mathbf{y}} \ \mathbf{f}) + \frac{h^2}{2!}\beta_{21}^2(\mathbf{f}_{\mathbf{yy}} \ \mathbf{f} \ \mathbf{f}) + \frac{h^3}{3!}\beta_{21}^3 \ (\mathbf{f}_{\mathbf{yyy}} \ \mathbf{f} \ \mathbf{f} \ \mathbf{f}) + \cdots,\end{aligned}$$

$$\begin{aligned}\mathbf{y}'_{n,3} &= \mathbf{f}(\mathbf{y}_n + h(\beta_{31}\mathbf{y}'_{n,1} + \beta_{32}\mathbf{y}'_{n,2})) \\ &= \mathbf{f}(\mathbf{y}_n + h(\beta_{31} + \beta_{32})\mathbf{f} \\ &\quad + h^2 \beta_{32}\beta_{21}(\mathbf{f}_{\mathbf{y}} \ \mathbf{f}) + \frac{h^3}{2!}\beta_{32}\beta_{21}^2(\mathbf{f}_{\mathbf{yy}} \ \mathbf{f} \ \mathbf{f}) + \cdots)\end{aligned}$$

$$= \mathbf{f} + (\mathbf{f_y} \; (h(\beta_{31} + \beta_{32})\mathbf{f}$$
$$+ h^2 \beta_{32}\beta_{21}(\mathbf{f_y} \; \mathbf{f}) + \frac{h^3}{2!}\beta_{32}\beta_{21}^2(\mathbf{f_{yy}} \; \mathbf{f} \; \mathbf{f}) + \cdots))$$
$$+ (\frac{\mathbf{f_{yy}}}{2!} \; (h(\beta_{31} + \beta_{32})\mathbf{f} + h^2\beta_{32}\beta_{21}(\mathbf{f_y} \; \mathbf{f}) + \cdots)^2)$$
$$+ (\frac{\mathbf{f_{yyy}}}{3!} \; (h(\beta_{31} + \beta_{32})\mathbf{f} + \cdots)^3) + \cdots$$
$$= \mathbf{f} + h(\beta_{31} + \beta_{32})(\mathbf{f_y} \; \mathbf{f}) + h^2 \beta_{32}\beta_{21}(\mathbf{f_y} \; (\mathbf{f_y} \; \mathbf{f}))$$
$$+ \frac{h^3}{2!}\beta_{32}\beta_{21}^2(\mathbf{f_y} \; (\mathbf{f_{yy}} \; \mathbf{f} \; \mathbf{f})) + \cdots$$
$$+ \frac{h^2}{2!}(\beta_{31} + \beta_{32})^2(\mathbf{f_{yy}} \; \mathbf{f} \; \mathbf{f})$$
$$+ \frac{2h^3}{2!}(\beta_{31} + \beta_{32})\beta_{32}\beta_{21}(\mathbf{f_{yy}} \; (\mathbf{f_y} \; \mathbf{f}) \; \mathbf{f})$$
$$+ \frac{h^3}{3!}(\beta_{31} + \beta_{32})^3(\mathbf{f_{yyy}} \; \mathbf{f} \; \mathbf{f} \; \mathbf{f}) + \cdots,$$

$$\mathbf{y}'_{n,4} = \mathbf{f}(\mathbf{y}_n + h(\beta_{41}\mathbf{y}'_{n,1} + \beta_{42}\mathbf{y}'_{n,2} + \beta_{43}\mathbf{y}'_{n,3}))$$
$$= \mathbf{f}(\mathbf{y}_n + h(\beta_{41}+\beta_{42}+\beta_{43})\mathbf{f} + h^2(\beta_{42}\beta_{21} + \beta_{43}(\beta_{31} + \beta_{32}))(\mathbf{f_y} \; \mathbf{f})$$
$$+ \frac{h^3}{2!}(2\beta_{43}\beta_{32}\beta_{21}(\mathbf{f_y} \; (\mathbf{f_y} \; \mathbf{f}))$$
$$+ (\beta_{42}\beta_{21}^2 + \beta_{43}(\beta_{31} + \beta_{32})^2)(\mathbf{f_{yy}} \; \mathbf{f} \; \mathbf{f})) + \cdots)$$
$$= \mathbf{f} + (\mathbf{f_y}(h(\beta_{41}+\beta_{42}+\beta_{43})\mathbf{f}+h^2(\beta_{42}\beta_{21}+\beta_{43}(\beta_{31}+\beta_{32}))(\mathbf{f_y} \; \mathbf{f})$$
$$+ \frac{h^3}{2!}(2\beta_{43}\beta_{32}\beta_{21}(\mathbf{f_y} \; (\mathbf{f_y} \; \mathbf{f}))$$
$$+ (\beta_{42}\beta_{21}^2 + \beta_{43}(\beta_{31} + \beta_{32})^2)(\mathbf{f_{yy}} \; \mathbf{f} \; \mathbf{f})) + \cdots))$$
$$+ (\frac{\mathbf{f_{yy}}}{2!} \; (h(\beta_{41} + \beta_{42} + \beta_{43})\mathbf{f}$$
$$+ h^2(\beta_{42}\beta_{21} + \beta_{43}(\beta_{31} + \beta_{32}))(\mathbf{f_y} \; \mathbf{f}) + \cdots))^2$$
$$+ (\frac{\mathbf{f_{yyy}}}{3!} \; (h(\beta_{41} + \beta_{42} + \beta_{43})\mathbf{f} + \cdots))^3 + \cdots$$

$$= \mathbf{f} + h(\beta_{41} + \beta_{42} + \beta_{43})(\mathbf{f_y}\ \mathbf{f})$$
$$+ h^2(\beta_{42}\beta_{21} + \beta_{43}(\beta_{31} + \beta_{32}))(\mathbf{f_y}\ (\mathbf{f_y}\ \mathbf{f}))$$
$$+ \frac{h^3}{2!}(2\beta_{43}\beta_{32}\beta_{21}(\mathbf{f_y}\ (\mathbf{f_y}\ (\mathbf{f_y}\ \mathbf{f}))))$$
$$+ (\beta_{42}\beta_{21}^2 + \beta_{43}(\beta_{31} + \beta_{32})^2)(\mathbf{f_y}\ (\mathbf{f_{yy}}\ \mathbf{f}\ \mathbf{f}) + \cdots))$$
$$+ \frac{h^2}{2!}(\beta_{41} + \beta_{42} + \beta_{43})^2(\mathbf{f_{yy}}\ \mathbf{f}\ \mathbf{f})$$
$$+ \frac{2h^3}{2!}(\beta_{41} + \beta_{42} + \beta_{43})(\beta_{42}\beta_{21} + \beta_{43}(\beta_{31} + \beta_{32}))(\mathbf{f_{yy}}(\mathbf{f_y}\ \mathbf{f})\ \mathbf{f}) + \cdots$$
$$+ \frac{h^3}{3!}((\beta_{41} + \beta_{42} + \beta_{43})^3(\mathbf{f_{yyy}}\ \mathbf{f}\ \mathbf{f}\ \mathbf{f}) + \cdots).$$

Below is a summary of the elementary differential terms that appeared in the expansion, according to their order in h. The notation $R^i_{l,m}$ identifies the order h^i, the order l of the leading derivative of $\mathbf{f}$, and the index m among such terms. Next to this is the T^k_j of the corresponding term in the Taylor expansion, followed by the equation of coefficients from the corresponding expansions. Recall the notation $\alpha_i = \sum_{j=1}^{i-1} \beta_{ij}$ for expressions that appear repeatedly in the expansion.

h^k, Elementary Differential	$R^i_{l,m}$	T^k_j	RK coefficient $= T$ coefficient
$h^1(\mathbf{f})$	$R^1_{1,1}$	T^1_1	$\gamma_1 + \gamma_2 + \gamma_3 + \gamma_4 = \frac{1}{1!}$
$h^2(\mathbf{f_y}\ \mathbf{f})$	$R^2_{1,1}$	T^2_1	$\gamma_2\alpha_2 + \gamma_3\alpha_3 + \gamma_4\alpha_4 = \frac{1}{2!}$
$h^3(\mathbf{f_{yy}}\ \mathbf{f}\ \mathbf{f})$	$R^3_{2,1}$	T^3_1	$\frac{1}{2!}(\gamma_2\alpha_2^2 + \gamma_3\alpha_3^2 + \gamma_4\alpha_4^2) = \frac{1}{3!}$
$h^3(\mathbf{f_y}\ (\mathbf{f_y}\ \mathbf{f}))$	$R^3_{1,1}$	T^3_2	$\gamma_3\beta_{32}\alpha_2 + \gamma_4(\beta_{42}\alpha_2 + \beta_{43}\alpha_3) = \frac{1}{3!}$
$h^4(\mathbf{f_{yyy}}\ \mathbf{f}\ \mathbf{f}\ \mathbf{f})$	$R^4_{3,1}$	T^4_1	$\frac{1}{3!}(\gamma_2\alpha_2^3 + \gamma_3\alpha_3^3 + \gamma_4\alpha_4^3) = \frac{1}{4!}$
$h^4(\mathbf{f_y}\ (\mathbf{f_{yy}}\ \mathbf{f}\ \mathbf{f}))$	$R^4_{1,2}$	T^4_3	$\frac{1}{2!}(\gamma_3\beta_{32}\alpha_2^2 + \gamma_4(\beta_{42}\alpha_2^2 + \beta_{43}\alpha_3^2)) = \frac{1}{4!}$
$h^4(\mathbf{f_{yy}}\ (\mathbf{f_y}\ \mathbf{f})\ \mathbf{f})$	$R^4_{2,1}$	T^4_3	$\gamma_3\alpha_3\beta_{32}\alpha_2 + \gamma_4\alpha_4(\beta_{42}\alpha_2 + \beta_{43}\alpha_3) = \frac{3}{4!}$
$h^4(\mathbf{f_y}\ (\mathbf{f_y}\ (\mathbf{f_y}\ \mathbf{f}])$	$R^4_{1,1}$	T^4_4	$\gamma_4\beta_{43}\beta_{32}\alpha_2 = \frac{1}{4!}$

The rooted trees $R^i_{l,m}$ corresponding to elementary differentials up to order $k = 4$ are shown below in order of their occurrence in stages $i = 1, \ldots, 4$ of the Runge-Kutta approximation. Within a

H. Runge-Kutta Methods

stage, we have listed trees by the order l of the l-fold product that produces them, i.e., by how many previously existing trees are attached to a new root node in order to construct the tree.

$R^1_{1,1} : \sum_i \gamma_i = 1$

$R^2_{1,1} : \sum \gamma_i \alpha_i = \frac{1}{2}$

$R^2_{2,1} : \sum \gamma_i \alpha_i^2 = \frac{1}{3}$

$R^2_{3,1} : \sum_i \gamma_i \alpha_i^3 = \frac{1}{3}$

$R^3_{1,1} : \sum\sum \gamma_i \beta_{ij} \alpha_j = \frac{1}{6}$

$R^3_{1,2} : \sum\sum \gamma_i \beta_{ij} \alpha_j^2 = \frac{1}{12}$

$R^3_{2,1} : \sum\sum \gamma_i \alpha_i \beta_{ij} \alpha_j = \frac{1}{8}$

$R^4_{1,1} : \sum\sum\sum \gamma_i \beta_{ij} \beta_{jk} \alpha_k = \frac{1}{24}$

Just as for the first-order Taylor expansion of $\mathbf{y}(t+h)$, there is only one tree at the first stage of the Runge-Kutta expansion, the tree corresponding to $\mathbf{f}$ itself. However, at the second stage, there is already an infinite family of trees corresponding to the infinite series of terms $(\mathbf{f_{y\cdots y}} \, \mathbf{f} \cdots \mathbf{f})$ with l $\mathbf{y}$'s and operands $\mathbf{f}$ for $l = 0, 1, \ldots$. So while the third-order tree corresponding to $(\mathbf{f_{yy}} \, \mathbf{f} \, \mathbf{f})$ appears at the second stage, the other third-order tree corresponding to $(\mathbf{f_y} \, (\mathbf{f_y} \, \mathbf{f}))$ does not. This tree only appears in the first-order term of the expansion of this series at the third stage when $(\mathbf{f_y}$ is added in front of existing terms including $(\mathbf{f_y} \, \mathbf{f})$, or in tree form, the tree corresponding to $(\mathbf{f_y} \, \mathbf{f})$ is attached to a new root node. Carrying this further, we can see that the tree with r nodes having depth $r-1$ corresponding to the elementary differential of the form $(\mathbf{f_y} \, (\mathbf{f_y} \ldots (\mathbf{f_y} \, \mathbf{f}])$ does not occur until the rth stage of a Runge-Kutta expansion. Note that the tree corresponding to $(\mathbf{f_{yy}} \, \mathbf{f} \, \mathbf{f})$, and in fact every tree that occurs at the second stage, also recurs as the first term in each order at every subsequent stage.

Conversely, since every tree with r nodes arises by attaching some number of trees with strictly fewer nodes to its root, every tree with r nodes does occur by the rth stage. This shows that r stages are necessary for a Runge-Kutta expansion to match a Taylor expansion to order r, because at least one term is missing with fewer stages. It also shows that all of the terms necessary for matching are present at the rth stage, but sufficiency depends upon the relation between the number of parameters, $r(r+1)/2$, that define an r stage method and the number of coefficient equations corresponding to the elementary differential (or rooted tree) up to a given order. In particular, we have seen that there is one parameter for one-stage methods and one tree for first-order agreement, and one method, Euler's Method, satisfies the matching conditions. There are three parameters for two-stage methods, and only two trees of order two or less, resulting in a one-parameter family of explicit two-stage Runge-Kutta Methods of order two. There are three more parameters for three-stage methods, and with two more trees at order three with conditions to match, we reach a two-parameter family of three-stage methods of order three. With four more parameters for a four-stage method, but also four additional trees with four nodes, there will again be a two-parameter family of four-stage methods of order four. But since there are nine rooted trees having five nodes, two free fourth-stage parameters plus five new fifth-stage parameters are still deficient by two. Six stages are required to achieve a fifth-order method.

The conditions for matching the Runge-Kutta and Taylor expansion terms involving an elementary differential can be obtained directly from the structure of the corresponding rooted tree. For this purpose, we relabel the nonleaf nodes with index symbols for summation. Note that the earlier labeling with **y**-derivatives of **f** was helpful but not actually necessary to recover the elementary differential, and the same holds here for recovering coefficients. The coefficients developed in successive stages arose from summing over the β parameters of the method times corresponding coefficients of previous elementary differential terms. Therefore, each time we attach an existing node to a new root node, we contribute a sum of the corresponding β coefficients to the coefficient corresponding to the resulting tree. Each leaf

H. Runge-Kutta Methods

node contributes $\alpha_i = \beta_{ij}$ for the stage it represents and therefore does not need to be indexed. Though the purposes are slightly different, the form in which β coefficients appear is the same as that of the γ coefficients. The β's are used to construct evaluation points for $\mathbf{f}$, $\mathbf{y}'_{n,i} = \mathbf{f}(\mathbf{y}_{n,i})$, with $\mathbf{y}_{n,i} = \mathbf{y}_n + \beta_{n,1}\mathbf{y}'_{n,1} + \cdots + \beta_{n,i}\mathbf{y}'_{n,i-1}$. When we have obtained sufficiently many (i.e., r) of these evaluations, the γ's are used to obtain $\mathbf{y}_{n+1} = \mathbf{y}_n + \gamma_{n,1}\mathbf{y}'_{n,1} + \cdots + \gamma_{n,r}\mathbf{y}'_{n,r}$. Because of this, the form of the coefficient formula associated with a particular tree is the same regardless of the number of stages of the method in which it appears. In other words, the coefficient formulas and corresponding matching conditions for a four-stage method reduce to those for a three-stage method simply by eliminating all terms involving coefficients whose first index is ≥ 4. If we only retain terms involving coefficients whose first index is ≤ 2, we recover the two conditions we found for a two-stage method to be second-order.

The rooted trees corresponding to exactness of the solution of the polynomial accuracy model problems $y' = (t^P)'$ are the depth-one trees with P nodes, $R^2_{P-1,1} = T^P_1$, that occur at the second stage. These are the only rooted trees of order $P \leq 2$. Therefore, exactness on equations whose solutions are polynomials of degree P is necessary and sufficient for general Pth-order accuracy of an explicit Runge-Kutta Method when $P \leq 2$. Even though these trees occur at the second stage, for $P > 2$, P stages are required to match the Taylor coefficient. The rooted trees corresponding to Pth-order accuracy for the absolute stability model problem $y' = \lambda y$ are the maximal depth trees with P nodes, $R^P_{1,1} = T^P_P$, i.e., the trees with one edge leaving every node except the leaf. For this problem, all derivatives of $\mathbf{f}(\mathbf{y})$ beyond the first are zero. This tree does not occur until the Pth stage. Therefore, P stages are necessary for general Pth-order accuracy.

Below the trees representing eight elementary differentials of order ≤ 4, we have collected the factors from the two sides of the matching conditions in the table into one of the form $1/d(T)$. Here $d(T)$ arises as the ratio of the multiplicity of ways the tree can be constructed by addition of edges and leaves on the Taylor side and a combination of lth-order expansion factorials and multinomial coefficients on the Runge-Kutta side. These factors can also be computed directly from their trees using the following simple algorithm. The density of any leaf, a rooted tree of order one arising from evaluating **f** in the first stage, is 1. At every subsequent stage, at which we attach one or more trees to a new root node, the density of the resulting trees is the product of the densities of the trees being attached, times the order of the resulting tree. For example, the tree $R_{2,1}^3$ is a rooted tree with four nodes corresponding to the elementary differential (**f**$_{yy}$ (**f**$_y$ **f**) **f**). It is first obtained at the $i = $ 3rd stage of the Runge-Kutta expansion in the $l = $ 2nd-order term of the expansion, by joining the trees corresponding to (**f**$_y$ **f**) and **f**. The former has density $1 \cdot 2$ and the latter has density 1, so since the resulting tree has order 4, its density is $1 \cdot 2 \cdot 4 = 8$.

For $r = 4$ stages, the eight fourth-order matching conditions are

$\gamma_1 + \gamma_2 + \gamma_3 + \gamma_4 = 1$, $\qquad \gamma_2\alpha_2 + \gamma_3\alpha_3 + \gamma_4\alpha_4 = \frac{1}{2}$,
$\gamma_2\alpha_2^2 + \gamma_3\alpha_3^2 + \gamma_4\alpha_4^2 = \frac{1}{3}$, $\qquad \gamma_3\beta_{32}\alpha_2 + \gamma_4(\beta_{42}\alpha_2 + \beta_{43}\alpha_3) = \frac{1}{6}$,
$\gamma_2\alpha_2^3 + \gamma_3\alpha_3^3 + \gamma_4\alpha_4^3 = \frac{1}{4}$, $\qquad \gamma_3\beta_{32}\alpha_2^2 + \gamma_4(\beta_{42}\alpha_2^2 + \beta_{43}\alpha_3^2) = \frac{1}{12}$,
$\gamma_3\alpha_3\beta_{32}\alpha_2$
$+\gamma_4\alpha_4(\beta_{42}\alpha_2 + \beta_{43}\alpha_3) = \frac{1}{8}$, $\quad \gamma_4\beta_{43}\beta_{32}\alpha_2 = \frac{1}{24}$.

The recommended procedure for solving the equations is to choose $\alpha_2, \ldots, \alpha_r$ and then solve the r equations, $\sum_{i=1}^{r} \gamma_i \alpha_i^k$, $k = 0, \ldots, r-1$, for $\gamma_1, \ldots, \gamma_r$. Next, solve for the β_{ij} that are determined by linear equations. In this case, the fourth, sixth, and seventh equations above allow us to solve for β_{32}, β_{42}, and β_{43}. The classical fourth-order Runge-Kutta Method corresponds to the solution obtained by setting $\alpha_2 = \alpha_3 = \frac{1}{2}$, $\alpha_4 = 1$, given in tableau form as

0				
$\frac{1}{2}$	$\frac{1}{2}$			
$\frac{1}{2}$	0	$\frac{1}{2}$		
1	0	0	1	
	$\frac{1}{6}$	$\frac{2}{6}$	$\frac{2}{6}$	$\frac{1}{6}$

For the scalar ODE with $f_y = 0$, i.e., $y' = f(t)$, this method reduces to the Simpson-parabolic quadrature method.

H. Runge-Kutta Methods

Among the methods satisfying the matching equations of a given order, optimal methods can be obtained by minimizing local truncation error bounds. A well-known example of this is the two-stage method of order 2 known as *Ralston's Method*, given in tableau form as

$$
\begin{array}{c|cc}
0 & & \\
\frac{3}{4} & \frac{3}{4} & \\
\hline
 & \frac{1}{3} & \frac{2}{3}
\end{array}
$$

Pairs of closely related Runge-Kutta Methods can be used for automatic step-size control in the same manner that pairs of multistep methods are used for error estimation and step-size modification. A well-known example of this technique is the Runge-Kutta-Fehlberg pair consisting of a five-stage and six-stage method of orders 4 and 5, respectively. Further details on these and other topics, including special cases for scalar, autonomous, and constant coefficient systems of equations, methods based on extrapolation, methods to treat second- and higher-order equations directly, etc., can be found in [BJ].

The region of absolute stability for an explicit r-stage method is determined by one step of the method applied to the absolute stability model problem, $y' = \lambda y$. For $r \leq 4$, we know that the coefficients can be chosen so that the method has order of accuracy $P = 4$. In this case the region of absolute stability is $\{w \in \mathbf{C} \mid |p_r(w)| \leq 1\}$ where $p_r(z) = \sum_{k=0}^{r} z^k/k!$, the truncation to degree r of the exact exponential series solution of the model problem (Figure 5.14). For $r > 4$, we must replace p_r by some polynomial of degree $\leq r$ that depends on the specifics of the method.

Implicit Runge-Kutta Methods can be employed if larger stability regions are required. See [IA1] for a discussion of these methods and their relation to Gauss-Legendre quadrature and collocation methods. Of particular relevance to the topic of this appendix are several publications on Runge-Kutta Methods for Hamiltonian systems. Explicit symplectic Runge-Kutta Methods only exist for general Hamiltonians that are separable, but the implicit Gauss-Legendre Runge-Kutta Methods are symplectic and they are optimal for general Hamiltonians. See [IA2], [SJM], [CS], [HLW], [YH], [CP].

Appendix I

Multistep Methods

In this appendix, we provide a brief introduction to the derivation of linear multistep methods and analyze the conditions on the coefficients that are necessary and sufficient to guarantee convergence of order P.

Among the seven basic examples in Chapter 5, one was a two-step method, the leapfrog method. Multistep methods potentially obtain more accurate approximations from fewer costly evaluations of a vector field if prior evaluations and values of the solution are stored for later use. With each additional stored value comes the need for an additional value for initialization not provided by the analytical problem. Each of these carries an additional degree of potential instability and adds to the cost of changing the step-size. In spite of this, some multistep methods have desirable absolute stability properties. We will also describe some relationships between the accuracy and stability of these methods.

Recall that we are considering methods for approximating solutions of the IVP

$$\mathbf{y}' = \mathbf{f}(t, \mathbf{y}), \quad \mathbf{y}(t_o) = \mathbf{y}_o, \quad t \in [t_o, t_o + T], \quad \mathbf{y} \in \mathbf{R}^D, \qquad (\text{I.1})$$

satisfying a Lipschitz condition in some norm on $\mathbf{R}^D$,

$$||\mathbf{f}(t, \mathbf{y}_1) - \mathbf{f}(t, \mathbf{y}_2)|| \leq L ||\mathbf{y}_1 - \mathbf{y}_2||. \qquad (\text{I.2})$$

For simplicity of exposition, we will henceforth use notation for the scalar case, but unless otherwise noted, generalizations to the case of systems are typically straightforward. In scalar notation, linear

m-step methods take the form

$$y_{n+1} = \sum_{j=0}^{m-1} a_j y_{n-j} + h \sum_{j=-1}^{m-1} b_j y'_{n-j}, \qquad (\text{I.3})$$

for all integers n satisfying $0 \leq nh \leq T$. Here, $t_0 = t_o$, $t_{n+1} = t_n + h$, $y'_j = f(t_j, y_j)$, and $y_0, \ldots, y_{m-1}$ are initial values obtained by companion methods discussed below. When the meaning is unambiguous, we will leave the dependence of y_n on h implicit.

There are several strategies that may be used to obtain families of linear m-step methods with higher-order accuracy. The first such family we will consider is the m-step *backward difference formula* methods, BDFm. These methods are derived by replacing the derivative on the left of $y'(t_{n+1}) = f(t_{n+1}, y(t_{n+1}))$ with the approximation obtained by differentiating the polynomial $p_m(t)$ of degree m that interpolates $y(t)$ at $t_{n+1}, t_n, \ldots, t_{n+1-m}$ and then discretizing. For $m = 1$, since

$$p_1(t) = y(t_{n+1}) + (t - t_{n+1}) \frac{y(t_{n+1}) - y(t_n)}{h},$$

we discretize

$$\frac{y(t_{n+1}) - y(t_n)}{h} = f(t_{n+1}, y(t_{n+1}))$$

and find that BDF1 is the Backward Euler Method,

$$y_{n+1} = y_n + h f(t_{n+1}, y_{n+1}).$$

For $m = 2$, iterative interpolation, described in Appendix J, yields

$$p_2(t) = y(t_{n+1}) + (t - t_{n+1})[\frac{y(t_{n+1}) - y(t_n)}{h}$$
$$+ (t - t_n) \frac{y(t_{n+1}) - 2y(t_n) + y(t_{n-1})}{2h^2}]$$

and

$$p'_2(t_{n+1}) = \frac{3y(t_{n+1}) - 4y(t_n) + y(t_{n-1})}{2h}.$$

Discretizing, we find that BDF2 is given by

$$y_{n+1} = \frac{4}{3} y_n - \frac{1}{3} y_{n-1} + \frac{2h}{3} f(t_{n+1}, y_{n+1}).$$

I. Multistep Methods

The Adams family of methods arises when we approximate the integral on the right of $y(t_{n+1}) - y(t_n) = \int_{t_n}^{t_{n+1}} y'(s)\, ds$ with

$$\int_{t_n}^{t_{n+1}} P_m^{A\cdot}(s)\, ds.$$

where $P^{A\cdot}$ interpolates $y'(s)$ at a prescribed set of time-steps, and then discretize. For the explicit *Adams-Bashforth* Methods, ABm, $P_m^{AB}(s)$ is the polynomial of degree m that interpolates $y'(s)$ at $t_n, \ldots, t_{n+1-m}$. For the implicit *Adams-Moulton* Methods, AMm, $P_m^{AM}(s)$ is the polynomial of degree $m+1$ that interpolates $y'(s)$ at $t_{n+1}, t_n, \ldots, t_{n+1-m}$.

For $m = 1$, $P_1^{AB}(s) = y'(t_n)$, and we find that AB1 is Euler's Method. Moreover,

$$P_1^{AM}(s) = y'(t_{n+1}) + (s - t_{n+1})\frac{y'(t_{n+1}) - y'(t_n)}{h}$$

and

$$\int_{t_n}^{t_{n+1}} P_1^{AM}(s)\, ds = h\frac{y'(t_{n+1}) + y'(t_n)}{2},$$

so we find that AM1 is the trapezoidal method. For $m = 2$,

$$P_2^{AB}(s) = y'(t_n) + (s - t_n)\frac{y'(t_n) - y'(t_{n-1})}{h}$$

and

$$\int_{t_n}^{t_{n+1}} P_2^{AB}(s)\, ds = h\frac{3y'(t_n) - y'(t_{n-1})}{2},$$

so AB2 is given by

$$y_{n+1} = y_n + h(\frac{3}{2}y'_n - \frac{1}{2}y'_{n-1}).$$

Again iterative interpolation yields

$$P_2^{AM}(s) = y'(t_{n+1}) + (s - t_{n+1})[\frac{y'(t_{n+1}) - y'(t_n)}{h}$$
$$+ (s - t_n)\frac{y'(t_{n+1}) - 2y'(t_n) + y'(t_{n-1})}{2h^2}].$$

Using

$$\int_{t_n}^{t_{n+1}} (s - t_{n+1})(s - t_n)\, ds = \int_0^h (u^2 - uh)\, du = -h^3/6$$

reduces this to

$$\int_{t_n}^{t_{n+1}} P_2^{AM}(s)\, ds = h[\frac{y'(t_{n+1}) + y'(t_n)}{2}$$
$$- \frac{y'(t_{n+1}) - 2y'(t_n) + y'(t_{n-1})}{12}].$$

Discretizing, we find that AM2 is given by

$$y_{n+1} = y_n + h(\frac{5}{12}y'_p + \frac{8}{12}y'_n - \frac{1}{12}y'_{n-1}).$$

Another strategy for deriving multistep methods obtains the coefficients a_j, b_j as solutions of linear equations that guarantee the method is formally accurate of order P. These conditions are related to the order of accuracy of a convergent method by the *local truncation error* of the method, ϵ_n. This quantity measures by how much a solution of the differential equation fails to satisfy the difference equation, in the sense

$$y(t_{n+1}) = \sum_{j=0}^{m-1} a_j y(t_{n-j})$$
$$+ h \sum_{j=-1}^{m-1} b_j y'(t_{n-j}) + \epsilon_n, \quad \text{where } y'_j = f(t_j, y(t_j)). \tag{I.4}$$

In its top row, Table I.1 contains the first few terms of the Taylor expansion of the left-hand side of (I.4), $y(t_{n+1})$, about t_n, in powers of h. Below the line, the rows contain the Taylor expansions of the terms $y(t_n - jh)$ and $y'(t_n - jh)$ on the right of (I.4), where we have placed terms of the same order in the same column and set $q = m = 1$ for compactness of notation.

Algebraic conditions that determine a bound on the order of ϵ_n are obtained by comparing the collective expansions of both sides. The terms in each column are multiples of $h^p y_n^{(p)}$. If we form a common denominator by multiplying the b terms by p/p, the right-hand

I. Multistep Methods

sides of order $p \geq 1$ have the form

$$\sum_{j=0}^{m-1} \frac{1}{p!}(-j)^p a_j h^p y_n^{(p)} + \sum_{j=-1}^{m-1} \frac{1}{p!} p(-j)^{p-1} b_j h^p y_n^{(p)},$$

so to make these terms equal, we must have

$$\sum_{j=0}^{m-1} (-j)^p a_j + \sum_{j=-1}^{m-1} p(-j)^{p-1} b_j = 1. \qquad (\text{I.5})$$

It can be shown using the linearity of the local truncation error with respect to solutions $y(t)$ and a basis of functions of the form $(t - t_n)^k$ that

$$\epsilon_n \leq C h^{P+1} \qquad (\text{I.6})$$

if and only if conditions (I.5) are satisfied for each order h^p, $p = 0, 1, \ldots, P$. The conditions (I.5) are equivalent to requiring that the numerical method is exact on polynomials of degree P, assuming that it is initialized with exact values.

Table I.1: Taylor expansions of $y(t_{n+1})$
and terms in multistep methods

	$y_{n+1} =$	$y_n +$	$y_n' h +$	$\frac{1}{2} y_n'' h^2 +$	$\frac{1}{6} y_n''' h^3 + \cdots$
	$b_{-1} h y_{n+1}' =$		$b_{-1} y_n' h +$	$b_{-1} y_n'' h^2 +$	$\frac{1}{2} b_{-1} y_n''' h^3 + \cdots$
$+$	$a_0 y_n = a_0 y_n$				
$+$	$b_0 h y_n' =$		$b_0 y_n' h$		
$+$	$a_1 y_{n-1} = a_1 y_n -$		$a_1 y_n' h +$	$\frac{1}{2} a_1 y_n'' h^2 -$	$\frac{1}{6} a_1 y_n''' h^3 + \cdots$
$+$	$b_1 h y_{n-1}' =$		$b_1 y_n' h -$	$b_1 y_n'' h^2 +$	$\frac{1}{2} b_1 y_n''' h^3 + \cdots$
	$\vdots$	$\vdots$	$\vdots$	$\vdots$	$\vdots$
$+$	$a_q y_{n-q} = a_q y_n -$		$a_q y_n' q h +$	$\frac{1}{2} a_q y_n'' q^2 h^2 -$	$\frac{1}{6} a_q y_{n-q}''' q^3 h^3 + \cdots$
$+$	$b_q h y_{n-q}' =$		$+ b_q y_n' h -$	$b_q y_n'' q h^2$	$-\frac{1}{2} b_q y_{n-q}''' q^2 h^3 + \cdots$

We say a method is *consistent* if conditions (I.5) are satisfied for $p = 0$ and $p = 1$, i.e., if

$$\sum_{j=0}^{m-1} a_j = 1 \quad \text{and} \quad \sum_{j=0}^{m-1} -j a_j + \sum_{j=-1}^{m-1} b_j = 1. \tag{I.7}$$

In this case we know that the method formally approximates the differential equation. This guarantees that the approximated equation is the one that we intended. The more subtle issue of convergence of a numerical method involves determining whether solutions of the approximating equation (in this case the multistep method) do indeed approximate solutions of the approximated equation as the discretization parameter tends to zero. The root condition for 0-stability discussed in Chapter 5 together with consistency are necessary and sufficient for a multistep method to be convergent. If, in addition, (I.5) is satisfied for all $p \leq P$, then the convergence is with global order of accuracy P.

Four of our working example methods of Chapter 5 and three additional methods discussed above fit into the linear m-step framework with $m \leq 2$. Table I.2 summarizes the nonzero coefficients defining these methods and identifies the value of P for which the matching conditions up to order P are satisfied, but not the conditions of order $P + 1$. For reference, the conditions for $p = 2, 3,$ and 4 are

$$\sum_{j=0}^{m-1} j^2 a_j - \sum_{j=-1}^{m-1} 2j b_j = 1,$$

$$-\sum_{j=0}^{m-1} j^3 a_j + \sum_{j=-1}^{m-1} 3j^2 b_j = 1,$$

and

$$\sum_{j=0}^{m-1} j^4 a_j - \sum_{j=-1}^{m-1} 4j^3 b_j = 1,$$

respectively.

I. Multistep Methods

Table I.2: Coefficients and order of accuracy
of example multistep methods

Method (P)	Coefficients
Euler (1)	$a_0 = b_0 = 1$
Backward Euler*(1)	$a_0 = 1$, $b_{-1} = 1$
Trapezoidal*(2)	$a_0 = 1$, $b_{-1} = b_0 = \frac{1}{2}$
Leapfrog (2)	$a_1 = 1$, $b_0 = 2$
BDF2*(2)	$a_0 = \frac{4}{3}$, $a_1 = -\frac{1}{3}$, $b_{-1} = \frac{2}{3}$
AB2(2)	$a_0 = 1$, $b_0 = \frac{3}{2}$, $b_1 = -\frac{1}{2}$
AM2*(3)	$a_0 = 1$, $b_{-1} = \frac{5}{12}$, $b_0 = \frac{8}{12}$, $b_1 = -\frac{1}{12}$

*Implicit method

Since explicit linear m-step methods are determined by $2m$ coefficients and implicit linear m-step methods are determined by $2m+1$ coefficients, we can obtain multistep methods not belonging to the BDF or Adams families by requiring that a method satisfy as many of the matching conditions of linear equations as possible. The first $2m$ or $2m+1$ conditions, respectively, form nonsingular systems of linear equations in those coefficients whose solution maximizes the order of the local truncation error.

Any linear m-step method with $m = 1$ that satisfies the consistency conditions $a_0 = 1$ and $b_{-1} + b_0 = 1$ is among the family of θ-methods:

$$y_{n+1} = y_n + h((1-\theta)y'_n + \theta y'_{n+1}). \tag{I.8}$$

This family includes one explicit method, Euler's Method, for $\theta = 0$. Second-order accuracy requires $2b_{-1} = 1$, corresponding to the trapezoidal method with $\theta = \frac{1}{2}$. Since the order 3 condition $3b_{-1} = 1$ is not satisfied, the maximal order of an implicit method with $m = 1$ is 2, attained by the trapezoidal method. The θ-method family also includes the Backward Euler Method ($\theta = 1$). The restriction $\theta \in [0,1]$ is not required for consistency, but since the amplification factor is

$$a(w) = \frac{(1+(1-\theta)w)}{(1-\theta w)},$$

it is standard to assure greater stability for λ in the left half-plane,

To obtain an explicit two-step method with local truncation error of order 4 in this way, we look for a method of the form

$$y_{n+1} = a_0 y_n + a_1 y_{n-1} + h(b_0 y'_n + b_1 y'_{n-1})$$

whose coefficients satisfy the four linear conditions $a_0 + a_1 = 1$, $-a_1 + b_0 + b_1 = 1$, $a_1 - 2b_1 = 1$, $-a_1 + 3b_1 = 1$. The method corresponding to the unique solution of this system is

$$y_{n+1} = -4y_n + 5y_{n-1} + h(4y'_n + 2y'_{n-1}). \tag{I.9}$$

We showed in Section 5.3 that (I.9) is not 0-stable. Another method that is consistent, but not 0-stable, is

$$y_{n+1} = 3y_n - 2y_{n-1} - hy'_n. \tag{I.9$'$}$$

We can confirm the instability by considering the roots of its characteristic polynomial for $w = 0$, $p_0(r) = \rho(r) = r^2 - 3r + 2 = (r-1)(r-2)$. Though this method does not satisfy the second-order accuracy conditions, keeping the same $a_0 = 3$, $a_1 = -2$ and modifying the derivative coefficients to $b_0 = \frac{1}{2}$ and $b_1 = -\frac{3}{2}$ yields a method that would be second-order accurate were it not for the same instability.

The connection between the truly unstable behavior of the method (I.10), $y_{n+1} = 3y_n - 2y_{n-1} - hy'_n$, and the roots of its characteristic polynomial for $w = 0$, $p_0(r) = \rho(r) = r^2 - 3r + 2 = (r-1)(r-2)$, is apparent. This also make it clear that we could extend the example to have higher-order truncation error while retaining the same unstable behavior by keeping the same $a_0 = 3$, $a_1 = -2$ but modifying the derivative coefficients to $b_0 = \frac{1}{2}$ and $b_1 = -\frac{3}{2}$.

To obtain an implicit two-step method with local truncation error of order 5 by solving order conditions, we look for a method of the form

$$y_{n+1} = a_0 y_n + a_1 y_{n-1} + h(b_{-1} y'_{n+1} + b_0 y'_n + b_1 y'_{n-1})$$

whose coefficients satisfy the five linear conditions $a_0 + a_1 = 1$, $-a_1 + b_{-1} + b_0 + b_1 = 1$, $a_1 + 2b_{-1} - 2b_1 = 1$, $-a_1 + 3b_{-1} + 3b_1 = 1$,

I. Multistep Methods

$a_1 + 4b_{-1} - 4b_1 = 1$. The method corresponding to the unique solution of this system is

$$y_{n+1} = y_{n-1} + 2h\left(\frac{1}{6}y'_{n+1} + \frac{4}{6}y'_n + \frac{1}{6}y'_{n-1}\right), \quad (\text{I}.10)$$

known as *Milne's corrector*. We can also interpret this as integrating quadratic interpolation of y' at t_{n+1}, t_n, t_{n-1} (the Simpson-parabolic rule) to approximate the integral in

$$y_{n+1} - y_{n-1} = \int_{t_{n-1}}^{t_{n+1}} y'(s)\, ds.$$

Additional families of methods may be obtained using approximations of the integral in

$$y_{n+1} - y_{n-j} = \int_{t_{n-j}}^{t_{n+1}} y'(s)\, ds$$

for larger values of j.

Seeking higher-order accuracy to improve efficiency does not assure convergence. It can actually hinder it by compromising 0-stability. This is the case even for the implicit methods, i.e., they do not always have stability properties that are superior to those of explicit methods. In fact, for $m > 6$, the backward difference methods, BDFm, are implicit methods with arbitrarily high formal accuracy, but they are not even 0-stable.

If a general multistep method is applied to the model problem $y' = \lambda y$ and we set $w = \lambda h$, it takes the form of a homogeneous linear difference equation

$$(1 - b_{-1}w)y_{n+1} = \sum_{j=0}^{m-1}(a_j + b_j w)y_{n-j}. \quad (\text{I}.11)$$

We call the polynomial

$$p_w(r) = (1 - b_{-1}w)r^m - \sum_{j=0}^{m-1}(a_j + b_j w)r^{m-(j+1)}$$

the *characteristic polynomial* of the multistep method (I.3). We also define $\rho(r)$ and $\sigma(r)$ by $p_w(r) = \rho(r) + w\sigma(r)$, so in particular,

$$\rho(r) = p_0(r) = r^m - \sum_{j=0}^{m-1} a_j r^{m-(j+1)}.$$

When $p_w(r)$ has distinct roots $r_j(w)$, $j = 0, \ldots, m-1$, the general solution of (I.11) is a linear combination

$$y_n = \sum_{j=0}^{m-1} c_j r_j^n. \tag{I.12}$$

If $p_w(r)$ has some multiple roots, we can index any set of them consecutively, $r_j(w) = \cdots = r_{j+s}(w)$, in which case we replace the corresponding terms in (I.12) by terms of the form $c_{j+k} n^k r_j^n$, $k = 0, \ldots, s$.

As $w \to 0$, the roots of $r_j(w)$ approach corresponding roots of $\rho(r)$. We can use the fact that some root $r(w)$ must approximate $e^w = 1 + 1w$ as $w \to 0$ as another derivation of the consistency conditions (I.7). Since $e^0 = 1$ must be a root of p_0, $p_0(1) = 1 - \sum_j a_j = 0$, which is the zeroth-order consistency condition. Treating $r(w)$ as a curve defined implicitly by the relation $P(r, w) = p_w(r) = 0$ and differentiate implicitly with respect to w at $(r, w) = (1, 0)$, we obtain

$$-\sum_{j=-1}^{m-1} b_j + r'(w) \left(m - \sum_{j=0}^{m-1} a_j (m - (j+1)) \right) = 0.$$

Employing the zeroth-order consistency condition, factoring m from the second term, and setting $r'(w) = 1$ yields the first-order consistency condition of (I.7). This approach can be continued to any order. Alternatively, we may consider to what degree $r = e^w$ is a solution of the characteristic equation $\rho(r) + w\sigma(r) = 0$. The equations (I.5) for $p = 0, \ldots, P$ are equivalent to $\rho(e^w) + w\sigma(e^w) = O(w^{P+1})$ as $w \to 0$. If we use $w = \ln(r)$ to write this in the form $\rho(r) + \ln(r)\sigma(r) = 0$, they are also equivalent to

$$\rho(r) + \ln(r)\sigma(r) = C|r-1|^{P+1} + O(|r-1|^{P+2}), \tag{I.13}$$

I. Multistep Methods

as $r \to 1$ (so $w \to 0$). It is convenient to expand $\ln(r)$ in powers of $u = r - 1$ near $u = 0$, in which case (I.13) becomes

$$\rho(1+u) + \ln(1+u)\sigma(1+u) = C|u|^{P+1} + O(|u|^{P+2}).$$

In terms of the coefficients a_j and b_j of the numerical method, and using $q = m - 1$ as before, this becomes

$$(1+u)^m - (a_0(1+u)^q + \cdots + a_q) - (u - \frac{u^2}{2} + \frac{u^3}{3} - \cdots)$$
$$\times (b_{-1}(1+u)^m + b_0(1+u)^q + \cdots + b_q) = C|u|^{P+1} + O(|u|^{P+2}). \tag{I.14}$$

The condition that the coefficient of the u^p term on the left-hand side vanishes is equivalent to the order p matching condition we have given above.

The competition between accuracy and stability is explained in part by two results of Dahlquist that describe barriers to the order of accuracy of multistep methods that satisfy certain stability conditions. The first barrier gives the maximum order of a stable m-step method. Specifying $m - 1$ nonprincipal roots of $\rho(r)$ that satisfy the root condition is equivalent to specifying $m - 1$ real parameters that describe some combination of real roots and complex conjugate pairs. Along with $r_0 = 1$, these determine the real coefficients a_j through $\rho(r) = \Pi(r - r_j)$. Depending on whether the method is explicit or implicit, this leaves m or $m+1$ coefficients b_j with which to satisfy the accuracy conditions of order $p = 1, \ldots, P$. If the method is explicit, one would expect that this is possible through $P = m$, and through $P = m + 1$ if the method is implicit. We know that these are attainable from the examples of AB2 and AM2, stable two-step methods of order 2 and 3, respectively. In the explicit case, this bound turns out to be correct in general, and also in the implicit case if m is odd. However, if m is even, it is possible to satisfy one more additional equation, i.e., there are stable implicit m-step methods with order $m + 2$, but none higher. Milne's corrector satisfies the root condition, so it is 0-stable and convergent. But, for arbitrarily small w the magnitude of the root of $\rho_w(r)$ that approaches -1 as $w \to 0$ can exceed that of the principal root that approaches $+1$. Because of this, it lacks a desirable property called relative stability, but it is still

0-stable and convergent. It is suggestive that this method contains a form of the Simpson-parabolic integration method, an example of the Newton-Cotes quadrature methods based on an odd number of nodes. Due to symmetry, these quadrature methods attain an additional degree of accuracy over the number of nodes when the number of nodes is odd.

The second barrier refers to methods that are *A-stable*, which means that their region of absolute stability contains the entire left half-plane, i.e., all $w \in \mathbf{C}$ such that $\text{Re}(w) \leq 0$. Dahlquist showed that any A-stable linear multistep method has order of accuracy less than or equal to 2. Because of the usefulness of methods with large regions of absolute stability, considerable effort has gone into finding higher-order $A(\alpha)$-stable methods whose regions of absolute stability contain large wedges symmetric about the negative real axis in the left half-plane.

The analysis of propagation of errors for linear multistep methods involves issues arising from multiple initial values and modes of amplification that are not present in one-step methods. When we analyzed the error propagation of Euler's Method, we saw that the global error is bounded in terms of a sum of contributions arising from initial error and local truncation error, interacting with the amplification associated with the method. The portion of the bound resulting from the local truncation error has order one less than that of the local truncation error itself, while the portion resulting from the initialization error has the same order of the initialization error. The heuristic explanation is that the number of steps in which truncation errors are introduced grows in inverse proportion to the step size, contributing a factor of h^{-1}. Initialization errors may be amplified by some bounded constant, but they are introduced in a fixed number of steps that are independent of h. So the global order of accuracy of one-step and 0-stable linear multistep methods is at most one less than the order of the local truncation error. But initialization errors are only introduced in a fixed number of steps that is independent of h, so their contribution to the global error has the same magnitude as that of the initialization errors themselves. For the global order to be as small as possible, the initial values must also be one less than the

I. Multistep Methods

order of the local truncation error; any more accuracy is wasted. For one step methods, the initial value can be considered exact, since it is given in the IVP, though even this value may include experimental or computational errors. But for m-step methods with $m > 1$, we must use one-step methods to generate one or more additional values. Once we have a second initial value, we could also use a two-step method to generate a third, then a three-step method to generate a fourth, and so on. No matter how we choose to do this, it is just the order of the (local truncation) error of the initial values that limits the global error of the solution. For this reason, it is sufficient to initialize a method whose local truncation error has order $P + 1$ using a method whose local truncation error has order P. For example, the local truncation error of the leapfrog method has order 3. If $y_0 = y_o$, the exact initial value, and we use Euler's Method, whose local truncation error has order 2, to obtain y_1 from y_0, the resulting method has global order of accuracy 2. If we use the midpoint method or Heun's Method, whose local truncation errors both have order 3, the global order of accuracy of the resulting methods is still 2, no more accurate than if we use Euler's Method to initialize. But if we use a lower-order approximation, $y_1 = y_0$, a method whose local truncation error has order 1 and is not even consistent, the savings of one evaluation of f degrades the convergence of all subsequent steps to global order 1. As another example, the two-step implicit Adams-Moulton Method, AM2, has local truncation error of order 4. If we initialize it with the midpoint method or Heun's Method, we achieve the greatest possible global order of accuracy, 3. Initializing with RK4 will not improve this behavior, and initializing with Euler's Method degrades the order to 2. So the reason for including initial errors in the analysis of error propagation for one-step methods is clarified when we consider multistep methods.

When y_{n+1} is only defined implicitly, the ease with which we can determine its value from y_n (and previous values in the case of a multistep method) is significant from both practical and theoretical points of view. In the first place, a solution might not even exist for all values of $h > 0$. For a simple one-step method such as the Backward Euler Method, it can fail to have a solution even for the linear equation

$y' = \lambda y$, $y(0) = y_o$, where it reduces to $y_{n+1}(1 - \lambda h) = y_n$, which clearly has no solution if $\lambda h = 1$ and $y_n \neq 0$.

When $b_{-1} \neq 0$, (I.3) can be considered as a family of fixed-point equations $y_{n+1} = T(y_{n+1}, h)$ depending on the parameter h. If we let

$$y_{n+1}{}^* = \sum_{j=0}^{m-1} a_j y_{n-j}, \qquad (I.15)$$

then $y_{n+1}{}^* = T(y_{n+1}{}^*, 0)$. Using the Lipschitz continuity of f with respect to y and the linearity of T with respect to h, we can show that for sufficiently small h, $T(\cdot, h)$ is a contraction that maps an interval I containing $y_{n+1}{}^*$ into itself. By the contraction mapping principle, for any $y_{n+1}{}^{(0)}$ in this interval, the iteration

$$y_{n+1}{}^{(k+1)} = T(y_{n+1}{}^{(k)}, h) \qquad (I.16)$$

converges linearly to a unique fixed point y_{n+1}, satisfying $y_{n+1} = T(y_{n+1}, h)$, with rate

$$b_{-1} h |\frac{\partial f}{\partial y}(y_{n+1})| \leq |b_{-1} h L|. \qquad (I.17)$$

The situation for implicit Runge-Kutta Methods is more involved, since each step requires the solution of a nonlinear system of equations, but the same principles can be extended to derive existence, smooth dependence, and a convergence rate proportional to h, when h is sufficiently small.

It is a key principle in the design, analysis, and implementation of predictor-corrector methods that the convergence rate of the fixed-point iteration (I.16) is proportional to h. If we perform a single step of an implicit method of order P by iterating (I.16) to convergence, the resulting $y_{n+1}{}^\infty$ has a local truncation error that behaves like $|y(t_{n+1}) - y_{n+1}{}^\infty| \approx C h^{P+1}$ as $h \to 0$. If we only iterate (I.16) so far that $|y_{n+1}{}^{(k)} - y_{n+1}{}^\infty| \approx C' h^{P+1}$ as $h \to 0$ as well, then using $y_{n+1}{}^{(k)}$ instead of $y_{n+1}{}^\infty$ should still result in a method with the same order.

We may do this in a variety of ways, but it is common to initialize the iteration with an explicit method, called a *predictor*, whose order is the same or one less than that of the implicit method, P.

I. Multistep Methods

Each iteration of (I.16) is called a *corrector* step, and if our predictor has global order $P - 1$, its local truncation error will behave like $|y(t_{n+1}) - y_{n+1}^{(0)}| \approx C_P h^P$. Since this dominates the local truncation error of the corrector, $|y_{n+1}^\infty - y_{n+1}^{(0)}| \approx C_P h^P$, it makes sense to perform one corrector iteration. Due to the h dependence of the rate of convergence, $|y_{n+1}^\infty - y_{n+1}^{(1)}| \approx C_C h^{P+1}$ and therefore $|y(t_{n+1}) - y_{n+1}^{(1)}| \approx C_C' h^{P+1}$, and further iterations do not increase the order of the local truncation error. If a predictor is already as accurate as the implicit method, $y(t_{n+1}) - y_{n+1}^{(0)} \approx C_P' h^{P+1}$, it would seem pointless to iterate, since one iteration provides no improvement in overall accuracy. At the opposite extreme, we could even initialize with the *constant method*, $y_{n+1} = y_n$, for which the local truncation error $|y(t_{n+1}) - y_{n+1}^{(0)}| \approx y_n' h$, and perform P corrector iterations.

We now consider two simple concrete examples. The implicit method we will use in the first example is the Backward Euler Method, and in the second example we will use the trapezoidal method. We will analyze both accuracy and stability for the model problem $y' = \lambda y$ in order to understand why it makes sense to correct to—or even beyond—the maximal achievable accuracy of the method. One reason is improvement in the region of absolute stability. The region of absolute stability of the explicit method corresponding to $y_{n+1} = y_{n+1}^{(0)}$ gets deformed step by step into that of the implicit method corresponding to $y_{n+1} = y_{n+1}^\infty$. A second reason is that the difference between $y_{n+1}^{(0)}$ and $y_{n+1}^{(1)}$, obtained from a corrector of the same order, can be used to estimate local errors with very little additional computation, and this can be used to adjust the step-size automatically and even change on the fly to an appropriate higher- or lower-order method.

For the purpose of analyzing the local truncation error in both examples, we let $y(t_{n+1})$ be the exact solution passing through (t_n, y_n), evaluated at t_{n+1}, so $y(t_{n+1}) = y_n e^{\lambda h} = \sum_{j=0}^\infty \frac{(\lambda h)^j}{j!}$. The iteration (I.16) corresponding to the Backward Euler Method is

$$y_{n+1}^{(k+1)} = y_n + h f(t_{n+1}, y_{n+1}^{(k)}).$$

For the model problem $f(t, y) = \lambda y$, if we initialize the iteration with $y_{n+1}^{(0)} = y_n$ and perform no iterations, the local truncation error behaves like $y'(t_n)h = \lambda h y_n$ as $h \to 0$. If we iterate once, $y_{n+1}^{(1)} = (1 + \lambda h)y_n$, and the result is no different than if we had applied one step of Euler's Method, which has local truncation error $y(t_{n+1}) - y_{n+1}^{(1)} \approx y''(t_n)\frac{h^2}{2} = \frac{(\lambda h)^2}{2}y_n$ as $h \to 0$. Another iteration gives $y_{n+1}^{(2)} = (1 + \lambda h + (\lambda h)^2)y_n$, and we may also think of this as using an Euler's Method predictor followed by one Backward Euler corrector step. The local truncation error $y(t_{n+1}) - y_{n+1}^{(2)} \approx -y''(t_n)\frac{h^2}{2} = -\frac{(\lambda h)^2}{2}y_n$. This is to be expected since after this iteration, $y_{n+1}^{(2)}$ is an $O(h^3)$ approximation of the approximation $y_{n+1}^{(\infty)} = (1 - \lambda h)^{-1}y_n = \sum_{j=0}^{\infty}(\lambda h)^j$ whose terms to order h^2 agree with those of $y_{n+1}^{(2)}$ above and only agree to order h with $y(t_{n+1})$. Therefore, $y_{n+1}^{(2)}$ shares the same error behavior as $y(t_{n+1}) - y_{n+1}^{(\infty)} \approx -y''(t_n)\frac{h^2}{2} = -\frac{(\lambda h)^2}{2}y_n$ as $h \to 0$. Better approximations of $y_{n+1}^{(\infty)}$ are not better approximations of $y(t_{n+1})$.

The benefits of these iterations are increased stability and error estimation. Euler's Method is never absolutely stable for $w = \lambda h$ on the imaginary axis, since $|1 + w| > 1$ for $w = ai$, $a \neq 0$. The amplification factor corresponding to $y_{n+1}^{(2)}$, $a(w) = 1 + w + w^2$, satisfies $|1+w+w^2| \leq 1$ for $w = ai$, $a \in [-1, 1]$. Also, we can subtract $y(t_{n+1}) - y_{n+1}^{(2)} \approx -y''(t_n)\frac{h^2}{2}$ from $y_{n+1}^{(2)} - y_{n+1}^{(1)} \approx y''(t_n)h^2$, to obtain an estimate of the local truncation error in terms of computed quantities, $y(t_{n+1}) - y_{n+1}^{(2)} \approx \frac{1}{2}(y_{n+1}^{(2)} - y_{n+1}^{(1)})$ as $h \to 0$. If this error exceeds a certain bound, we may decide to reduce the step size, or we can use these quantities once more to increase the order of our method by canceling the leading terms in their errors (a process known as extrapolation). In this case, we can define a new predictor from their mean, $y_{n+1}^0 = \frac{1}{2}(y_{n+1}^{(2)} - y_{n+1}^{(1)})$, and expect its local truncation error to behave as h^3 as $h \to 0$. This is indeed the case, as this is just Heun's Method. In conjunction with an implicit method with second-order accuracy, we could then continue the process. For this reason, we briefly perform a similar analysis of a trapezoidal corrector.

The iteration (I.16) corresponding to the trapezoidal method is

$$y_{n+1}^{(k+1)} = y_n + \frac{h}{2}(f(t_n, y_n) + f(t_{n+1}, y_{n+1}^{(k)})).$$

I. Multistep Methods

For the model problem $f(t,y) = \lambda y$, if we initialize the iteration with $y_{n+1}^{(0)} = y_n$ and perform no iterations, the local truncation error behaves like $y'(t_n)h = \lambda h y_n$ as $h \to 0$. If we iterate once, $y_{n+1}^{(1)} = (1+\lambda h)y_n$ and the result is *still* no different than if we had applied one step of Euler's Method, which has local truncation error $y(t_{n+1}) - y_{n+1}^{(1)} \approx y''(t_n)\frac{h^2}{2} = \frac{(\lambda h)^2}{2}y_n$ as $h \to 0$. But this time, another iteration gives $y_{n+1}^{(2)} = (1 + \lambda h + \frac{(\lambda h)^2}{2})y_n$, and we may also think of this as using an Euler's Method predictor followed by one trapezoidal corrector step, i.e., Heun's Method. The local truncation error $y(t_{n+1}) - y_{n+1}^{(2)} \approx -y''(t_n)\frac{h^3}{6} = \frac{(\lambda h)^3}{6}y_n$. Beyond the first iteration, $y_{n+1}^{(k)}$ attains the same accuracy of $y_{n+1}^{(\infty)}$; better approximations of $y_{n+1}^{(\infty)}$ are not better approximations of $y(t_{n+1})$. The asymptotic form of the third-order local error terms for Heun's Method, a Runge-Kutta Method, depends on the problem to which it is applied. For the model problem, the asymptotic form of the local error is identical to that of the trapezoidal method, so they cannot be used together even in this situation for local error estimation with the Milne device.

For any $m > 0$, the ABm predictor, AM$m-1$ corrector pair has good stability and convergence properties and does satisfy the common order requirements of the Milne device.

Using a predictor with one iteration of a corrector is sometimes denoted PC or PC^1, with n corrector iterations, PC^n, and corrector iterations to convergence, PC^∞. Redefining $\tilde{P} = PC^{n-1}$ to be a new predictor turns PC^n into $\tilde{P}C$. The evaluations of $f(t,y)$ in the definition of the method are sometimes denoted as a distinct step with the letter E, especially when expensive updates can be omitted and iterations can be usefully performed with prior evaluations. This is somewhat analogous to using a single Jacobian for multiple iterations of Newton's Method for the reduction of operations from N^3 to N^2. When all evaluations are performed, the method is denoted $PE(CE)^n$, though there are many variations possible. If absolute stability of the implicit method is essential for the particular problem at hand, e.g., when an A-stable method is required, more efficient approaches to computing y_{n+1} (e.g., Newton or quasi-Newton Methods) should be implemented for quadratic rather than linear convergence.

This highlights a perspective that the predictor-corrector idea is *not* only about finding the solution of the implicit stepping method—that may be found by more efficient means. Rather, it is about designing efficient intermediate explicit methods with enhanced stability and error estimation properties. The use of two methods of common order that share a substantial portion of their computational effort for automatic step-size control is not limited to explicit-implicit pairs. The same idea is used in one-step explicit methods, e.g., the Runge-Kutta-Fehlberg pair.

We conclude with some general remarks on the theoretical and practical effects of stability and instability for multistep methods. The growth of errors in numerical approximations obtained from linear multistep methods is governed by a difference inequality consisting of a homogeneous error amplification term correlated with the stability properties of the method and an inhomogeneous error-forcing term correlated with the local truncation error. The contribution from error amplification is governed by the same recurrence that determines the stability of the method and in particular by the behavior of the method on the model problem $y' = \lambda y$ where $\lambda = L$, the Lipschitz constant for f.

The most fundamental result in the theory of linear multistep methods is known as *Dahlquist's Equivalence Theorem*. This theorem relates the root condition and convergence as follows. Let (I.3) be a consistent m-step method applied to the well-posed IVP (I.1), with initial values $y_0, \ldots, y_{m-1}$ approaching y_o as $h \to 0$. The method is convergent if and only if $\rho(r)$ satisfies the root condition. Furthermore, if the local truncation error defined in (I.4) satisfies $|\epsilon_n| \leq C_T h^{P+1}$ and the initial errors $e_{j,h} = y_{j,h} - y(t_j)$, $j = 0, \ldots, m-1$, satisfy
$$\max_{j=0,\ldots,m-1} |e_{j,h}| \leq C_I h^P,$$
then
$$\max_{0 \leq nh \leq T} |y_{n,h} - y(t_n)| \leq C_G h^P$$
as $h \to 0$. In other words, if the local truncation error has order $P+1$, the initial error only needs to have order P for the global convergence

I. Multistep Methods

to have order P. A proof that applies to an even more general class of methods may be found in [IK]. In the much greater generality of linear finite difference methods for partial differential equations, the fact that stability and consistency together are both necessary and sufficient for convergence is the content of the important Lax-Richtmyer Equivalence Theorem [LR].

There are several other more stringent conditions that have been developed to distinguish the behavior observed in convergent linear multistep methods. The *strong root condition* says that except for r_0, the roots of $\rho(r)$ are all inside the open unit disc, a condition that, as the name suggests, clearly implies the root condition. By continuity, for sufficiently small h, the nonprincipal roots $r_j(w)$ of $\rho(r) + w\sigma(r)$ will also have magnitude less than 1. If the coefficients of an m-step method with $m > 1$ satisfy $a_j = 0$ for $j < m - 1$, so $a_{m-1} = 1$ for consistency, it cannot satisfy the strong root condition, since the roots of $\rho(r) = r^m - 1$ are all of the form $r_j = e^{2\pi j/m}$, $j = 0, \ldots, m-1$. This class includes the leapfrog method and Milne's corrector. Even if the strong root condition is not satisfied, we can require that for sufficiently small h, the parasitic roots of $p_w(r)$ have magnitudes less than or equal to the magnitude of the principal root, a condition called *relative stability*. In this case, parasitic roots can only grow exponentially when the principal root is growing faster exponentially, making them less of a concern. The term *weak stability* is used to describe a method that is stable but not relatively stable. Since the leapfrog method satisfies the root condition and we have shown that as $h \to 0$, the parasitic root of the leapfrog method has magnitude greater than its principal root, the leapfrog method is weakly stable and demonstrates that the root condition cannot imply relative stability. However, the continuity argument that shows that the strong root condition implies the root condition can be used just as easily to show that the strong root condition implies relative stability. By the observation above, the consistent two-step method $y_{n+1} = y_{n-1} + 2hy'_{n-1}$ cannot satisfy the strong root condition. But since its characteristic polynomial is

$$r^2 - (1 + 2w) = (r - \sqrt{1 + 2w})(r + \sqrt{1 + 2w}),$$

its principal and parasitic roots have the same magnitude and it is relatively stable. This shows that relative stability is strictly weaker than the strong root condition.

For Euler's Method, the Backward Euler Method, and the trapezoidal method, $\rho(r) = r - 1$. Since there are no nonprincipal roots, they satisfy the strong root condition, the root condition, and the relative stability condition by default. Both the explicit and the implicit m-step Adams Methods, ABm and AMm, are designed to have $\rho(r) = r^m - r^{m-1} = (r-1)r^{m-1}$, so that all parasitic roots are zero! These methods satisfy the strong root condition, as nicely as possible. For BDF2, $\rho(r) = r^2 - \frac{4}{3}r + \frac{1}{3} = (r-1)(r - \frac{1}{3})$ satisfies the strong root condition. For higher m, BDFm is designed to have order of accuracy m if the method is convergent. However, these methods are only 0-stable for $m \leq 6$, so BDFm is not convergent for $m \geq 7$.

If we apply Milne's corrector, the implicit two-step method having maximal local truncation error, to the model problem $y' = \lambda y$, it takes the form

$$y_{n+1} = y_{n-1} + (\frac{w}{3}y_{n+1} + \frac{4w}{3}y_n + \frac{w}{3}y_{n-1}).$$

Solutions are linear combinations $y_n = c_+ r_+ + c_- r_-$ where $r_\pm$ are the roots of

$$p_w(r) = (1 - w/3)r^2 - (4w/3)r - (1 + w/3).$$

By setting $u = w/3$ and multiplying by $1/(1-u) = 1 + u + \cdots$, to first order in u, these roots satisfy

$$r^2 - 4u(1 + \cdots)r - (1 + 2u + \cdots) = 0$$

or

$$r_\pm = 2u \pm \sqrt{4u^2 + 1 + 2u}.$$

Using the binomial expansion $(1+2u)^{1/2} \approx 1 + u + \cdots$, to first order in u, $r_+ \approx 1 + 3u$ and $r_- \approx -1 + u$. The root $\approx 1 + 3u = 1 + \lambda h$ approximates the solution of the model problem $y' = \lambda y$. As $\lambda h \to 0$ in a way that u in a neighborhood of the negative real axis near the origin, the other root $\approx -1 + u$ has magnitude greater than 1, showing that Milne's Method corrector is not relatively stable. Like the leapfrog method, it satisfies the root condition, so it is stable, but

I. Multistep Methods

only weakly stable. So even a convergent implicit method can have worse stability properties than a convergent explicit method.

We can summarize the relationship among various types of stability for multistep methods and their consequences as follows:

strong root condition ⇒ relative stability ⇒ root condition,

absolute stability on $M_S(0)$ ⇒ root condition on $\rho(r)$ ⇔ 0-stability,

consistency + root condition ⇔ convergence,

Pth-order formal accuracy+root condition ⇔ Pth-order convergence.

We conclude by cautioning the reader that many examples in Chapter 5 show that the behavior of a method is not determined by either the order of its local truncation error or by the rather loose bounds provided by 0-stability. 0-stability is certainly important, helping us to avoid nonconvergent methods, a caveat that cannot be overemphasized. Convergence and stability are a minimal but important requirement for a useful method. Along with the order of accuracy, the actual performance of a method is more closely correlated with its absolute stability with respect to modes present in the ODE.

▷ **Exercise I–1.** Determine the regions of absolute stability for BDF2, AM2, and AM2. For implicit methods, it may be advantageous to consider the fact that if $r \neq 0$ is a root of $ar^2 + br + c$, then $s = \frac{1}{r}$ is a root of $cs^2 + bs + a$.

Appendix J

Iterative Interpolation and Its Error

In this appendix we give a brief review of iterative polynomial interpolation and corresponding error estimates used in the development and analysis of numerical methods for differential equations.

The unique polynomial of degree n,

$$p_{x_0,\ldots,x_n}(x) = \sum_{j=0}^{n} a_j x^j, \qquad (J.1)$$

that interpolates a function $f(x)$ at $n+1$ points,

$$p_{x_0,\ldots,x_n}(x_i) = y_i = f(x_i), \quad 0 \le i \le n, \qquad (J.2)$$

can be found by solving simultaneously the $(n+1) \times (n+1)$ linear system of equations for the $n+1$ unknown coefficients a_j given by (J.2). It can also be found using Lagrange polynomials

$$p_{x_0,\ldots,x_n}(x) = \sum_{i=0}^{n} y_i L_{i,x_0,\ldots,x_n}(x) \qquad (J.3)$$

where

$$L_{i,x_0,\ldots,x_n}(x) = \prod_{0 \le j \le n, j \ne i} \frac{(x - x_j)}{(x_i - x_j)}. \qquad (J.4)$$

Here, we develop $p_{x_0,\ldots,x_n}(x)$ inductively, starting from $p_{x_0}(x) = y_0$ and letting

$$p_{x_0,\ldots,x_{j+1}}(x)$$
$$= p_{x_0,\ldots,x_j}(x) + c_{j+1}(x - x_0)\cdots(x - x_j), \quad j = 0,\ldots,n-1 \qquad (J.5)$$

(so that each successive term does not disturb the correctness of the prior interpolation) and defining c_{j+1} so that $p_{x_0,\ldots,x_{j+1}}(x_{j+1}) = y_{j+1}$, i.e.,

$$c_{j+1} = \frac{y_{j+1} - p_{x_0,\ldots,x_j}(x_{j+1})}{(x_{j+1} - x_0)} = f[x_0,\ldots,x_{j+1}]. \quad (J.6)$$

Comparing (J.6) with (J.3), (J.4) gives an alternate explicit expression for $f[x_0,\ldots,x_n]$, the leading coefficient of the polynomial of degree n that interpolates f at $x_0,\ldots,x_n$:

$$f[x_0,\ldots,x_n] = \sum_{i=0}^{n} \frac{f(x_i)}{\prod_{j \neq i}(x_i - x_j)} \quad (J.7)$$

from which follows the divided difference relation

$$f[x_0,\ldots,x_n] = \frac{f[x_0,\ldots,\hat{x}_j,\ldots x_n] - f[x_0,\ldots,\hat{x}_i \ldots, x_n]}{x_i - x_j} \quad (J.8)$$

(where $\hat{}$ indicates omission).

For our purposes, we want to estimate $p_{x_0,\ldots,x_n}(t) - f(t)$, and to do so, we simply treat t as the next point at which we wish to interpolate f in (J.5):

$$p_{x_0,\ldots,x_n}(t) + f[x_0,\ldots,x_n,t](t - x_0)\cdots(t - x_n) = f(t)$$

or

$$p_{x_0,\ldots,x_n}(t) - f(t) = f[x_0,\ldots,x_n,t](t - x_0)\cdots(t - x_n). \quad (J.9)$$

Finally, we estimate the coefficient $f[x_0,\ldots,x_n,t]$ using several applications of Rolle's Theorem. Since $p_{x_0,\ldots,x_n,t}(x) = f(x)$ or $p_{x_0,\ldots,x_n,t}(x) - f(x) = 0$ at $n+2$ points $x_0,\ldots,x_n,t$, Rolle's Theorem says that $p'_{x_0,\ldots,x_n,t}(x) - f'(x) = 0$ at $n+1$ points, one in each open interval between consecutive distinct points of $x_0,\ldots,x_n,t$. Repeating this argument, $p''_{x_0,\ldots,x_n,t}(x) - f''(x) = 0$ at n points on the intervals between the points described in the previous stage, and repeating this $n-1$ more times, there is one point ξ in the interior of the minimal closed interval containing all of the original points $x_0,\ldots,x_n,t$ at which

$$p^{(n+1)}_{x_0,\ldots,x_n,t}(\xi) - f^{(n+1)}(\xi) = 0. \quad (J.10)$$

J. Iterative Interpolation and Its Error

But because $f[x_0, \ldots, x_n, t]$ is the leading coefficient of the polynomial $p_{x_0,\ldots,x_n,t}(x)$ of degree $n+1$ that interpolates f at the $n+2$ points $x_0, \ldots, x_n, t$, if we take $n+1$ derivatives, we are left with a constant, that leading coefficient times $(n+1)!$:

$$p^{(n+1)}_{x_0,\ldots,x_n,t}(x) = (n+1)! f[x_0, \ldots, x_n, t]. \qquad (J.11)$$

Combining this with (J.10) gives

$$f[x_0, \ldots, x_n, t] = \frac{f^{(n+1)}(\xi)}{(n+1)!} \qquad (J.12)$$

where ξ in the interior of the minimal closed interval containing all of the original points $x_0, \ldots, x_n, t$, and substituting into (J.9) yields the basic interpolation error estimate:

$$p_{x_0,\ldots,x_n}(t) - f(t) = \frac{f^{(n+1)}(\xi)}{(n+1)!}(t - x_0) \cdots (t - x_n). \qquad (J.13)$$

For $n = 0$ this recovers the mean value theorem

$$\frac{f(t) - f(x_0)}{t - x_0} = f'(\xi) \qquad (J.14)$$

for some $\xi \in (x_0, t)$.

Since many multistep methods involve simultaneous interpolation of y and y' at $t_n, \ldots, t_{n-m+1}$, to treat these, we would want to have the corresponding estimates for osculatory interpolation that can be obtained by letting pairs of interpolation points coalesce. In the simplest cases, for two points, this process recovers the tangent line approximation and estimate

$$f[x_0] + f[x_0, x_0](x - x_0) = f(x_0) + f'(x_0)(x - x_0).$$

For four points, it recovers the cubic spline interpolation approximating a function and its derivative at two points.

Bibliography

[BB] Bernoff, A.J., and Bertozzi, A. L, *Singularities in a modified Kuramoto-Sivashinsky equation describing interface motion for phase transition*, Physica D 85 (1995), 375–404.

[BH] Briggs, W. L., and Henson, V. E., *The DFT: An Owner's Manual for the Discrete Fourier Transform*, SIAM, Philadelphia, PA, 1995.

[BJ] Butcher, J., *Numerical Methods for Ordinary Differential Equations*, Second Edition, Wiley, Hoboken, NJ, 2008.

[CP] Cartwright, J. H. E., and Piro, O., *The Dynamics of Runge–Kutta Methods*, Int. J. Bifurcation and Chaos 2 (1992), 427–449..

[CS] Channell, P. J., and Scovel, C., *Symplectic integration of Hamiltonian systems*, Nonlinearity 3 (1990), 231.

[CFL] Courant, R., Friedrichs, K., and Lewy, H., *On the partial difference equations of mathematical physics*, IBM Journal, March 1967, pp. 215–234, English translation of the 1928 German original, Mathematische Annalen 100 (1928), 32–74.

[DR] Devaney, R., *First Course in Chaotic Dynamical System: Theory and Experiment*, Westview Press, Reading, MA, 1992.

[DK] Devaney, R., and Keen, L., Editors, *Chaos and Fractals: The Mathematics Behind the Computer Graphics*, American Mathematical Society, Providence, RI, 1989.

[EWH] Enright, W. H., *Continuous numerical methods for ODEs with defect control*, Journal of Computational and Applied Mathematics 125 (2000), n. 1-2, 159–170.

[FB] Fornberg, B., *Practical Guide to Pseudospectral Methods*, Cambridge University Press, New York, NY, 1996.

[FP] Friedlander, S., and Pavlović, S., *Blowup in a three-dimensional vector model for the Euler equations*, Comm. Pure Appl. Math. 57 (6) (2004), 705–725.

[GCW] Gear, C. W., *Numerical Initial Value Problems in Ordinary Differential Equations*, Prentice Hall, Englewood Cliffs, NJ, 1971.

[GS] Gill, S., *A process for the step-by-step integration of differential equations in an automatic computing machine*, Proc. Cambridge Philos. Soc. 47 (1951), 96–108.

[GJ] Gleick, J., *Chaos: Making a New Science*, Penguin, London, 2008.

[GH] Goldstein, H., *Classical Mechanics*, Addison Wesley, San Francisco, 2002.

[GBL] Gowers, T., Barrow-Green, J., and Leader, I., Editors, *The Princeton Companion to Mathematics*, Princeton University Press, Princeton, NJ, 2008.

[HLW] Hairer, E., Lubich, C., and Wanner, G., *Geometric Numerical Integration. Structure-Preserving Algorithms for Ordinary Differential Equations*, Springer Series in Comput. Mathematics, Vol. 31, Springer-Verlag, 2002.

[HP] Henrici, P., *Discrete Variable Methods in Ordinary Differential Equations*, John Wiley & Sons, New York, NY, 1962.

[HDJ] Higham, D. J., *Robust defect control with Runge-Kutta schemes*, SIAM Journal on Numerical Analysis 26 (1989) n. 5, 1175–1183.

[HS] Hirsch, M. W., and Smale, S., *Differential Equations, Dynamical Systems, and Linear Algebra*, Academic Press, New York, 1974.

[HW] Hubbard, J., and West, B., *Differential Equations, Dynamical Systems, and Linear Algebra*, Springer, New York, 1991.

[IK] Isaacson, E., and Keller, H. B., *Analysis of Numerical Methods*, Reprint of 1966 Edition, Dover, New York, 1994.

[IA1] Iserles, A., *A First Course in the Numerical Analysis of Differential Equations*, Cambridge University Press, Cambridge, 1996.

[IA2] Iserles, A., *Efficient Runge-Kutta methods for Hamiltonian equations*, Bull. Hellenic Math. Soc. 32 (1991), 3–20.

[KH] Karcher, H., *A Gronwall argument for error estimates of ODE integration schemes*, available online at http://www.math.uni-bonn.de/people/karcher/ODEerrorViaGronwall.pdf.

[LG] Lamb, G. L., *Elements of Soliton Theory*, Wiley, New York, NY, 1980.

[LL] Landau, L. D., and Lifshitz, E. M., *Mechanics*, Addison-Wesley, Reading, MA, 1974.

[LR] Lax, P. D., and Richtmyer, R. D., *Survey of the stability of linear finite difference equations*, Comm. Pure Appl. Math. 9 (1956), 267–293.

Bibliography

[LE] Lorenz, E., *Deterministic non-periodic flow*, Journal of the Atmospheric Sciences 20 (1963), 130–141.

[MRH] Merson, R. H., *An operational method for the study of integration processes*, Proc. Symposium Data Processing, Weapons Research Establishment, Salisbury, S. Australia, 1957.

[MT] Mullins, T., *The Nature of Chaos*, Oxford University Press, USA, 1993.

[PB] Palais, B., *Blowup for nonlinear equations using a comparison principle in Fourier space*, Commun. Pure Appl. Math 41 (1988), 165–196.

[PDB] Polking, J., Arnold, D., and Boggess, A., *Differential Equations*, Second Edition, Prentice Hall, 2005.

[DR] Ruelle, D., *Chance and Chaos*, Penguin, London, 1993.

[SJM] Sanz-Serna, J. M., *Runge–Kutta schemes for Hamiltonian systems*, BIT 28 (1988), 877.

[SLF] Shampine, L. F., *Numerical Solution of Ordinary Differential Equations*, Chapman and Hall, New York, NY, 1994.

[SG] Strang, G., *On the Construction and Comparison of Difference Schemes*, SIAM J. Numer. Anal. 5 (1968) (3), 506–517.

[SS] Strogatz, S., *Nonlinear Dynamics and Chaos*, Perseus, 2000.

[TF] Tappert, F., *Numerical Solutions of the Korteweg-de Vries Equations and Its Generalizations by the Split-Step Fourier Method*, in Nonlinear Wave Motion, Lectures in Applied Math., vol. 15, American Mathematical Society, Providence, RI, pp. 215–216, 1974.

[YH] Yoshida, H., *Construction of higher order symplectic integrators*, Phys. Lett. A 150 (1990), 262.

Index

0-stability, 137
$2h$ Backward Euler Method, 168
$2h$ trapezoidal method, 168
Pth-order accurate, 136
r-stage one-step methods, 134
y-midpoint method, 169

A-stable, 162, 292
absolute stability, 137
Adams-Bashforth, 283
Adams-Moulton, 283
advection equation, 197
aliasing, 210
amplification factor, 140, 213, 220
analytic ODE, 31
analytical solution, 141
Asymptotic Stability Theorem, 47
asymptotically stable, 47
autonomous, 12

backward difference formula, 282
Backward Euler Method, 165
Banach Contraction Principle, 233, 235, 237
Banach space, 226
Butcher tableau, 266
butterfly, 25

Calculus of Variations, 63, 68
canonical coordinates, 64
canonical lifting, 65

Cartesian, 248
CFL condition, 200
chaos, 25
characteristic polynomial, 157, 290
characteristics, 199, 211
Classical Fourth-Order Runge-Kutta (Classical RK4), 176
clock, 22
closed orbit, 20
closed system, 91
configuration space, 96
conjugate momentum, 73
Conservation Laws, 73
Conservation of Angular Momentum, 120
Conservation of Energy, 74
Conservation of Linear Momentum, 118
consistent, 138, 286
constant method, 295
constant of the motion, 33
continuity of solutions w.r.t. V, 24
continuity with respect to initial conditions, 22
continuous extension, 216, 221
convergent, 136
convergent algorithm, 18
Coriolis force, 104
corrector, 295

311

coupled harmonic oscillators, 50
critical point, 20

Dahlquist root condition, 172
Dahlquist's Equivalence Theorem, 298
defect, 221
dense output, 221
differential equation, 1, 5
discretization methods, 133
discretization parameter, 134
dynamical systems theory, 15

E. Noether's Principle, 81
ecological models, 56
eigenvalue, 260
elementary differentials, 268
embedded, 165
embedded methods, 180
equation of exponential growth, 37
equation of simple harmonic motion, 37
equations of evolution, 1
equilibrium point, 20
Euler's Method, 16, 17, 144
Euler-Lagrange Equations, 68, 70
exercises, 2
Existence Theorem, 8
explicit method, 134, 159
exponential growth, 15
extrapolation, 151
extremal, 72

fixed point, 20
fixed-point iteration, 160
flow, 15
flow-box coordinates, 252
forced harmonic oscillator, 55
formal accuracy, 142
Foucault pendulum, 103

geodesics, 86
global error, 18, 145, 219
Gram-Schmidt Procedure, 231
Green's operator, 56
Gronwall's Inequality, 22

Hamilton's Equations, 86
harmonic oscillator, 10, 112, 181

heat equation, 187
Heun's Method, 162, 265
homoclinic tangle, 31
Hooke's Law, 38
horseshoe map, 31
hyperbolic, 46, 197

ignorable coordinate, 73
implicit method, 134, 159
implicit midpoint method, 169
improved Euler Method, 162
inhomogeneous linear ODE, 52
initial value problem, 1, 7
inner-product space, 228
invariance properties of flows, 33
Inverse Function Theorem, 240
isolated system, 91
IVP, 7

Jordan canonical form, 44

KdV equation, 198

Lagrange's Equations, 86
law of evolution, 1
leapfrog method, 154
linear m-step method, 135
linearized Korteweg-deVries equation, 198
local constant of the motion, 253
local error, 146, 214
local truncation error, 213, 215, 284
local truncation errors, 146
logistic equation, 6, 59
Lorenz attractor, 28
Lyapounov exponent, 31

Maximal Solution Theorem, 13
method of small vibrations, 128
method of successive approximations, 10
metric space, 225
midpoint method, 151, 265
Milne device, 178
Milne's corrector, 289
model parameter, 140
modified trapezoidal method, 162
multistage, 134
multistep, 134

Index

Newton's equations, 86
Newton's Laws of Motion, 92
Newton's Method, 160
No Bounded Escape Theorem, 14
nonautonomous, 12
norm, 226
normed vector space, 226
numerical methods, 133

orbit, 14
order of accuracy, 136
ordinary differential equation (ODE), 1

parabolic, 197
partial differential equation (PDE), 1
pendulum, 113
Pendulum Equation, 111
period, 20
periodic solution, 20
Pierre Simon de Laplace, 25
polynomial accuracy, 142
potential function, 106
precession, 121
predator-prey model, 59
predictor, 294
predictor-corrector, 163
prerequisites, 2
prime period, 21
propagator, 49

Ralston's Method, 279
region of absolute stability, 148
relative stability, 299
residual, 216, 221
residual error, 214
resonant, 56
rest point, 20
Rikitake Two-Disk Dynamo, 29
root condition, 172
rooted trees, 267
Runge-Kutta, 134
Runge-Kutta expansion, 267

second-order ODE, 66
sensitive dependence on initial conditions, 27
singularity, 20

small oscillations about equilibrium, 126
smoke particle, 6
smoothness w.r.t. initial conditions, 23
smoothness w.r.t. parameters, 24
solution (of an IVP), 8
spectral method, 141
Spectral Theorem, 259
splitting method, 208
stability, 137
stable equilibrium, 47
stable subspace, 45
stationary point, 20
step-size, 134
stiff, 139
Straightening Theorem, 252
Strang splitting, 209
strange attractor, 27, 31
strong root condition, 299
successive approximations, 39

Taylor Methods, 263
Theorem on Smoothness w.r.t. Initial Conditions, 258
time-steps, 134
total energy function, 74
trapezoidal method, 159

Uniqueness Theorem, 8
unstable subspace, 45

variation of parameters, 52
variational equation, 257
vector field, 5
visual aids, 3
Volterra-Lotka equations, 59

weak stability, 299
Web Companion, 3
well-posed, 26

DATE DUE

QA 371 .P34 2009

Palais, Richard S.

Differential equations,
  mechanics, and computation